Springer Series in
SOLID-STATE SCIENCES 6

Springer
*Berlin
Heidelberg
New York
Barcelona
Hong Kong
London
Milan
Paris
Singapore
Tokyo*

Springer Series in
SOLID-STATE SCIENCES

Series Editors:
M. Cardona   P. Fulde   K. von Klitzing   R. Merlin   H.-J. Queisser   H. Störmer

The Springer Series in Solid-State Sciences consists of fundamental scientific books prepared by leading researchers in the field. They strive to communicate, in a systematic and comprehensive way, the basic principles as well as new developments in theoretical and experimental solid-state physics.

126 **Physical Properties of Quasicrystals**
    Editor: Z.M. Stadnik
127 **Positron Annihilation in Semiconductors**
    Defect Studies
    By R. Krause-Rehberg and H.S. Leipner
128 **Magneto-Optics**
    Editors: S. Sugano and N. Kojima
129 **Computational Materials Science**
    From Ab Initio to Monte Carlo Methods
    By K. Ohno, K. Esfarjani, and Y. Kawazoe
130 **Contact, Adhesion and Rupture of Elastic Solids**
    By D. Maugis
131 **Field Theories for Low-Dimensional Condensed Matter Systems**
    Spin Systems and Strongly Correlated Electrons
    By G. Morandi, P. Sodano, A. Tagliacozzo, and V. Tognetti

Series homepage – http://www.springer.de/phys/books/sss/

Volumes 1–125 are listed at the end of the book.

R. P. Huebener

Magnetic Flux Structures in Superconductors

Extended Reprint of a Classic Text

Second Edition
With 111 Figures

 Springer

Prof. Dr. Rudolf Peter Huebener
Physikalisches Institut
Lehrstuhl für Experimentalphysik II
Auf der Morgenstelle 14
72076 Tübingen
Germany

Series Editors:

Professor Dr., Dres. h. c. Manuel Cardona
Professor Dr., Dres. h. c. Peter Fulde*
Professor Dr., Dres. h. c. Klaus von Klitzing
Professor Dr., Dres. h. c. Hans-Joachim Queisser
Max-Planck-Institut für Festkörperforschung, Heisenbergstrasse 1, D-70569 Stuttgart, Germany
* Max-Planck-Institut für Physik komplexer Systeme, Nöthnitzer Strasse 38
  D-01187 Dresden, Germany

Professor Dr. Roberto Merlin
Department of Physics, 5000 East University, University of Michigan
Ann Arbor, MI 48109-1120, USA

Professor Dr. Horst Störmer
Dept. Phys. and Dept. Appl. Physics, Columbia University, New York, NY 10023 and
Bell Labs., Lucent Technologies, Murray Hill, NJ 07974, USA

ISSN 0171-1873

ISBN 3-540-67953-7 Springer-Verlag Berlin Heidelberg New York

ISBN 3-540-09213-7 Springer-Verlag Berlin Heidelberg New York

Library of Congress Cataloging-in-Publication Data. Huebener, Rudolf Peter, 1931- . Magnetic flux structures in superconductors: extended reprint of a classic text / R.P. Huebener.--[2nd ed.]. p. cm. -- (Springer series in solid-state sciences, ISSN 0171-1873; 6) Includes bibliographical references and index. ISBN 3540679537 (alk. paper) 1. Superconductors. 2. Magnetic flux. I. Title. II. Series. QC611.95.H84 2000   537.6'23--dc21   00-061911

This work is subject to copyright. All rights are reserved, whether the whole or part of the material is concerned, specifically the rights of translation, reprinting, reuse of illustrations, recitation, broadcasting, reproduction on microfilm or in any other way, and storage in data banks. Duplication of this publication or parts thereof is permitted only under the provisions of the German Copyright Law of September 9, 1965, in its current version, and permission for use must always be obtained from Springer-Verlag. Violations are liable for prosecution under the German Copyright Law.

Springer-Verlag Berlin Heidelberg New York
a member of BertelsmannSpringer Science+Business Media GmbH

© Springer-Verlag Berlin Heidelberg 1979, 2001
Printed in Germany

The use of general descriptive names, registered names, trademarks, etc. in this publication does not imply, even in the absence of a specific statement, that such names are exempt from the relevant protective laws and regulations and therefore free for general use.

Typesetting: Camera copy of the First Edition;
             new parts, data conversion by LE-TeX Jelonek, Schmidt & Voeckler GbR, Leipzig
Cover concept: eStudio Calamar Steinen
Cover production: *design & production* GmbH, Heidelberg

Printed on acid-free paper   SPIN: 10778922   57/3141/YL - 5 4 3 2 1 0

To Gerda,
Ingrid, Christoph, and Monika

Preface to the Second Edition

The first edition of this book provided an introduction to the many static and dynamic features of magnetic flux structures in what are now called classical or low-temperature superconductors. It went out of print not long after the discovery of high-temperature superconductors in 1986 by J.G. Bednorz and K.A. Müller, a discovery which resulted worldwide in an explosive growth of research and development in the field of superconductivity. Because of this upsurge of activities, a strong demand for this book clearly continued.

Since the contents of the fourteen chapters of the first edition are still valid and continue to represent a useful introduction into the various subjects, it was felt that a reprinting of these chapters in this second edition would be highly attractive. In this way, the reader is also able to trace the earlier scientific developments, themselves constituting important ideas sometimes forgotten by the new community dealing with high-temperature superconductivity. However, because of the exciting and important recent progress in the field of high-temperature superconductivity, an extensive chapter has been added in this second edition. It provides a summary of the new developments and a discussion of the highlights. Here keywords such as vortex matter, vortex imaging, and half-integer magnetic flux quanta describe surprising new issues.

Like the first edition, this text is intended for researchers and graduate students, and it may serve as supplementary material for a graduate course on low-temperature solid-state physics. The static and dynamic properties of magnetic flux structures in low-temperature and high-temperature superconductors continue to play a crucial role in many technological applications of these materials. Hence, in addition to researchers and graduate students, this new edition may also be useful to electronic engineers.

In preparing this new edition, I benefitted from helpful comments by many people, of which I would like to mention in particular A.A. Abrikosov, E.H. Brandt, U.R. Fischer, R. Kümmel, Yu.N. Ovchinnikov, N. Schopohl, O.M. Stoll, A. Tonomura, and C. C. Tsuei.

Tübingen,
November 2000   *R.P. Huebener*

Preface to the First Edition

The idea for this book originated from an International Conference on Magnetic Structures in Superconductors organized by John R. Clem and the author at Argonne National Laboratory (ANL) in September of 1973. Large parts of the book evolved from lectures given to graduate students at the University of Tübingen during the past three years.

It is the purpose of this book to provide an introduction to the many features of magnetic flux structures in superconductors and to discuss the recent developments in this field. Here, in addition to the static properties of magnetic flux structures, the time-dependent phenomena represent an important subject including flux flow and the transport effects in superconductors. Throughout the book the emphasis is placed on the physical phenomena and the experimental results. We do not attempt a general introduction to superconductivity. Except for a brief discussion of the Ginzburg-Landau theory, with respect to the theoretical developments we only give an outline and refer to the original papers or other reviews for the detail of the calculations.

The book is intended for researchers and graduate students interested in the subject of magnetic flux structures in superconductors. It may serve as supplementary material for a graduate course on low-temperature solid-state physics. During recent years technological applications of superconducting materials are becoming increasingly important. Here the static and dynamic behavior of magnetic flux structures play a distinguished role. The book may be helpful for people involved in these engineering aspects of superconductivity.

A short remark on notation: different symbols are used for different physical quantities as much as possible. For distinction, different types of the same letter, subscripts, underlines, etc. are employed. In some cases where the possibility for confusion hardly exists the same symbol refers to different quantities. A list of all important symbols used throughout the book is added at the end.

During recent years I benefitted greatly from discussions with many people, in particular with my former colleagues at ANL and my present associates an the University of Tübingen. Here I wish to mention W. Buck, D.E. Chimenti, John R. Clem, K.E. Gray, R.T. Kampwirth, K. Noto, M.C.L. Orlowski, and K.-P. Selig. I am grateful to D.E. Chimenti, U. Essmann, A. Kiendl, B. Obst, L. Rinderer, P. Thorel, and H. Ullmaier for providing photographs or drawings for some of the figures. Thanks are due to John R. Clem for critically reading various sections of the manuscript and to H.-G. Wener for helping with the preparation of the photographs and drawings. Finally I am pleased to thank my secretary, Miss M.-L. Weisschuh, for her expert and always cheerful assistance in preparing the material for this book.

Tübingen,
December 1978

R.P. Huebener

Contents

1. **Introduction** 1

2. **Magnetic Properties of Type-I Superconductors** 4
   2.1 Meissner Effect, Penetration Depth, and Coherence Length.. 4
   2.2 Intermediate State and Wall Energy 13
   2.3 Landau Domain Theory 16
   2.4 Domain Patterns 21

3. **Ginzburg-Landau Theory** 33
   3.1 Free Energy and the Ginzburg-Landau Equations 34
   3.2 The Two Characteristic Lengths $\xi(T)$, $\lambda(T)$
       and the Wall-Energy Parameter 37
   3.3 Critical Current in a Thin Film or Wire 41
   3.4 Quantization of the Fluxoid 44
   3.5 Nucleation of Superconductivity in Bulk Samples 46
   3.6 Nucleation of Superconductivity at Surfaces 48
   3.7 Abrikosov Vortex State Near H_{c2} 50

4. **Magnetic Properties of Type-II Superconductors** 58
   4.1 Mixed State 58
   4.2 London Model 61
       4.2.1 Isolated Vortex Line 61
       4.2.2 Lower Critical Field H_{c1} 64
       4.2.3 Interaction Between Vortex Lines 65
       4.2.4 Magnetization Near H_{c1} 67
   4.3 Clem Model 69
   4.4 Theory of the Static Vortex Structure 72
   4.5 Flux-Line Lattices 75
       4.5.1 Correlation Between the Vortex Lattice
             and the Crystal Lattice 75
       4.5.2 Defects in the Vortex Lattice 78
   4.6 Surface Effects 82
       4.6.1 Energy Barrier Near a Surface 83
       4.6.2 Vortex Nucleation at the Surface 86

	4.7	Attractive Vortex Interaction 87
		4.7.1 Experiments 88
		4.7.2 Theory 92

5. Thin Films 94

6. Experimental Techniques 100
   6.1 Bitter Method 100
   6.2 Magneto Optics 102
   6.3 Micro Field Probes 110
   6.4 Neutron Diffraction 110
   6.5 Magnetization 114
   6.6 Miscellaneous 118

7. Lorentz Force and Flux Motion 121
   7.1 Motion of Magnetic Flux Structures 121
   7.2 Lorentz Force 122
   7.3 Flux Flow Resistance 124
   7.4 Flux Penetration into a Superconductor 130
   7.5 Hall Effect 134
   7.6 Ettinghausen and Peltier Effect 136
   7.7 Josephson Relation 138
   7.8 Instabilities 141
   7.9 Force-Free Configurations 143

8. Special Experiments 144
   8.1 Magneto Optics 144
   8.2 Nuclear Magnetic Resonance 145
   8.3 Neutron Diffraction 145
   8.4 Sharvin Point Contact 147
   8.5 Magnetic Coupling 148
   8.6 Micro Field Probes 152
   8.7 Simulation Experiments 154

9. Thermal Force and Flux Motion 155
   9.1 Thermal Force 155
   9.2 Nernst and Seebeck Effect 156
   9.3 Transport Entropy 160

10. Time-Dependent Theories 165
   10.1 Phenomenological Theories 165
   10.2 Time-Dependent Ginzburg-Landau Theory 170
        Flux-Flow Resistivity 174
        Hall Effect 177
        Transport Entropy 177

11. Flux Pinning 179
11.1 Critical State.. 179
11.2 Summation of Pinning Forces 184
11.3 Fundamental Pinning Interactions 186
   11.3.1 Core Interaction 186
   11.3.2 Magnetic Interaction 187
   11.3.3 Elastic Interaction 189
   11.3.4 Ginzburg-Landau Free Energy 189
11.4 Some Model Experiments 191

12. Flux Creep and Flux Jumps 195

13. Electrical Noise Power 200
13.1 Autocorrelation Function and Power Spectrum 200
13.2 Influence of the Geometry of the Contacts 205
13.3 Experiments 209

14. Current-Induced Resistive State 211
14.1 Wire Geometry 211
14.2 Thin Film Geometry 215
   14.2.1 Dynamic Model: Nucleation of Flux-Tube Trains 215
   14.2.2 Gibbs Free-Energy Barrier 220
   14.2.3 Constricted Geometry 221
14.3 Microbridges 228
14.4 Thermal Effects 231

15. High-Temperature Superconductors: Summary of Recent Developments 235
15.1 Overview 235
15.2 Static Single Vortex 236
   15.2.1 Pancake Vortices 236
   15.2.2 Electronic Vortex Structure........................ 238
   15.2.3 Half-Integer Magnetic Flux Quanta 240
15.3 Static Vortex Lattice 241
   15.3.1 Vortex Matter 241
   15.3.2 Electronic Structure 243
15.4 Lorentz Force and Vortex Motion 245
   15.4.1 Thermally Assisted Flux Flow 245
   15.4.2 Broadening of the Resistive Transition 249
   15.4.3 Damping of the Vortex Motion and Hall Effect 251
   15.4.4 Flux-Flow Instabilities at High Velocities 254
   15.4.5 Dynamic Correlation 260
   15.4.6 Quantum Tunneling 261
15.5 Thermal Force and Vortex Motion 261
   15.5.1 Thermal Diffusion of the Quasiparticles 262

		15.5.2 Thermal Diffusion of the Magnetic Flux Quanta 265
		15.5.3 Hall Angle 267
15.6	Flux Pinning 268
		15.6.1 Vortex Glass and Universal Scaling 269
		15.6.2 Nanostructuring and Confined Geometries 270
15.7	Vortex Imaging 271

List of Symbols 275

Bibliography 285

References 287

Index 307

1. Introduction

Following the discovery of the disappearance of electrical resistance in a "superconducting" material at low temperatures by Kamerlingh Onnes in 1911, a perhaps more fundamental aspect of superconductivity has been uncovered by MEISSNER and OCHSENFELD in 1933, when they observed that a superconductor, placed in a weak magnetic field, completeley expels the field from the superconducting material except for a thin layer at the surface. This expulsion of a magnetic field is generally referred to as the Meissner effect [1.1], and, at first sight, may appear to be the end of the story on magnetic flux structures in superconductors. However, as has been shown during the subsequent years, in an applied magnetic field superconductors display a rich variety of phenomena. Here, the Meissner effect only represents the beginning rather than the end.

Complete flux expulsion in the simple form of the Meissner effect only takes place when the dimensions of the superconductor perpendicular to the applied magnetic field are relatively small such that demagnetizing effects can be neglected. If the applied field is sufficiently strong and if demagnetization becomes appreciable, for example, in a thin superconducting plate oriented perpendicular to the field, then magnetic flux penetrates through portions of the superconductor, and the material splits up into normal and superconducting domains. The geometrical detail of this domain structure depends, of course, sensitively on the wall energy associated with the interface between the normal and superconducting domains. This wall energy is directly proportional to the difference between the coherence length and the penetration depth of the superconductor.

In type-I superconductors the wall energy is positive, and the amount of flux contained in a single normal domain can be many flux quanta. This domain configuration is called the intermediate state. The first theory of the domain structure in the intermediate state has been developed by LANDAU [1.2] about 40 years ago. In the normal domains the magnetic field is equal to the thermodynamic critical field H_c, whereas in the superconducting domains it is zero. The intermediate state is attained in the magnetic field range $H_c(1-D) < H < H_c$, where D is the demagnetization coefficient.

Type-II superconductors are characterized by a negative wall energy. In this case magnetic flux can be dispersed through the material in the form of single flux quanta of flux $\varphi_0 = hc/2e = 2.07 \times 10^{-7}$ G cm^2, the smallest possible unit of magnetic flux. This distribution of flux is referred to as the mixed state, attained in the field range $H_{c1}(1-D) < H < H_{c2}$. Here, H_{c1} and H_{c2} are the lower and upper critical field, respectively. Type-II superconductivity has been theoretically predicted first by ABRIKOSOV [1.3] from the general phenomenological theory of GINZBURG and LANDAU [1.4] of the superconducting phase transition. Subsequently, GOR'KOV [1.5,6] reformulated the BCS (Bardeen-Cooper-Schrieffer) theory in terms of Green's functions and showed that, for superconductors (of arbitrary electron mean free path), the Ginzburg-Landau-Abrikosov theory follows from the microscopic theory for the region near the transition temperature T_c, provided that the order parameter is small, and that the spatial variation of both the order parameter and the magnetic field is slow. The content of this formalism, based on the work of Ginzburg, Landau, Abrikosov, and Gor'kov, is now often referred to as the GLAG theory. In recent years the restricted regime of validity of the GLAG theory has been extended by numerous authors in various ways, the generalizations being applicable to different regimes of temperature, magnetic field, and mean free path of the electrons [1.7].

Besides the observation of the intermediate and the mixed state, a third type of magnetic flux structure has been discovered, consisting of a mixture of flux-free domains (Meissner phase) and domains containing a flux-line lattice (Shubnikov phase) with constant lattice parameter. This behavior is accompanied by a first-order phase transition near H_{c1}, and is explained in terms of a long-range attractive interaction between vortices.

For some time now it is clear, that magnetic flux structures in superconductors possess interesting dynamic properties. Flux flow, under the influence of the Lorentz force of an electrical current or of the thermal force of a temperature gradient, contributes importantly to the numerous transport properties of superconductors. It is in particular this dynamic aspect of magnetic flux structures, which plays a strong role in a variety of technological applications of superconductivity. The dynamics of vortex structures has been treated in a number of phenomenological theories. A more fundamental theoretical approach is based on time-dependent extensions of the Ginzburg-Landau theory. Here the theoretical treatment encounters a serious difficulty arising from the existence of a gap in the energy spectrum and the interconversion between normal excitations and superfluid in nonequilibrium situations. Presently the time-dependent Ginzburg-Landau theory provides much insight into vortex dynamics, and semiquantitatively accounts for the various transport phenomena associated with flux flow in type-II superconductors. However, compared with the theory of the static vortex structure, the time-dependent theory is still in an early stage, and considerably more work needs to be done.

The rapid progress achieved in recent years in the field of magnetic flux structures in both type-I and type-II superconductors is characterized by a strong and fruitful interplay between experimental and theoretical developments. Among the experimental advances we note the application of novel experimental techniques or of a more sophisticated version of some traditional methods. In this book we attempt to summarize these developments.

In the following we do not present a general introduction into superconductivity, since a number of excellent texts are available, which are listed in the bibliography at the end of this book. We restrict ourselves to the subject indicated in the title and shall emphasize the basic aspects. Although in recent years our subject has become increasingly important for numerous technological applications, we include these aspects only in a minor way. We do not present an independent complete treatise on magnetic flux structures in superconductors and often refer to the original papers or review articles when appropriate.

2. Magnetic Properties of Type-I Superconductors

2.1 Meissner Effect, Penetration Depth, and Coherence Length

The expulsion of magnetic flux from the interior of a superconductor, as displayed in the Meissner effect, represents an important aspect of superconductivity in addition to the infinite electrical conductivity. In the absence of the Meissner effect, a material with infinite conductivity would have the peculiar property, that its state for a given value of temperature and magnetic field depends on the path along which the values of these two thermodynamic variables have been approached. If this material is cooled through its "superconducting" transition temperature T_c in the presence of an external applied magnetic field (less than the critical field H_c), the magnetic field distribution within the material will be unaffected by the appearance of perfect electrical conductivity. On the other hand, if the external field is raised from zero to the same value *after* the material had been cooled below T_c, shielding currents flowing with zero electrical resistance will be generated at the surface, resulting in complete flux expulsion from the interior. It is the Meissner effect, which removes this dependence of the state of the material upon the thermodynamic path, and which allows the application of equilibrium thermodynamics to the superconducting phase transition.

We denote the free-energy density in zero magnetic field in the normal and the superconducting phase by $f_n(T)$ and $f_s(T)$, respectively. For the difference $f_n(T)-f_s(T)$ one obtains [2.1]

$$f_n(T) - f_s(T) = \frac{H_c^2(T)}{8\pi} \quad . \tag{2.1}$$

This represents the condensation energy per volume gained through the superconducting phase transition. For a value H_c = 100 gauss this energy amounts to 398 erg/cm^3.

The perfect diamagnetism shown by a superconducting material can be expressed in the diagram given in Fig.2.1a, where the magnetic flux density B is plotted versus the magnetic field H. For a perfect diamagnet the magnetization M is

$$M = -\frac{1}{4\pi} H \quad , \tag{2.2}$$

resulting in the M(H) diagram shown in Fig.2.1b. The critical field H_c, above which superconductivity disappears, has been found to follow in good approximation the empirical relation

$$H_c(T) \approx H_c(0)\left[1 - \left(\frac{T}{T_c}\right)^2\right] \quad . \tag{2.3}$$

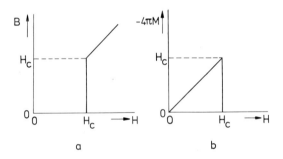

<u>Fig.2.1a and b.</u> Magnetic flux density (a) and magnetization (b) versus applied magnetic field for a bulk type-I superconductor

Looking rather closely at the surface of a superconductor, one finds that the Meissner effect does not occur exactly, and that the magnetic field penetrates a rather small but finite distance into the superconducting metal. This distance is called penetration depth and usually amounts to a few hundred angströms. It is within this surface layer, that the shielding supercurrents are flowing. Effects due to the finite penetration depth become more pronounced, of course, the smaller the overall sample dimensions. The first phenomenological theory of the Meissner effect has been put forth by the LONDON brothers [2.2].

For an outline of the *London theory*, we start with the case of infinite electrical conductivity. The microscopic force equation is then

$$m\dot{\underline{v}}_s = e\underline{E} \; , \tag{2.4}$$

without including a dissipative term. Here m and e are the electronic mass and charge, respectively; $\underline{v}_s$ is the electron velocity and $\underline{E}$ the electric field. The density of the supercurrent is

$$\underline{J}_s = n_s e \underline{v}_s \; , \tag{2.5}$$

n_s being the number density of superconducting electrons. From (2.4,5) one finds

$$\underline{E} = \frac{4\pi}{c^2} \lambda_L^2 \underline{\dot{J}}_s \; , \tag{2.6}$$

using the definition

$$\lambda_L^2 = \frac{c^2}{4\pi} \frac{m}{n_s e^2} \; . \tag{2.7}$$

The quantity λ_L represents a length of order 100 Å. Inserting the Maxwell equation

$$\text{curl } \underline{E} = -\frac{1}{c} \dot{\underline{h}} \tag{2.8}$$

into (2.6) we have

$$\frac{4\pi}{c} \lambda_L^2 \text{ curl } \underline{\dot{J}}_s + \dot{\underline{h}} = 0 \; . \tag{2.9}$$

Here and in the following we use the notation $\underline{h}$ for the local value of the magnetic flux density (on a microscopic scale), reserving $\underline{B}$ for its macroscopic average. From the Maxwell equation

$$\text{curl } \underline{h} = \frac{4\pi}{c} \underline{J} \tag{2.10}$$

we then obtain

$$\lambda_L^2 \text{ curl curl } \dot{\underline{h}} + \dot{\underline{h}} = 0 \; , \tag{2.11}$$

or

$$\nabla^2 \dot{\underline{h}} = \frac{1}{\lambda_L^2} \dot{\underline{h}} \quad . \tag{2.12}$$

Equation (2.12) has the solution

$$\dot{\underline{h}}(x) = \dot{\underline{h}}(0) \exp(-x/\lambda_L) \quad , \tag{2.13}$$

considering a superconductor filling the half-space with $x > 0$ and letting the coordinate x run from the surface (at $x = 0$) into the interior of the superconductor. From (2.13) we note that $\dot{\underline{h}} = 0$ in the interior, as is characteristic for any material with infinite electrical conductivity.

It was the central idea of F. and H. LONDON, to extend (2.9) by removing the time derivative and thereby postulating the new equation

$$\frac{4\pi}{c} \lambda_L^2 \, \text{curl} \, \underline{J}_s + \underline{h} = 0 \quad . \tag{2.14}$$

With the Maxwell equation (2.10), we then obtain

$$\nabla^2 \underline{h} = \frac{1}{\lambda_L^2} \underline{h} \quad , \tag{2.15}$$

yielding the solution

$$\underline{h}(x) = \underline{h}(0) \exp(-x/\lambda_L) \tag{2.16}$$

for the same geometrical conditions as before. Equations (2.6,14) are referred to as the *London equations*. In addition to Maxwell's equations, they apply to superconducting materials and distinguish these substances from other materials. Equation (2.16) indicates that the magnetic field is exponentially screened from the interior of a superconductor, the screening taking place within a surface layer of thickness λ_L. The quantity λ_L, defined in (2.7), is generally referred to as the London penetration depth.

From the appearance of the density n_s of the superconducting electrons in the denominator of expression (2.7) we expect the London penetration depth to be temperature dependent, with $\lambda_L \to \infty$ for $T \to T_c$. The experimental results on the temperature dependence of the penetration depth λ can be described in good approximation by the empirical relation

$$\frac{\lambda(T)}{\lambda(0)} = \left[1 - \left(\frac{T}{T_c}\right)^4\right]^{-\frac{1}{2}} \equiv y\left(\frac{T}{T_c}\right) \quad . \tag{2.17}$$

Deviations from this relation are displayed with high sensitivity in a plot of the derivative $\partial\lambda/\partial y$ versus the function $y(T/T_c)$. A typical example of such a plot is shown in Fig.2.2. Apparently the function $\lambda(T)$ deviates slightly from relation (2.17), the deviation becoming more pronounced at low temperatures. From the BCS theory [2.3] a temperature dependence of the penetration depth, slightly different from that of relation (2.17), is expected. The solid line in Fig.2.2 is a theoretical curve calculated on the basis of the BCS theory.

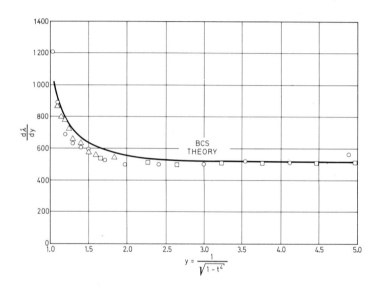

Fig.2.2. $\partial\lambda/\partial y$ versus y for a tin sample. Solid line: theoretical curve from the BCS theory [2.4]

The experimental techniques for measuring the penetration depth require the determination of the rather small length of λ with reasonable accuracy. The early experiments measured the magnetic susceptibility and relied on small samples (colloidal particles, thin wires or thin films) with dimensions comparable to λ, such that appreciable effects could be expected. Macroscopic specimens have been investigated using various methods including ac susceptibility, microwave surface impedance, and ac resonance techniques. More recently superconducting quantum-interference devices have been used in dc flux measurements with a sensitivity comparable to the high-frequency methods. We note that the highest precision is obtained only in measurements of *changes* of λ with temperature or magnetic field, whereas the absolute values of λ, and thereby the quoted values of $\lambda(0)$, are known less accurately. Results of the earlier work are summarized in the book by SHOENBERG [2.5]. The various high-frequency measurements have been reviewed by WALDRAM [2.6]. A compilation of some experimental results, including the proper references, is given in Table 2.1.

The experimental values of the penetration depth at zero temperature, $\lambda(0)$, have been found to be up to 5 times larger than the quantity $\lambda_L(0)$ obtained from (2.7). An extension of the theory has been given by PIPPARD [2.7], explaining this discrepancy in terms of the nonlocality of the superconducting properties. The nonlocal aspects arise from the fact that spatial changes in a superconductor, say, of the quantity n_s, cannot occur over arbitrarily small distances and are possible only within a certain distance, ξ_0, the coherence length of the superconductor. The importance of this coherence length is an indication of the rigidity of the functions describing the superconducting state. The characteristic length ξ_0 can be estimated from the uncertainty principle as follows. The gap Δ in the energy spectrum of the superconductor corresponds to the momentum range δp, given by

$$\Delta = \frac{(p+\delta p)^2}{2m} - \frac{p^2}{2m} \approx \frac{p\delta p}{m} = v_F \, \delta p \quad . \tag{2.18}$$

Here p and v_F are the electron momentum and the Fermi velocity, respectively. A particle whose momentum is uncertain within δp has a minimum spatial extent $\delta x \approx \hbar/\delta p$. From this we estimate

$$\xi_0 \approx \frac{\hbar v_F}{\Delta} \approx \frac{\hbar v_F}{k_B T_c} \quad . \tag{2.19}$$

In this derivation we have used the fact that $\Delta \ll E_F$, the Fermi energy, and that $\Delta \approx k_B T_c$, k_B being the Boltzmann constant. For type-I superconductors such as indium, tin, and aluminum, one finds $\xi_0 \gg \lambda_L(0)$. We note that the BCS theory yields the more accurate result

---

$\hbar = h/2\pi$ (normalized Planck's constant).

Table 2.1. Experimental values of the penetration depth λ(0), the coherence length ξ_0, and the wall-energy parameter δ(0) at zero temperature for some type-I superconductors. Following each value the reference is indicated

Material	λ(0) (Å)	ξ_0 (Å)	δ(0) (Å)
Pb	630 [a] 390 [b] 453 ± 8 [c] 418 ± 9 [d]	960 [d] 510 [m]	553 [s] 560 ± 110 [t] 600 – 800 [u]
Sn	520 ± 30 [e] 510 [f] 510 [b] 750 [a]	940 [m]	2300 [v] 2300 [w] 2300 [x] 2130 [dd]
In	640 [b] 395 ± 25 [g] 430 ± 20 [h]	3540 [n] 2400 [o] 2700 [p] 2600 [q] 2500 [r]	3400 [x] 5300 [y] 4700 – 6800 [z] 3300 [aa] 10000 [bb]
Al	490 [i] 515 [j] 500 [k] 530 [l]	13600 [m]	18000 [w]
Hg	380–450 [f]	---	2500 [cc]
Ta	---	---	480–540 [ee]

References to Table 2.1

[a] G.E. Peabody, R. Meservey: Phys. Rev. B *6*, 2579 (1972)
[b] J.M. Lock: Proc. Roy. Soc. A *208*, 391 (1951)
[c] H.R. Kerchner, D.M. Ginsberg: Phys. Rev. B *10*, 1916 (1974)
[d] R.F. Gasparovic, W.L. McLean: Phys. Rev. B *2*, 2519 (1970)
[e] H. Parr: Phys. Rev. B *12*, 4886 (1975)
[f] E. Lauerman, D. Shoenberg: Proc. Roy. Soc. A *198*, 560 (1949)
[g] H. Parr: Phys. Rev. B *14*, 2842 (1976)
[h] P.N. Dheer: Proc. Roy. Soc. A *260*, 333 (1961)
[i] T.E. Faber, A.B. Pippard: Proc. Roy. Soc. (London) A *231*, 336 (1955)
[j] M.A. Biondi, M.P. Garfunkel: Phys. Rev. *116*, 862 (1959)
[k] P.M. Tedrow, G. Farazi, R. Meservey: Phys. Rev. B *4*, 74 (1971)
[l] W.L. McLean: Proc. Phys. Soc. (London) *79*, 572 (1962)
[m] J.J. Hauser: Phys. Rev. B *10*, 2792 (1974)
[n] L.V. Del Vecchio, P. Lindenfeld: Phys. Rev. B *1*, 1097 (1970)

References to Table 2.1 (continued)

o E. Guyon, F. Meunier, R.S. Thompson: Phys. Rev. *156*, 452 (1967)
p R.S. Thompson, A. Baratoff: Phys. Rev. Lett. *15*, 971 (1965)
q A.M. Toxen: Phys. Rev. *127*, 382 (1962)
r R.D. Chaudhari, J.B. Brown: Phys. Rev. *139*, A1482 (1965)
s U. Krägeloh: Phys. Status Solidi *42*, 559 (1970)
t A. Kiendl, H. Kirchner: J. Low Temp. Phys. *14*, 349 (1974)
u R.P. Huebener, R.T. Kampwirth: J. Low Temp. Phys. *15*, 47 (1974)
v Yu.V. Sharvin: Zh. Eksp. Teor. Fiz. *33*, 1341 (1957)
  [Sov. Phys. JETP *6*, 1031 (1958)]
w T.E. Faber: Proc. Roy. Soc. A *248*, 460 (1958)
x E.A. Davies: Proc. Roy. Soc. A *255*, 407 (1960)
y R.P. Huebener, R.T. Kampwirth: Phys. Status Solidi (a) *13*, 255 (1972)
z F. Haenssler, L. Rinderer: Helv. Phys. Acta *40*, 659 (1967)
aa Yu V. Sharvin: Zh. Eksp. Teor. Fiz. *38*, 298 (1960)
   [Sov. Phys. JETP *11*, 216 (1960)]
bb R.N. Goren, M. Tinkham: J. Low Temp. Phys. *5*, 465 (1971)
cc D.E. Farrell, R.P. Huebener, R.T. Kampwirth: J. Low Temp. Phys. *19*, 99 (1975)
dd W. Rodewald: Private Communication
ee U. Essmann, W. Wiethaup, H.U. Habermeier: Phys. Status Solidi (a) *43*, 151 (1977)

$$\xi_0 = 0.18 \frac{\hbar v_F}{k_B T_c} \quad , \tag{2.20}$$

which is not too different from (2.19). In Table 2.1 the coherence length ξ_0 is listed for a number of pure metals.

From (2.16) we see that, according to the London theory, the quantities $\underline{h}$ and $\underline{J}_s$ vary on a scale λ_L. Therefore, the derivation of the London equations only holds for $\lambda_L \gg \xi_0$, a condition contrary to our estimate from (2.19). The required extension of the London theory to a nonlocal situation is analogous to the development of nonlocal electrodynamics of normal metals (anomalous skin effect). Here the nonlocal aspects arise from the finite extent of the electron mean free path. We proceed by introducing the vector potential $\underline{A}$

$$\underline{h} = \mathrm{curl}\, \underline{A} \tag{2.21}$$

and rewriting the London equation (2.14) in the form

$$\underline{J}_s = -\frac{c}{4\pi \lambda_L^2} \underline{A} \quad . \tag{2.22}$$

Relation (2.22) between the supercurrent density and the vector potential only holds when $\underline{A}$ is nearly constant over the distance ξ_0. However, we know that $\underline{A}$ decreases to zero approximately within the distance λ from the surface of the superconductor. Obviously, for $\lambda \ll \xi_0$ not all electrons contained in a region within the distance ξ_0 from the surface contribute to the screening current at the surface. The effective electrons are reduced by a factor of about λ/ξ_0, such that (2.22) must be replaced by

$$\underline{J}_s = - \frac{c}{4\pi \lambda_L^2} \frac{\lambda}{\xi_0} \underline{A} \quad , \quad (\lambda \ll \xi_0) \quad . \tag{2.23}$$

Using Maxwell's equation (2.10) together with (2.23) one obtains

$$\nabla^2 \underline{h} = \frac{1}{\lambda_L^2} \cdot \frac{\lambda}{\xi_0} \underline{h} \quad , \tag{2.24}$$

and the solution $\underline{h}(x) = \underline{h}(0) \exp(-x/\lambda)$ finally yields

$$\lambda = (\lambda_L^2 \xi_0)^{1/3} \quad , \quad (\lambda \ll \xi_0) \quad . \tag{2.25}$$

We note, that for $\lambda \ll \xi_0$ the actual penetration depth exceeds λ_L by a factor of about $(\xi_0/\lambda_L)^{1/3}$.

So far we have tacitly assumed that we are dealing with a pure metal in which spatial variations of the superconducting properties are not limited by the electron mean free path. Extending our discussion now to impure systems such as alloys, the finite length of the electron mean free path must be taken into account. We expect the influence of the electron mean free path ℓ on the coherence length ξ to be such, that $\xi(\ell)$ decreases with decreasing ℓ and approaches the electron mean free path itself for $\ell \ll \xi_0$. A function $\xi(\ell)$ having this property is given by

$$\frac{1}{\xi(\ell)} = \frac{1}{\xi_0} + \frac{1}{\ell} \quad . \tag{2.26}$$

Analogous to (2.23) the effective number of electrons for screening is now reduced by the factor ξ/ξ_0, leading to the equation

$$\underline{J}_s = - \frac{c}{4\pi \lambda_L^2} \frac{\xi}{\xi_0} \underline{A} \quad , \tag{2.27}$$

and further to the penetration depth

$$\lambda(\ell,T) = \lambda_L(T) \left(\frac{\xi_0}{\xi}\right)^{\frac{1}{2}} = \lambda_L(T)\left(1 + \frac{\xi_0}{\ell}\right)^{\frac{1}{2}} \quad . \tag{2.28}$$

A dependence of the penetration depth on the electron mean free path such as shown in (2.28) has been observed experimentally by PIPPARD [2.8] in tin-indium alloys.

2.2 Intermediate State and Wall Energy

In our previous discussion of the Meissner effect resulting in complete flux expulsion from the superconductor, we have assumed that the dimensions of the superconductor perpendicular to the applied field are relatively small and that the distortion of the magnetic field due to the field expulsion from the superconductor can be neglected. In this case the demagnetization coefficient D of the sample is approximately zero. Such a situation may be realized for a thin superconducting wire or plate oriented everywhere parallel to the field. Now we consider a superconducting specimen with a finite demagnetization coefficient placed in an external magnetic field H. As a consequence of the Meissner effect, the field is expelled from the superconductor. The demagnetization results in a field, H_{edge}, at the lateral edges of the sample given by

$$H_{edge} = H/1-D \quad . \tag{2.29}$$

We see from (2.29) that in the range

$$H_c(1-D) < H < H_c \tag{2.30}$$

we have $H_{edge} > H_c$, and superconductivity must break down. In the field range given by relation (2.30) the superconductor splits up into an arrangement of normal and superconducting domains. In the normal domains magnetic flux passes through the sample, and the magnetic field is equal to H_c. In the superconducting domains the magnetic field is zero. This situation is shown schematically in Fig.2.3. Such a configuration consisting of a mixture of normal and superconducting domains is called the *intermediate state*.

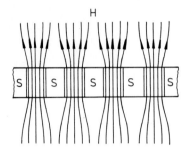

Fig.2.3. Intermediate state

In Table 2.2 we list the values of the demagnetization coefficient D for a series of sample geometries.

Table 2.2. Demagnetization coefficient D for various sample geometries

Geometry	D
Thin plate oriented parallel to H_e	≈ 0
Narrow cylinder oriented parallel to H_e	≈ 0
Sphere	1/3
Cylinder with circular cross section oriented perpendicular to H_e	1/2
Thin plate oriented perpendicular to H_e	≈ 1.0

An important quantity in determining the scale of the domain pattern of the intermediate state is the wall energy associated with the interface separating a normal from a superconducting domain. This energy is closely related to the two characteristic lengths, penetration depth and coherence length, discussed in the preceding section, and can be estimated in the following way. We consider a region of superconducting phase (S) located next to a region of normal phase (N), as depicted in Fig.2.4. The superconducting behavior, as represented by the density n_s of the superconducting electrons, cannot vanish abruptly at the interface and must approach zero rather gradually within a certain distance from the interphase separating the two regions. This distance, arising from the rigidity of the superconducting state, is expected to be approximately given by the coherence length ξ. As a consequence, due to the presence of the normal phase, we encounter a loss of condensation energy in the superconducting region, which, per unit area of interface, is approximately equal to $(H_c^2/8\pi)\xi$. This loss in condensation energy is equivalent to a positive contribution to

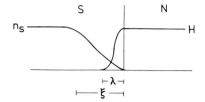

Fig.2.4. Variation of the density of the superconducting electrons, n_s, and of the magnetic field, H, near a S/N interface

the wall energy. A second contribution arises from the fact that a magnetic field penetrates a distance λ into the superconducting phase. This results in a *reduction* of condensation energy per unit area of interface of about $(H_c^2/8\pi)\lambda$. In this way, we finally have for the wall energy α per unit area

$$\alpha = \frac{H_c^2}{8\pi}(\xi - \lambda) \quad . \tag{2.31}$$

The difference

$$\delta \equiv \xi - \lambda \tag{2.32}$$

is often referred to as the wall-energy parameter.

From expression (2.31) we draw the important conclusion that for $\xi > \lambda$ the wall energy is positive. The class of materials with $\alpha > 0$ is called *type-I superconductors*. In an applied magnetic field satisfying relation (2.30), type-I superconductors reside in the intermediate state where large normal domains can be formed, each containing many flux quanta. On the other hand, we see that for $\xi < \lambda$ the wall energy will be negative. Materials with $\alpha < 0$ are called *type-II superconductors*. In this case in an applied magnetic field the domain configuration with the maximum amount of interface between normal and superconducting phase will be energetically most favorable. This leads to the dispersion of magnetic flux through these materials in the form of single flux quanta.

$$\varphi_0 = hc/2e = 2.07 \times 10^{-7} \text{ G cm}^2 \quad , \tag{2.33}$$

the smallest possible unit of magnetic flux. We note, that so far, the essential feature of the boundary between superconducting and normal phase, contained in (2.31), has been found from a rather qualitative argument. A more accurate treatment of this boundary problem, based on the Ginzburg-

2.3 Landau Domain Theory

A theoretical treatment of the domain structure in the intermediate state has first been given by LANDAU [1.2]. He assumed an infinite superconducting plate of thickness $d \gg \lambda$ placed in a perpendicular magnetic field H. The intermediate state is assumed to consist of a periodic arrangement of straight superconducting and normal domains (laminar model). The quantity to be calculated is the *periodicity length* a, which is the distance from center to center between two neighboring normal or superconducting laminae (see Fig.2.5). The periodicity length a is the sum of the width of the superconducting and normal domains, which we denote by a_s and a_n, respectively

$$a = a_s + a_n \quad . \tag{2.34}$$

Since the normal domains carry a field equal to H_c, flux conservation requires

$$a_n/a = H/H_c \quad . \tag{2.35}$$

For calculating the length a, we must collect the different contributions to the energy of the intermediate state and then find its minimum through a variation of a.

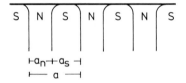

Fig.2.5. Scheme of the Landau domain structure. a_n and a_s are the width of the normal and superconducting domains, respectively. Orientation of the applied magnetic field similar to Fig.2.3

In the following we determine these different energies per unit area of the interface between the superconducting plate and the surrounding space. The first contribution, F_1, arises from the wall energy associated with the interfaces within the plate between superconducting (S) and normal (N) domains. From (2.31,32) F_1 is given by

$$F_1 = \frac{H_c^2}{8\pi} \delta \cdot \frac{2d}{a} ,\qquad(2.36)$$

where the geometrical factor $2d/a$ indicates the total area of domain walls per unit area of the plate. The contribution F_1 will become small, when the periodicity length a is rather large (assuming that the wall-energy parameter δ is positive). The second contribution, F_2, to the intermediate-state energy results from the excess energy of the nonuniform magnetic field outside the superconductor relative to a uniform magnetic field distribution. This energy will be proportional to the volume outside the superconductor occupied by the inhomogeneous distribution of magnetic field. The field inhomogeneity extends into the outer space over a distance of the order of the periodicity length a, such that

$$F_2 = \frac{H_c^2}{8\pi} 2a \cdot f_1(\tilde{h}) .\qquad(2.37)$$

Here $f_1(\tilde{h})$ represents a numerical function of the reduced field $\tilde{h} = H/H_c$. Finally, a third contribution, F_3, comes from the excess energy due to the broadening of the normal laminae at the surface of the superconducting plate (see Fig.2.5). Again, this contribution is expected to increase proportional to the periodicity length a, and we have

$$F_3 = \frac{H_c^2}{8\pi} 2a \cdot f_2(\tilde{h}) ,\qquad(2.38)$$

where $f_2(\tilde{h})$ is another numerical function. The sum of all three contributions yields

$$F_1 + F_2 + F_3 = \frac{H_c^2}{4\pi}\left[\frac{\delta d}{a} + a \cdot f(\tilde{h})\right] ,\qquad(2.39)$$

with $f(\tilde{h}) = f_1(\tilde{h}) + f_2(\tilde{h})$. A more detailed expression for the numerical function $f(\tilde{h})$ was given by LANDAU and LIFSHITZ [2.9]. By differentiating expression (2.39) with respect to a, we obtain the periodicity length for minimum energy

$$a = [\delta \cdot d/f(\tilde{h})]^{\frac{1}{2}} .\qquad(2.40)$$

Tabulated values of the numerical function $f(\tilde{h})$ have been published by LIFSHITZ and SHARVIN [2.10], and by HAENSSLER and RINDERER [2.11]. A plot of this function $f(\tilde{h})$ is presented in Fig.2.6.

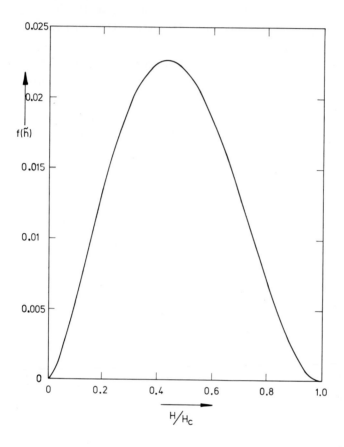

Fig.2.6. Numerical function $f(\tilde{h})$ of (2.40) versus reduced magnetic field $\tilde{h} = H/H_c$ [2.11]

The Landau model discussed so far, consisting of simply connected, straight laminae of normal material imbedded between two superconducting domains, is referred to as the *nonbranching model*. A refinement of this model has been carried out by LANDAU assuming that near the surface of the plate branching of the normal laminae into smaller portions takes place, the branches being separated from each other by inclusions of superconducting phase [2.12,13]. The Landau *branching model* is shown schematically in Fig.2.7. For the periodicity length a this model yields

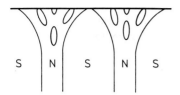

Fig.2.7. Scheme of the Landau branching model. Orientation of the applied magnetic field similar to Fig.2.3

$$a = \left[(2\sqrt{2}-2)^2 \frac{d^2 \delta}{\tilde{h}(1-\tilde{h}^2)} \right]^{1/3} . \qquad (2.41)$$

An unbranched laminar model of the intermediate state has also been discussed by KUPER [2.14], yielding a result similar to (2.40). ANDREW [2.14a] considered a model consisting of branching threads near the surface of the plate. For the periodicity length his model also yields the functional dependence $a \sim (d^2 \cdot \delta)^{1/3}$ of (2.41), with a slightly different variation with the reduced field $\tilde{h}$ than given in this equation. Branching of the normal domains near the surface of a type-I superconductor is expected to take place in rather thick specimens. HUBERT [2.15] has theoretically analyzed the transition from a nonbranching to a branching domain structure. He concluded that above a *critical thickness* d_s of the sample plate the branching domain structure is energetically more stable. According to his model, for $\tilde{h} = 0.5$ this critical thickness is approximately

$$d_s \approx 800 \, \delta , \quad (\tilde{h} = 0.5) \qquad (2.42)$$

yielding, say, for lead at 4.2 K the value $d_s \approx 100$ μm. Experiments on lead samples with different thickness agree reasonably with HUBERT's prediction [2.16]. Recent experiments on the intermediate state in mercury also indicate, that the transition from a nonbranching to a branching domain structure occurs at a sample thickness approximately given by (2.42) [2.17]. We note that the concepts of the Landau theory relating to the intermediate-state structure in type-I superconductors are quite similar to those encountered in other domain systems, like ferromagnets. In each case we are dealing with the competition between the field energy and the energy of the domain walls as the mechanism determining the domain configuration which is energetically most favorable. Experimental studies of the periodicity length in laminar domain configurations, in conjunction with (2.40,41), have served to determine the wall-energy parameter δ in various type-I superconductors. These experimental values of δ are listed in the last column of Table 2.1.

In addition to the nonbranching and branching laminar model, LANDAU already has pointed out that at low flux densities *normal tubes or flux spots* with approximately circular cross section should be energetically more favorable than extended laminae. Such flux spots are, indeed, frequently observed. GOREN and TINKHAM [2.18] have treated the flux-tube configuration in some detail and obtained for the equilibrium spacing of the normal spots

$$a_{spot} = [2\delta d/\tilde{h}(1-\tilde{h})(1-\tilde{h}^{\frac{1}{2}})]^{\frac{1}{2}} \quad . \tag{2.43}$$

Here, again d is the thickness of the film or plate. For the flux-spot diameter these authors found

$$D_{spot} = [2\delta d/(1-\tilde{h})(1-\tilde{h}^{\frac{1}{2}})]^{\frac{1}{2}} \quad . \tag{2.44}$$

We note that the free energy difference between the laminar and the flux-spot configuration is rather small. Because of the limited accuracy of the calculation of the various energy contributions, a theoretical prediction, which of the two configurations is more stable, still appears impossible at present, and experiments must be consulted for a final answer. The experimentally observed domain patterns will be discussed in Sect.2.4.

From (2.40,43) we see that the scale of the domain pattern (periodicity length a) in type-I superconductors becomes smaller and smaller, and the dispersion of magnetic flux finer and finer, with decreasing sample thickness d. With increasing subdivision of the normal domains the wall energy contributes more and more to the free energy of the superconductor. Eventually, when the film thickness becomes about equal to the wall-energy parameter δ, the wall energy would compensate the superconducting condensation energy completely, and the critical field perpendicular to the superconducting film would become zero [2.3]. However, thin films of type-I superconductors with a thickness less than δ assume a vortex state in a perpendicular magnetic field and behave like type-II superconductors. We will discuss this behavior in thin films of type-I materials in more detail in Sect.5.

2.4 Domain Patterns

In recent years magnetic flux structures in thin films or plates of type-I superconductors have been investigated by many authors. These experiments covered a range of different materials and included sample thicknesses from about 1 cm down to $10^{-5} - 10^{-6}$ cm, where type-I materials display the Abrikosov vortex state. Domain patterns in type-I superconductors can be studied utilizing a number of experimental techniques, which will be described in Sect.6. A review of the domain configurations in the intermediate state has been given some time ago by LIVINGSTON and DeSORBO [2.19].

In a *perpendicular* magnetic field, films or plates of type-I usually have an intermediate-state structure where the orientation of the normal and superconducting domains is highly irregular. In this case the absence of any preferred overall domain orientation directly results from the absence of any preferred orientation of a field component within the plane of the specimen. Locally, the detailed domain configuration will be influenced by the metallurgical microstructure of the sample. A typical example is presented in Fig.2.8 showing the intermediate-state structure in a lead film for different values of the applied magnetic field. These photographs have been obtained utilizing a magneto-optical method (see Sect.6.2). The bright and dark regions indicate the normal and superconducting phase, respectively. The series of photographs in Fig.2.8 was taken by increasing the applied magnetic field monotonically from zero. At low fields we note predominantly flux tubes. As the field is increased, the flux tubes gradually grow into extended normal laminae. As one approaches H_c, the superconducting domains constitute the distinct feature and eventually shrink into small inclusions of superconducting phase.

From photographs such as shown in Fig.2.8 the periodicity length a of the Landau model can often be obtained over a restricted field range with reasonable accuracy by selecting local regions where a few fairly straight laminae of normal and superconducting phase can be found parallel to each other. In this way for sample thicknesses below d_s the validity of (2.40) and the Landau nonbranching model has often been confirmed, and experimental values of the wall-energy parameter have been derived [2.17,20-25]. Whereas these experiments have been performed mostly in the thickness range below d_s where the nonbranching model appears to be valid, the earlier work to a large extent involved much thicker samples, their thickness ranging up to several mm [2.11,16,26-35].

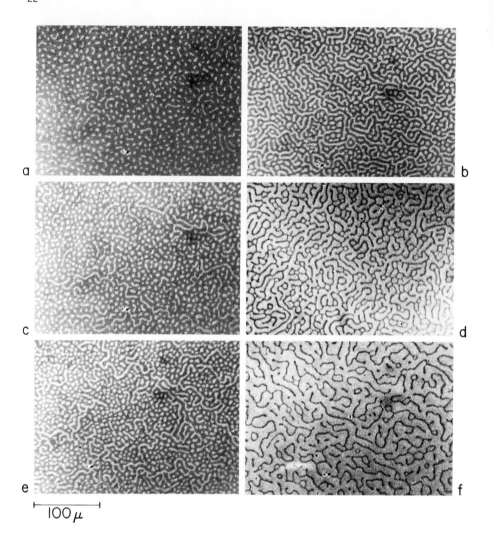

Fig.2.8a-f. Intermediate-state structure of a Pb film in perpendicular magnetic field for increasing values of H. (a) 95 G, (b) 132 G, (c) 178 G, (d) 218 G, (e) 348 G, (f) 409 G (normal domains are bright; T = 4.2 K; film thickness = 9.3 μm)

Fig.2.9a-e. Intermediate-state structure showing branching observed in a Pb disk of 12 mm diameter and 0.68 mm thickness in perpendicular magnetic field for increasing values of $\tilde{h} = H/H_c$. (a) $\tilde{h}$ = 0.29, (b) $\tilde{h}$ = 0.37, (c) $\tilde{h}$ = 0.53, (d) $\tilde{h}$ = 0.74, (e) $\tilde{h}$ = 0.84. Normal domains are bright, T = 4.2 K, $(SI)_T$ transition.(Courtesy of A. Kiendl)

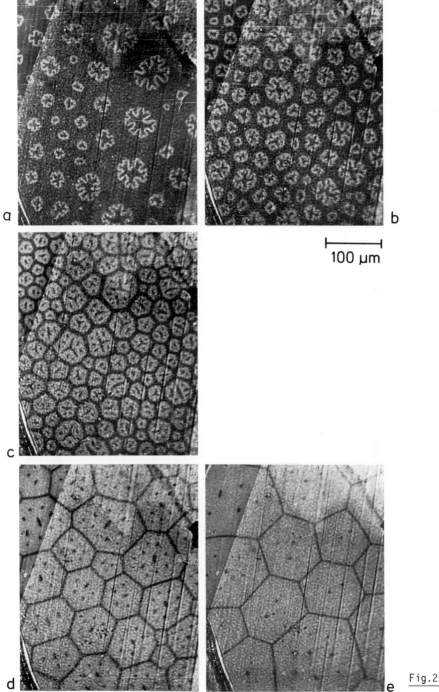

Fig.2.9a-e

A typical example of a domain structure showing branching near the sample surface is presented in Fig.2.9. The photographs were obtained magneto-optically and refer to a lead disk of 12 mm diameter and 0.68 mm thickness. Again, the bright and dark regions indicate the normal and superconducting phase, respectively. The series of photographs was taken by monotonically increasing the perpendicular magnetic field from zero. Apparently, at low magnetic fields flux enters the sample in the form of flux rings surrounding a superconducting region. The flux rings display a meandering shape. Above approximately $\tilde{h} = 0.4$ the superconducting region within a flux ring is not any more simply connected and consists of several superconducting inclusions separated from each other. Above about $\tilde{h} = 0.5$ the flux rings develop into polygonal normal domains, separated from each other by a "honeycomb" structure of the superconducting phase. With increasing field this honeycomb structure grows, and the superconducting regions shrink, approaching zero at $\tilde{h} = 1.0$. We note that the superconducting inclusions within the flux rings can be interpreted in terms of the branching model of Landau.

SHARVIN [2.29] has been the first to experimentally establish a nearly ideal, straight laminar domain structure by applying an *inclined magnetic field* to a superconducting plate of type I ("Sharvin geometry"). Because of the field component $H_\parallel$ parallel to the plate, the domain orientation is not any more isotropically distributed, but rather directed preferentially along $H_\parallel$. An example of the Landau domain structure observed magneto-optically in the Sharvin geometry is shown in Fig.2.10 for a lead single crystal of 1 mm thickness [2.36]. The black laminae indicate the superconducting phase. This domain pattern has been obtained for a 7° angle between the applied field and the large sample faces and by slowly reducing the magnetic field from $\tilde{h} > 1$ to $\tilde{h} = 0.92$. The straight Landau domains were always oriented precisely in the direction of the field component parallel to the large sample faces. Rotating this field component relative to the specimen resulted in a domain structure rotated by exactly the same amount. The direction of the sample edges did not seem to have any influence on the domain orientation. SHARVIN [2.29] has extended the laminar Landau model and derived for the periodicity length a for the case of an inclined field

$$a = [\delta d/f(\tilde{h})(1 - \tilde{h}^2 \cos^2\beta)]^{\frac{1}{2}} \ , \tag{2.45}$$

using the same notation as in (2.40). Here, β is the angle between the applied field and the large faces of the superconducting plate. Of course,

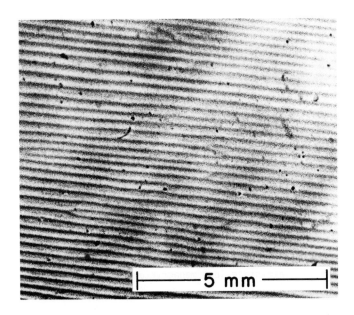

Fig.2.10. Landau domain structure observed in a lead single crystal of 1 mm thickness for 7° angle between applied field and large sample face; $(NI)_T$ transition, $h = 0.92$, $T = 4.2$ K, superconducting laminae are black [2.36]

a regular domain structure such as shown in Fig.2.10 serves rather well for determining the periodicity length a and subsequently the wall-energy parameter using (2.45).

As we can see from Fig.2.8, at low magnetic fields films or plates of type I display a domain structure consisting of simply connected *flux tubes*, if the specimen thickness does not become very large. Such flux tubes have been observed by many authors [2.18,20,22,24,25,37-40]. A quantitative comparison of the flux-tube diameter obtained from the Goren-Tinkham model and given in (2.44) with the experimental values has been performed in some cases. In mercury there appears to be reasonable agreement [2.17]. For lead the experimental results scatter within a factor 2 around the value from (2.44) [2.24,38]. In tin and indium the flux-spot densities have been measured using a high-resolution Hall probe, and the correct order of magnitude expected from the Goren-Tinkham model has been observed [2.18]. However, it is important to note that the observation of flux-spot densities is rather sensitive to pinning effects and deviations from thermodynamic equilibrium. Such deviations of the field within the sample from its equilibrium value can be minimized by cooling the specimen through the transition temperature while the applied field is kept constant.

Most of the experimental results discussed so far refer to *polycrystalline* specimens. Now we address ourselves to those additional features which appear in *monocrystalline* samples and which are based on the anisotropic properties of the superconducting crystal. Here a rather early experiment has been performed by SHARVIN and GANTMAKHER [2.31] who studied the anisotropy of the domain structure in single crystals of tin in a perpendicular magnetic field using a Bitter method. From these experiments they obtained information on the dependence of the wall-energy parameter upon the crystallographic direction. Examples of anisotropic domain patterns observed in monocrystalline foils or plates are shown in Figs.2.11-15. In all cases the external magnetic field has been applied perpendicular to the large sample faces. In the photographs of Figs.2.11-13, which were obtained using a Bitter method, the normal domains are dark. Figures 2.14,15 were taken from magneto-optic experiments, and here the normal domains are indicated by the bright regions. Figure 2.11 shows the intermediate state in a monocrystalline tin foil of 29 μm thickness at $\tilde{h}$ = 0.61. The normal domains are seen to be aligned parallel to the [100] directions. In similar tin samples flux tubes were found to assume a nearly quadratic shape with the interfaces parallel to the [100] directions [2.21]. Figure 2.12 shows the intermediate state in a monocrystalline, (111)-oriented lead foil of 6 μm thickness at $\tilde{h}$ = 0.15. A triangular flux-tube lattice can be seen, and in addition some normal laminae oriented parallel to the [$\bar{1}\bar{1}2$] directions. Apparently, in Fig.2.12 there is no correlation between the orientation of the flux-tube lattice and the crystal lattice. Figure 2.13 presents some results obtained with a single-crystalline lead foil with 23-27 μm thickness and (357) orientation at $\tilde{h} \approx 0.35$. The normal laminae are parallel to the projections of the [111] directions into the plane of the foil. In Fig.2.14 we see the anisotropic domain structure in a monocrystalline mercury plate of 106 μm thickness and with a roughly elliptical area at $\tilde{h}$ = 0.67. Figure 2.15 refers to the same sample as Fig.2.14 and shows a configuration of flux tubes and extended normal laminae near the sample edge at a lower field ($\tilde{h}$ = 0.29). The examples given in Figs.2.11-15 clearly demonstrate a pronounced anisotropy in the intermediate-state structure resulting from the anisotropy of the crystal lattice.

So far in our discussion we have ignored the fact that the detail of the intermediate-state structure often depends on the thermodynamic path along which the state of the sample has been established. Therefore, the intermediate state does not necessarily represent a thermodynamic equilibrium configuration. This *path-dependence of the domain pattern* has been stressed

Fig.2.11. Intermediate state of a monocrystalline tin foil of 29 μm thickness at $\bar{h} = 0.61$ and $T = 1.2$ K, $(NI)_T$ transition, normal regions are dark [2.21]

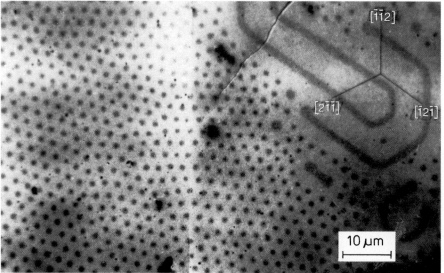

Fig.2.12. Intermediate state of a monocrystalline lead foil of 6 μm thickness at $\bar{h} = 0.15$ and $T = 1.2$ K, $(SI)_T$ transition, normal regions are dark [2.21]

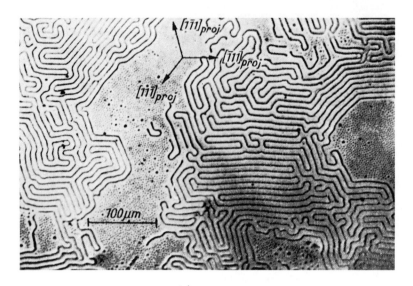

Fig.2.13. Intermediate state of a monocrystalline (357)-oriented lead foil of 23-27 μm thickness at $\bar{h} \approx 0.35$ and T = 1.2 K, (SI)$_T$ transition, normal regions are dark [2.24]

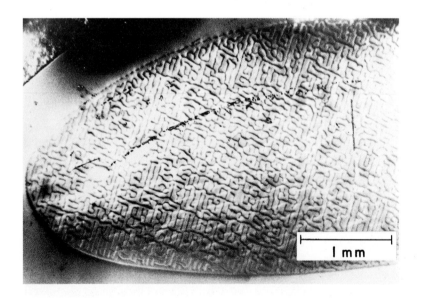

Fig.2.14. Intermediate state of a monocrystalline mercury plate of 106 μm thickness at $\bar{h}$ = 0.67 and T = 1.7 K; H was decreased monotonically from just below H_c; superconducting phase is dark [2.17]

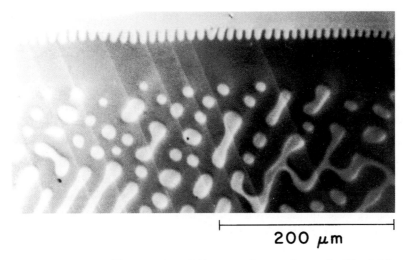

Fig.2.15. Intermediate state of the same Hg sample as in Fig.2.14 near the specimen edge at $\bar{h} = 0.29$ and $T = 1.7$ K, $(SI)_T$ transition, normal phase is bright [2.17]

in particular by HAENSSLER and RINDERER [2.11], and subsequently has been discussed by various authors [2.16,17,20,22,23,40,41]. The different thermodynamic paths used experimentally for establishing the intermediate state are shown schematically in Fig.2.16. Within the H(T) phase diagram we can reach a particular point, A, by raising the field from zero at constant temperature, by lowering the field from above H_c at constant temperature, or by reducing the temperature from above T_c at constant field. These three paths are generally referred to as $(SI)_T$, $(NI)_T$, and $(NI)_H$ transitions, respectively. Experimentally, only the first two transitions are usually investigated.

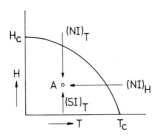

Fig.2.16. The different thermodynamic paths for establishing the intermediate state

The path-dependence of the domain structure can be seen from Fig.2.17, which represents a $(NI)_T$ transition, and which must be compared with the $(SI)_T$ transition in the same sample shown in Fig.2.9. The comparison of both figures demonstrates the rather general result that the $(SI)_T$ transition produces magnetic flux structures with a *closed topology*, where normal domains are completely surrounded by superconducting phase, often in the form of closed loops of superconducting laminae. On the other hand, the $(NI)_T$ transition nearly always results in an *open topology*, where the normal domains are open and directly connected to the sample edges. This topological difference can be understood from the way in which magnetic flux enters and leaves the specimen. In the $(SI)_T$ transition, flux enters the specimen from the edge in the form of flux tubes or flux rings which remain surrounded by superconducting phase. Increasing the field results in the replacement of the multiply connected superconducting region by normal phase, until finally closed loops of superconducting laminae remain. In the $(NI)_T$ transition, magnetic flux leaves the sample towards the edges, leaving behind only simply connected inclusions of superconducting phase. Only at very low fields normal domains (flux tubes or flux rings) completely surrounded by superconducting phase can sometimes be observed in a $(NI)_T$ transition. In Fig.2.17 we note the absence of the closed superconducting loops contained in the honeycomb structure of Fig.2.9. For the $(NI)_T$ transition the normal domains are seen to stretch continuously (and perhaps in a multiply connected configuration) across the whole field of view, except at the lowest magnetic field.

In addition to these topological differences arising from the different thermodynamic paths, irreversible behavior in the domain pattern can also result from *flux-pinning effects* which become more significant at lower magnetic fields. The various mechanisms for flux-pinning will be discussed in Sect.11.

As we have pointed out above, the dependence of the domain pattern upon the thermodynamic path essentially is associated with the *process of nucleation and growth* of the normal and superconducting phase. In the $(SI)_T$ transition individual normal domains (flux tubes or flux rings) are nucleated at the sample edges and propagate from there into the sample interior. In increasing magnetic field, the number of these normal domains increases at first, and at higher fields these domains gradually grow larger. In the $(NI)_T$ transition small superconducting domains are nucleated at first in the sample interior, growing into long laminae as the field is reduced. The growth process of the superconducting Landau domains for the Sharvin geo-

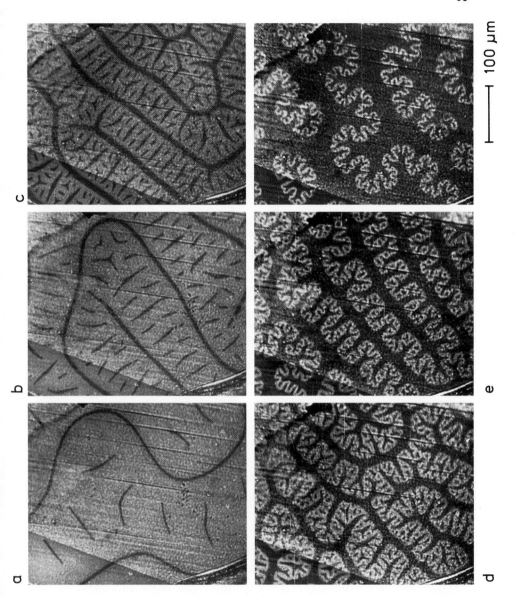

Fig.2.17a-f. Intermediate state of the same Pb disk as in Fig.2.9 in perpendicular magnetic field for decreasing values of $\tilde{h} = H/H_c$. (a) $\tilde{h} = 0.84$, (b) $\tilde{h} = 0.74$, (c) $\tilde{h} = 0.58$, (d) $\tilde{h} = 0.42$, (e) $\tilde{h} = 0.32$, (f) $\tilde{h} = 0.21$. Normal phase is bright, $T = 4.2$ K, $(NI)_T$ transition. (Courtesy of A. Kiendl)

metry shows rather interesting regularity, as shown in experiments by FARRELL et al. [2.36]. The essential features of this process are indicated schematically in Fig.2.18. As the magnetic field is reduced the straight superconducting domain (a) first becomes thicker in the center (b) and develops a small kink into the sample interior (c), from which a new domain grows in the direction of the field component parallel to the large sample faces (d). Near the branching point, the superconducting phase then grows also in opposite direction, resulting in a complete second domain, which is still linked with the original domain (e). Finally, the link between both domains disappears (f).

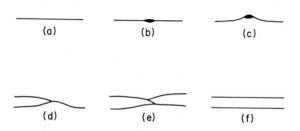

Fig.2.18. Growth of the superconducting Landau domains, schematically [2.36]

We note that in the growth processes the dynamic properties of the flux structures play an important role. These dynamic aspects will be discussed in Sect.7.

Summarizing the experimental results, the domain patterns in the intermediate state display rich topological variations depending on sample thickness (branching or nonbranching), the crystalline state of the superconductor (metallurgical microstructure, single- or polycrystal), and the thermodynamic path for establishing the intermediate state. The simplest domain structure consisting of straight, regular laminae is obtained at an inclined field angle (Sharvin geometry). The dominant features can be understood from the Landau domain theory and its extensions, which essentially rests on the wall-energy parameter δ as its most significant quantity. Experimental values of δ derived from observations of domain patterns agree reasonably with results obtained from theory or other experiments.

3. Ginzburg-Landau Theory

We have seen in Sect.2 that a superconductor often resides in a state of high spatial inhomogeneity. Here the intermediate state and the situation near a domain wall are just an example showing spatial variations of the order parameter. Spatial inhomogeneity is displayed perhaps even more importantly in the case of the vortex state in type-II superconductors. A phenomenological theory particularly suited for dealing with such inhomogeneous situations has been developed by GINZBURG and LANDAU [1.4]. The Ginzburg-Landau (GL) theory is based on LANDAU's [1.2] theory of second-order phase transitions, in which LANDAU introduced the important concept of the *order parameter*. This concept, originally developed for treating structural phase transitions, has since proved extremely useful in many systems where phase transitions take place. Depending on the physical system, the order parameter can have different dimensions. In order-disorder phase transitions the order parameter is a scalar. In a ferromagnetic and ferroelectric phase transition the order parameter is a vector, namely the magnetization and the polarization, respectively. In the superconducting phase transition the order parameter is the density of the superconducting electrons (pair wave function) or the energy-gap parameter. More recently, the concept of the order parameter has been extended to nonequilibrium phase transitions [3.1]. Here an example is the laser near threshold pumping power which can be treated using the light field or the photon number as the order parameter. In the original GL theory the order parameter is a complex quantity, namely a pseudo wave function $\Psi(\underline{r})$. The absolute value $|\Psi(\underline{r})|$ is connected with the local density of superconducting electrons, $|\Psi(\underline{r})|^2 = n_s(\underline{r})$. The phase of the order parameter is needed for describing supercurrents. The free-energy density is then expanded in powers of $|\Psi|^2$ and $|\nabla\Psi|^2$, assuming that Ψ and $\nabla\Psi$ are small. The minimum energy is found from a variational method leading to a pair of coupled differential equations for $\Psi(\underline{r})$ and the vector potential $\underline{A}(\underline{r})$. It is important to note that the GL theory is based on purely phenomenological concepts. However, GOR'KOV [1.5,6] has shown that, for both very pure and very impure superconductors, the GL theory follows rigorously from the BCS theory for the temperature region near T_c, provided that spatial variations of the order parameter and of the magnetic field are slow. In GOR'KOV's reformulation of the GL theory the quantity $\Psi(\underline{r})$ is denoted by $\Delta(\underline{r})$, which often has been called the gap parameter. However, in general there is no simple relationship between this function $\Delta(\underline{r})$ and a gap in the excitation spectrum.

3.1 Free Energy and the Ginzburg-Landau Equations

Starting with the simplest case, we assume the order parameter $\Psi(\underline{r})$ to be constant and the local magnetic flux density $\underline{h}$ to be zero throughout the superconductor. For small values of $\Psi(\underline{r})$, i.e., for $T \to T_c$, the free-energy density f can be expanded in the form

$$f = f_n + \alpha(T)|\Psi|^2 + \frac{\beta(T)}{2}|\Psi|^4 + \ldots \quad . \tag{3.1}$$

Stability of the system at the transition point (at which $\Psi = 0$) requires f to attain a minimum for $\Psi = 0$. Hence, in the expansion of (3.1) only even powers can appear. For the minimum in f to occur at finite values of $|\Psi|^2$, we must have $\beta > 0$. Otherwise the lowest value of f would be reached at arbitrarily large values of $|\Psi|^2$, where the expansion (3.1) is inadequate. As is shown in Fig.3.1, the variation of $f - f_n$ depends on whether α is positive or negative. For $\alpha > 0$ the minimum occurs at $|\Psi|^2 = 0$ corresponding to the normal state and the case $T > T_c$. On the other hand, for $\alpha < 0$ the minimum occurs at

$$|\Psi|^2 = |\Psi_0|^2 \equiv -\alpha/\beta \quad , \tag{3.2}$$

corresponding to $T < T_c$. We note that α must change its sign at $T = T_c$, and we can use the expansion

$$\alpha(T) = a \cdot (T - T_c) \quad , \quad a = \text{const} > 0 \quad . \tag{3.3}$$

With (3.2) we obtain

$$|\Psi_0|^2 = \frac{a}{\beta(T)} \cdot (T_c - T) \quad , \tag{3.4}$$

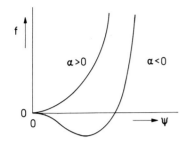

Fig.3.1. Free-energy density f versus the order parameter Ψ from (3.1)

representing a rather general result characteristic of a second-order phase transition. From (3.4) we can approximate very close to T_c

$$f = f_n + \alpha(T)|\Psi_0|^2 = f_n - \frac{a^2}{\beta(T)} \cdot (T_c - T)^2 \; , \tag{3.5}$$

yielding

$$\frac{\partial(f_n - f)}{\partial T} = -\frac{2a^2}{\beta} \cdot (T_c - T) \; . \tag{3.6}$$

We see that for $T \to T_c$ we have $\partial(f_n - f)/\partial T \to 0$, indicating a phase transition at least of second order.

Next we relax our assumptions, allowing spatial variations of the order parameter, however, still keeping $\underline{h} = 0$. To the free-energy expression of (3.1) we now must add terms of the form $(\partial\Psi/\partial x)^2$, $(\partial\Psi/\partial x)(\partial\Psi/\partial y)$, etc., the first significant terms being of second order, since in the absence of a magnetic field the equilibrium corresponds to Ψ = const. For cubic symmetry we simply have the expansion

$$f = f_n + \alpha(T)|\Psi|^2 + \frac{\beta(T)}{2}|\Psi|^4 + \gamma\left[\left(\frac{\partial\Psi}{\partial x}\right)^2 + \left(\frac{\partial\Psi}{\partial y}\right)^2 + \left(\frac{\partial\Psi}{\partial z}\right)^2\right] + \ldots \tag{3.7}$$

with $\gamma > 0$ for $T = T_c$. Equation (3.7) is the basis of Landau's general theory of second-order phase transitions.

Finally we include the presence of magnetic fields $\underline{h} = \mathrm{curl}(\underline{A})$. Then the free-energy density can be expanded in the form

$$f = f_n + \alpha|\Psi|^2 + \frac{\beta}{2}|\Psi|^4 + \frac{1}{2m^*}\left|\left(\frac{\hbar}{i}\nabla - \frac{e^*}{c}\underline{A}\right)\Psi\right|^2 + \frac{h^2}{8\pi} \tag{3.8}$$

representing the starting point of the GL theory. We note that for $\Psi = 0$ we have $f = f_n + \underline{h}^2/8\pi$, i.e., the free-energy density of the normal state. The fourth term in the expansion (3.8) becomes clearer by writing Ψ in the form

$$\Psi = |\Psi| \cdot e^{i\varphi} ; \tag{3.9}$$

it then becomes

$$\frac{1}{2m^*}\left[\hbar^2(\underline{\nabla}|\Psi|)^2 + (\hbar\,\underline{\nabla}\,\varphi - \frac{e^*}{c}\underline{A})^2 |\Psi|^2\right] . \tag{3.10}$$

The first contribution represents the additional energy arising from gradients in the magnitude of the order parameter. The second contribution contains the kinetic-energy density of the supercurrents, as we can see by identifying

$$|\Psi|^2 = n_s^* . \tag{3.11}$$

The kinetic-energy density is then $(1/2)m^* v_s^2 n_s^*$, where the supercurrent velocity v_s is given by

$$m^* \underline{v}_s = \underline{p}_s - \frac{e^*}{c}\underline{A} = \hbar\,\underline{\nabla}\,\varphi - \frac{e^*}{c}\underline{A} \tag{3.12}$$

($\underline{p}_s$: generalized particle momentum). It is interesting to note that when GINZBURG and LANDAU developed their theory, they already pointed out that the quantities m^* and e^* in (3.8) need not be identical with the mass and charge, respectively, of an electron. Indeed, it later turned out that the fundamental particles in superconductivity are the Cooper pairs, and we must take $e^* = 2e$ and $n_s^* = n_s/2$. It has been argued that the effective mass m^* can be chosen arbitrarily, since m^* is cancelled in the expression for the free-energy density [3.2]. However, ZIMMERMANN and MERCEREAU [3.3] obtained direct experimental evidence that $m^* = 2m$. They measured the maximum supercurrent I_{max} in a rotating superconducting interferometer as a function of rotation. I_{max} is proportional to $|\cos(\pi R_o^2\, m^*\, \omega/\hbar)|$, where R_o is the radius of the circular interferometer and ω the angular frequency of the rotation. ZIMMERMANN and MERCEREAU found the value $m^* = 2m$ with an uncertainty of ± 4%.

Having obtained the expression (3.8) for the free-energy density, we now must find its minimum with respect to spatial variations of the order parameter $\Psi(\underline{r})$ and the magnetic field distribution $\underline{A}(\underline{r})$. Following a standard variational procedure, one finds the two *GL differential equations*

$$\alpha\Psi + \beta|\Psi|^2\Psi + \frac{1}{2m^*}\left(\frac{\hbar}{i}\underline{\nabla} - \frac{e^*}{c}\underline{A}\right)^2 \Psi = 0 \tag{3.13a}$$

and

$$\underline{J}_s = \frac{e^*\hbar}{2m^*i}(\Psi^*\underline{\nabla}\Psi - \Psi\underline{\nabla}\Psi^*) - \frac{e^{*2}}{m^*c}\Psi^*\Psi\underline{A} \quad . \tag{3.13b}$$

Equation (3.13a) has the form of the Schrödinger equation with the energy eigenvalue $-\alpha$, the term $\beta|\Psi|^2\Psi$ acting like a repulsive potential. Equation (3.13b) represents the quantum-mechanical description of a current. Both equations refer to particles of mass m^*, charge e^*, and wave function $\Psi(\underline{r})$

The variational procedure requires the introduction of boundary conditions. GINZBURG and LANDAU assumed the boundary condition

$$\left(\frac{\hbar}{i}\underline{\nabla} - \frac{e^*}{c}\underline{A}\right)_n \Psi = 0 \quad , \tag{3.14}$$

where the subscript n denotes the component normal to the surface. Equation (3.14) indicates that no supercurrents pass through the surface. It implicitly assumes, that the expansion (3.8) of the free-energy density is valid directly up to the surface. However, this is only correct for a superconductor-insulator interface. For an interface between a superconductor and a normal metal the boundary condition (3.14) must be modified because of the proximity effect.

3.2 The Two Characteristic Lengths $\xi(T)$, $\lambda(T)$ and the Wall-Energy Parameter

From the GL equation (3.13a) we can derive a characteristic length in the following way. We assume currents and magnetic fields to be zero. Then in the one-dimensional case (3.13a) becomes

$$-\frac{\hbar^2}{2m^*}\frac{d^2\Psi}{dx^2} + \alpha\Psi + \beta\Psi^3 = 0 \quad . \tag{3.15}$$

This equation has two trivial solutions: $\Psi = 0$ and $\Psi = \Psi_0$ (with $|\Psi_0|^2 = -\alpha/\beta > 0$), describing the normal and superconducting state, respectively. However, we are also interested in more general solutions including spatial deformations of the order parameter. We introduce the normalized wave function $f = \Psi/\Psi_0$, with Ψ_0 given by (3.2). Equation (3.15) then becomes

$$\frac{\hbar^2}{2m^*|\alpha|}\frac{d^2f}{dx^2} + f - f^3 = 0 \quad . \tag{3.16}$$

Defining the characteristic length

$$\xi^2(T) = \frac{\hbar^2}{2m^*|\alpha|} \sim \left(1 - \frac{T}{T_c}\right)^{-1} \tag{3.17}$$

we obtain

$$\xi^2(T)\frac{d^2f}{dx^2} + f - f^3 = 0 \quad . \tag{3.18}$$

The length $\xi(T)$ is called the *temperature-dependent coherence length* and represents the natural length scale for spatial variations of Ψ. The quantity $\xi(T)$ is related to the PIPPARD coherence length ξ_0 introduced in (2.19,20). As shown in the BCS theory, we have near T_c in the pure and dirty limit, respectively,

$$\xi(T) = 0.74\, \xi_0/(1 - t)^{\frac{1}{2}} \qquad \text{(pure)} \tag{3.19}$$

and

$$\xi(T) = 0.855\, (\xi_0 \ell)^{\frac{1}{2}}/(1 - t)^{\frac{1}{2}} \qquad \text{(dirty)} \quad . \tag{3.20}$$

Here t is the reduced temperature, $t = T/T_c$. From (3.19,20) we see that near T_c the spatial changes of Ψ are slow compared to the length ξ_0.

A second characteristic length is associated with electromagnetic effects. For weak magnetic fields, to first order in $\underline{h}$ in the GL equation (3.13b) for the current, $|\Psi|^2$ can be replaced by its equilibrium value $|\Psi_0|^2$ from (3.2) valid in the absence of a field

$$\underline{J}_s = \frac{e^*\hbar}{2m^*i}(\Psi^*\underline{\nabla}\Psi - \Psi\underline{\nabla}\Psi^*) - \frac{e^{*2}}{m^*c}\Psi_0^2 \underline{A} \quad . \tag{3.21}$$

Taking the curl, we obtain the London-type equation

$$\text{curl } \underline{J}_s = -\frac{e^{*2}}{m^*c}\Psi_0^2 \cdot \underline{h} \quad , \tag{3.22}$$

yielding the *penetration depth*

$$\lambda(T) = \left[\frac{m^*c^2}{4\pi e^{*2}\Psi_0^2}\right]^{\frac{1}{2}} = \left[\frac{mc^2}{4\pi e^2 \cdot 2\Psi_0^2}\right]^{\frac{1}{2}} \quad . \tag{3.23}$$

With the identification $n_s = 2\Psi_0^2$ this agrees with the London value of (2.7). We see that $\lambda(T) \sim |\Psi_0|^{-1}$ and therefore proportional to $(1-t)^{-\frac{1}{2}}$. In the pure and dirty limit, near T_c the BCS theory yields

$$\lambda(T) = \frac{\lambda_L(0)}{\sqrt{2}(1-t)^{\frac{1}{2}}} \quad \text{(pure)} \quad , \tag{3.24}$$

and

$$\lambda(T) = \frac{\lambda_L(0)}{\sqrt{2}(1-t)^{\frac{1}{2}}}\left(\frac{\xi_0}{1.33\ell}\right)^{\frac{1}{2}} \quad \text{(dirty)} \quad , \tag{3.25}$$

respectively. Here $\lambda_L(0)$ is the London expression (2.7) at zero temperature.

We see that the two characteristic lengths $\xi(T)$ and $\lambda(T)$ show the same proportionality to $(1-t)^{-\frac{1}{2}}$ near the transition temperature. Therefore, it is useful to introduce the *Ginzburg-Landau parameter*

$$\kappa = \frac{\lambda(T)}{\xi(T)} \quad . \tag{3.26}$$

From (3.19,20,24,25) we have in the pure and dirty limit, respectively,

$$\kappa = 0.96 \frac{\lambda_L(0)}{\xi_0} \quad , \quad \text{(pure)} \tag{3.27}$$

and

$$\kappa = 0.715 \frac{\lambda_L(0)}{\ell} \quad \text{(dirty)} \quad . \tag{3.28}$$

Now we return to the wall energy and the wall-energy parameter δ, associated with a N-S interface, which has been discussed qualitatively in Sect.2.2. We recall the situation shown in Fig.2.4 where the order parameter (or the quantity n_s) falls to zero within the length ξ as one approaches the N-S interface. In the following we assume $\kappa \ll 1$, such that the magnetic field penetration into the superconducting region is negligible compared to the distance ξ. We choose our coordinate system such, that the N-S interface is placed at $x = 0$, and that the half-spaces $x > 0$ and $x < 0$ contain the superconducting and normal phase, respectively. Writing again $\Psi = f \cdot \Psi_0$, we have the boundary conditions

$$f = 0 \quad \text{for } x = 0, \quad \text{and} \tag{3.29a}$$

$$f = 1 \quad \text{for } x \to \infty . \tag{3.29b}$$

Multiplying (3.18) by df/dx and integrating we find

$$-\xi^2(T) \left(\frac{df}{dx}\right)^2 - f^2 + \frac{1}{2} f^4 = \text{const.} \tag{3.30}$$

Because of the condition (3.29b) the constant takes the value $-1/2$, and we have

$$\xi^2(T) \left(\frac{df}{dx}\right)^2 = \frac{1}{2} (1 - f^2)^2 . \tag{3.31}$$

Observing the condition (3.29a), this yields the solution

$$f = \tanh \frac{x}{\sqrt{2} \cdot \xi(T)} . \tag{3.32}$$

Using the expansion (3.8) for the case $\underline{h} = 0$, the condensation energy per unit area of interface is

$$\alpha_{\text{cond}} = \int_0^\infty dx \left[\alpha |\Psi|^2 + \frac{\beta}{2} |\Psi|^4 + \frac{\hbar^2}{2m^*} |\nabla \Psi|^2 \right] . \tag{3.33}$$

Writing the wall energy in the form (2.31), we have

$$\frac{H_c^2}{8\pi} \delta = \int_0^\infty \frac{H_c^2}{8\pi} dx - \left| \alpha_{\text{cond}} \right| . \tag{3.34}$$

Here the first term on the right represents the condensation energy in the case, when the half-space $x > 0$ remains fully superconducting up to the point $x = 0$. The difference in (3.34), therefore, yields the loss in condensation energy arising from the fact that Ψ gradually approaches zero for $x \to 0$, and thus can be identified with the wall energy. From (3.33,34) together with (3.17) and using the relation

$$f_n - f_s = \frac{H_c^2}{8\pi} = \frac{\alpha^2}{2\beta} \qquad (3.35)$$

[which follows from (3.1,2)], one obtains

$$\delta = \int_0^\infty dx \left[2\varepsilon^2(T)\left(\frac{df}{dx}\right)^2 + (1 - f^2)^2 \right] \qquad (3.36a)$$

By integration, following insertion of (3.31,32), we finally obtain

$$\delta = \frac{4}{3}\sqrt{2}\, \xi(T) = 1.89\, \xi(T) \qquad (3.36b)$$

We see that, neglecting the penetration depth $\lambda(T)$, the wall-energy parameter takes the form we had qualitatively expected in (2.32).

Generally, in the evaluation of δ the penetration depth must be taken into account by subtracting a term from the result (3.36b). Equation (3.34) must then be replaced by a more general expression based on the full expansion in (3.8) and where both $\Psi(x)$ and $\underline{h}(x)$ are spatially varying. Here, the case where both contributions to the wall-energy parameter exactly cancel each other, such that $\delta = 0$, is rather important. As shown in the original Ginzburg-Landau paper by numerical integration, cancellation of the positive and negative contribution to the wall energy occurs for $\kappa = 1/\sqrt{2}$. The ranges of the GL parameter $\kappa < 1/\sqrt{2}$ and $\kappa > 1/\sqrt{2}$ then characterize type-I and type-II superconductors, respectively.

3.3 Critical Current in a Thin Film or Wire

We now apply the GL equations for calculating the critical current in a thin superconducting sample (thin film or wire) at which superconductivity breaks down. For the sample dimension d perpendicular to the current direction we assume $d \ll \xi(T)$ and $d \ll \lambda(T)$, such that $|\Psi|$ and $\underline{J}_s$ are approximately constant over the sample cross section. We envision a situation in which no external magnetic field is applied. We can set $\Psi = |\Psi| e^{i\varphi(\underline{r})}$, where $|\Psi|$ is independent of $\underline{r}$. The second GL equation (3.13b) for the current then yields

$$\underline{J}_s = \frac{e^*}{m^*} |\Psi|^2 \left(\hbar\underline{\nabla}\varphi - \frac{e^*}{c} \underline{A}\right) = 2e|\Psi|^2 \underline{v}_s \quad , \tag{3.37}$$

and the free-energy density in (3.8) takes the form

$$f = f_n + |\Psi|^2 \left(\alpha + \frac{\beta}{2} |\Psi|^2 + \frac{1}{2} m^* v_s^2\right) + \frac{h^2}{8\pi} \quad . \tag{3.38}$$

We note that the total energy due to the magnetic field term $h^2/8\pi$ is smaller than the kinetic energy of the current by a factor approximately given by the ratio of the cross-sectional area of the conductor to λ^2, and, therefore, can be neglected in our case. For finding the optimum value of $|\Psi|$, we minimize f with respect to $|\Psi|$ and obtain

$$\alpha + \beta|\Psi|^2 + \frac{1}{2} m^* v_s^2 = 0 \quad . \tag{3.39}$$

Setting again $\Psi = f \cdot \Psi_0$ with $|\Psi_0|^2 = -\alpha/\beta$, we have

$$v_s^2 = \frac{2}{m^*} \alpha(f^2 - 1) \tag{3.40}$$

and

$$J_s = 2e|\Psi|^2 v_s = 2e|\Psi_0|^2 \left(\frac{2|\alpha|}{m^*}\right)^{\frac{1}{2}} f^2 (1 - f^2)^{\frac{1}{2}} \quad . \tag{3.41a}$$

In Fig.3.2 we show a plot of J_s versus f^2. For $J_s = 0$ we have $f = 1$, and f decreases with increasing current density. The quantity J_s reaches a maximum for $f^2 = 2/3$ corresponding to the *critical current density*

$$J_c = \frac{4}{3\sqrt{3}} e|\Psi_0|^2 \frac{\hbar}{m^*\xi} \quad . \tag{3.41b}$$

In (3.41a) we have replaced $|\alpha|$ by ξ using (3.17). For current densities larger than J_c there do not exist solutions with $f \neq 0$. As a consequence, for $J > J_c$ the superconductor becomes normal, and f changes abruptly from $(2/3)^{\frac{1}{2}}$ to zero.

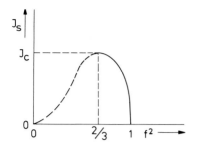

Fig.3.2. Supercurrent density J_s versus the normalized order parameter $f = \Psi/\Psi_0$ from (3.41)

Noting $|\Psi_0|^2 = |\alpha|/\beta \sim \xi^{-2}$ and the $\xi(T)$ relation indicated in (3.17) we find the characteristic temperature dependence of the GL critical current (valid close to the transition temperature)

$$J_c \sim \xi^{-3} \sim (1 - t)^{3/2} \quad . \tag{3.42}$$

Equation (3.41b) can be transformed into an expression containing more readily available quantities. Using (3.2,17,23,35) one finds

$$J_c = \frac{c}{3\sqrt{6}\pi} \frac{H_c(T)}{\lambda(T)} \quad . \tag{3.43}$$

A result similar to (3.43) can directly be obtained from an estimate of the shielding supercurrent densities near the surface of a specimen when the Meissner effect is established. In an applied magnetic field H we have from the London equation (2.14)

$$\mathrm{curl}\, J_s \approx -\frac{J_s}{\lambda} \approx -\frac{c}{4\pi\lambda^2} H \tag{3.44}$$

and

$$J_s \approx \frac{c}{4\pi} \frac{H}{\lambda} \quad . \tag{3.45}$$

In a type-I superconductor the maximum field for which (3.45) applies is H_c, yielding a maximum density of the shielding supercurrents close to the value of (3.43). For a typical value of H_c = 1000 G and λ = 100 Å, (3.43) yields a critical current density of about 10^9 A/cm^2.

It is interesting to compare the result of (3.41b,43) with the microscopic theory. As expected, near T_c the microscopic calculations agree with the results from the GL theory. A review of these calculations has been given by BARDEEN [3.4].

Experimentally, the $(1 - t)^{3/2}$ behavior of the critical current density near T_c has often been observed. Recent experiments also indicated reasonable agreement between the absolute magnitude of J_c and the expressions in (3.41b,43). In Sect.14 we shall return to the subject of the critical current for further discussion.

3.4 Quantization of the Fluxoid

In the last section we have seen how the flow of supercurrents causes a reduction of the order parameter, thereby establishing a critical current density as the upper limit at which superconductivity can still be maintained. Further, we have shown that the magnitude of the Meissner shielding currents can be understood from the same considerations. Here a multiply connected superconductor in the presence of a magnetic field represents an important configuration. In the simplest case we are dealing with a hole or normal inclusion extending throughout the superconductor. In analyzing this situation F. LONDON [3.5] introduced the concept of the *fluxoid* φ' associated with such a nonsuperconducting inclusion. According to LONDON the fluxoid is defined as

$$\varphi' = \varphi + \frac{4\pi}{c} \oint \lambda^2 \underline{J}_s \cdot d\underline{s} = \varphi + \left(\frac{m^*c}{e^*}\right) \oint \underline{v}_s \cdot d\underline{s} \quad . \tag{3.46}$$

Here

$$\varphi = \int \underline{h} \cdot d\underline{S} = \oint \underline{A} \cdot d\underline{s} \tag{3.47}$$

is the magnetic flux contained within the closed path of integration. From the London equation (2.14) we see that $\varphi' = 0$ for any path of integration enclosing only superconducting material. Consequently, the fluxoid is independent of the geometrical detail of the path surrounding the nonsuperconducting inclusion. On the other hand, for a time variation of $\underline{h}$ we have from the London equation (2.6) and Maxwell's equation (2.8)

$$\int \underline{\dot{h}} \cdot d\underline{S} + \frac{4\pi}{c} \oint \lambda^2 \underline{\dot{J}}_s \cdot d\underline{s} = 0 \quad , \tag{3.48}$$

i.e., a compensation of the changes $\dot{h}$ and $\dot{J}_s$ such that the fluxoid is held constant. We note that the fluxoid is characterized by attaining a distinct constant value for all contours surrounding the nonsuperconducting inclusion. Using (3.12) we can write

$$\varphi' = \frac{c}{e^*} \oint \left[m^* \underline{v}_s + \frac{e^*}{c} \underline{A} \right] \cdot d\underline{s} = \frac{c}{e^*} \oint \underline{p}_s \cdot d\underline{s} \quad . \tag{3.49}$$

Applying to the last integral the Bohr-Sommerfeld quantum condition we obtain

$$\varphi' = \frac{c}{e^*} n h \quad , \tag{3.50}$$

where n is an integer and h is Planck's constant. Setting $e^* = 2e$ we find the *fluxoid quantum*

$$\varphi_0 = \frac{hc}{2e} = 2.07 \times 10^{-7} \text{ G cm}^2 \quad . \tag{3.51}$$

The quantization of the fluxoid can also be deduced from the requirement that the complex pair wave function $\Psi(\underline{r})$ must be single-valued at any point in the superconductor. Therefore, the phase φ introduced in (3.9) must change by integral multiples of 2π following a complete turn along the path of integration, yielding

$$\oint \underline{\nabla}\varphi \cdot d\underline{s} = n \, 2\pi \quad . \tag{3.52}$$

Observing relation (3.12), this is identical with the result in (3.50). The quantization of the fluxoid clearly identifies a superconductor as a macroscopic quantum system, an aspect emphasized first by F. LONDON.

Applying our results to a superconducting ring or hollow cylinder placed in a magnetic field, we note that the second term in (3.46) can be neglected if the parts of the superconducting object are sufficiently thick such that $\underline{v}_s \to 0$. In this case the fluxoid and the total flux trapped in the superconductor are identical. This situation has been realized in the experiments which first demonstrated the validity of the fluxoid quantization [3.6,7]. On the other hand, if the superconducting walls are thin, the velocity $\underline{v}_s$ becomes appreciable, and the second term in (3.46) plays an important role. The periodic variation of the velocity $\underline{v}_s$ in a thin-walled superconducting cylinder with the magnetic field applied parallel to the cylinder axis, as required by the quantum condition (3.50), has been demonstrated in an ingeneous experiment by LITTLE and PARKS [3.8,9].

3.5 Nucleation of Superconductivity in Bulk Samples

In Sect.3.3 we discussed an example where $|\Psi|$ was approximately constant throughout the superconductor. In the following we turn to situations where the spatial variation of $|\Psi|$ becomes important.

First we consider the nucleation of superconductivity in the interior of a bulk specimen as an external magnetic field is gradually reduced. At what field value does spontaneous nucleation of superconducting regions start to occur? During the first appearance of superconductivity the quantity $|\Psi|$ will be small. Therefore, the GL equation can be *linearized* neglecting the term $\beta|\Psi|^2\Psi$ in (3.13a), and we have

$$\frac{1}{2m^*}\left(\frac{\hbar}{i}\underline{\nabla} - \frac{e^*}{c}\underline{A}\right)^2 \Psi = -\alpha\Psi \quad . \tag{3.53}$$

We note that in (3.53) the vector potential $\underline{A}$ is essentially given by the external magnetic field, since shielding supercurrents are proportional to $|\Psi|^2$ and can be neglected in the linearized approximation. The last equation resembles the Schrödinger equation of a free particle of mass m^* and charge e^* in a uniform magnetic field $\underline{h} = \text{curl } \underline{A}$, with the energy eigenvalue given by $-\alpha$.

In order to fix the geometry, we assume an infinite specimen with the magnetic field oriented in z-direction. Accordingly, the particle has a constant velocity component v_z and performs a circular motion in the x-y plane at the cyclotron frequency

$$\omega_c = \frac{e^*H}{m^*c} \quad . \tag{3.54}$$

The energy eigenvalues E_n of our problem correspond to the quantized states of the Landau levels and have the form

$$E_n = \frac{1}{2}m^*v_z^2 + (n + \frac{1}{2})\hbar\omega_c \quad , \tag{3.55}$$

where n is a positive integer. Inserting (3.54) and introducing the eigenvalue $-\alpha$, we have

$$-\alpha = |\alpha| = \frac{1}{2} m^* v_z^2 + (n + \frac{1}{2}) \frac{e^* \hbar H}{m^* c} \quad . \tag{3.56}$$

Obviously, the *highest field* satisfying this equation is attained for $v_z = 0$ and $n = 0$. We define this field as H_{c2} and obtain

$$|\alpha| = \frac{e\hbar}{m^*c} H_{c2} \quad . \tag{3.57}$$

On the other hand, using (3.2,17,23,26,35), $|\alpha|$ can be written as

$$|\alpha| = \sqrt{2} H_c \frac{e\hbar}{m^*c} \cdot \kappa \quad , \tag{3.58}$$

and we find from the last two equations the important relation

$$H_{c2} = \sqrt{2} \kappa H_c \quad . \tag{3.59}$$

Equation (3.59) again distinguishes the value $\kappa = 1/\sqrt{2}$ as the point which separates two kinds of superconductors. In type-II materials ($\kappa > 1/\sqrt{2}$) we have $H_{c2} > H_c$, and in decreasing field the *Shubnikov phase* will develop below H_{c2}. The onset of superconductivity below H_{c2} takes place in a second-order phase transition, with $|\Psi|$ increasing continuously from zero. The quantity H_{c2} is referred to as the upper critical field in type-II superconductors. This situation will be discussed in Sect.4. In type-I materials ($\kappa < 1/\sqrt{2}$) we have for the nucleation field $H_{c2} < H_c$. As a consequence, *supercooling* of the normal state below H_c is possible, in principle until the field value H_{c2} is reached. This theoretical limit of supercooling in type-I superconductors is difficult to observe since nucleation of the superconducting phase at sample inhomogeneities usually sets in above H_{c2}. However, with sufficient care it has been possible to observe suercooling down to the theoretical limit [3.10,11].

Equation (3.59) can be transformed into a different expression. Using (3.2,17,23,26,35) we find

$$H_{c2} = \frac{\varphi_0}{2\pi \xi^2} \quad . \tag{3.60}$$

3.6 Nucleation of Superconductivity at Surfaces

The results obtained in Sect.3.5 refer to an infinite superconductor, where the influence of the specimen surfaces can be neglected. A sample surface represents a severe inhomogeneity in the material, and during reduction of an external magnetic field we expect nucleation of superconductivity at the surface to set in at a higher field than in the interior of a bulk sample. Since this inhomogeneity arising from the surface of a specimen is rather well defined, a quantitative treatment of this nucleation problem is possible [3.12].

We assume that the superconductor occupies the half-space $x > 0$ and that its surface coincides with the y-z plane at $x = 0$. The surface is assumed to be electrically insulating. An external magnetic field is applied *parallel to the surface* along the z-direction. We start again with the linearized GL equation (3.53). Setting $A_x = A_z = 0$ and $A_y = Hx$ and noting the independence of z, we have

$$-\frac{\hbar^2}{2m^*}\frac{\partial^2 \Psi}{\partial x^2} + \frac{1}{2m^*}\left[\frac{\hbar}{i}\nabla_y - \frac{e^*}{c}Hx\right]^2 \Psi = -\alpha\Psi \quad . \tag{3.61}$$

We are looking for solutions of the form

$$\Psi = e^{iky}f(x) \quad . \tag{3.62}$$

The function $f(x)$ then satisfies the equation

$$-\frac{\hbar^2}{2m^*}\frac{d^2 f}{dx^2} + \frac{1}{2m^*}\left[\hbar k - \frac{e^*Hx}{c}\right]^2 f = -\alpha f \quad . \tag{3.63}$$

At the insulating surface we have the boundary condition (3.14), which becomes in our case

$$\left.\frac{d\Psi}{dx}\right|_{x=0} = \left.\frac{df}{dx}\right|_{x=0} = 0 \quad . \tag{3.64}$$

Equation (3.63) is analogous to the Schrödinger equation of a harmonic oscillator of the frequency $\omega = e^*H/m^*c$ with the equilibrium position given by

$$x_0 = \frac{\hbar kc}{e^* H} \quad . \tag{3.65}$$

The boundary condition (3.64) becomes important for an evaluation of the possible eigenvalues of (3.63). If the position $x_0 \gg \xi(T)$, the wave function Ψ will be localized near x_0, and its magnitude will be close to zero at the surface. The boundary condition (3.64) is then automatically satisfied. The solution of (3.63) has the form

$$f = \exp\left[-\frac{1}{2}\left(\frac{x-x_0}{\xi(T)}\right)^2\right] \quad , \tag{3.66}$$

yielding the same eigenvalue as in Sect.3.5 with the nucleation field equal to H_{c2}. The situation is the same for $x_0 = 0$, again resulting in the value H_{c2} for the nucleation field. These two cases are schematically shown in Fig.3.3a, where the potential $V(x)$ [by rearranging the terms in (3.63)] is

$$V(x) = \frac{e^{*2} H^2}{2m^* c^2}(x - x_0)^2 \quad . \tag{3.67}$$

However, a lower eigenvalue and correspondingly a higher nucleation field is obtained, when the position x_0 is placed approximately a distance ξ away from the surface, as we can see in the following way. We extend the potential $V(x)$ beyond the point $x = 0$ by its mirror image, as shown in Fig. 3.3b. The lowest eigenfunction of this potential is an even function of x, thus satisfying the condition (3.64). Obviously, the lowest eigenvalue of this potential is lower than for $V(x)$ of (3.67). The exact calculation yields for the lowest eigenvalue

$$|\alpha| = 0.59 \frac{e\hbar}{m^* c} H \tag{3.68}$$

which must be compared with the result in (3.57). Hence, for the nucleation field in the presence of a surface we find

$$H_{c3} = \frac{1}{0.59} H_{c2} = 1.695 \, H_{c2} \quad , \tag{3.69}$$

i.e., a value somewhat larger than H_{c2} as we had expected. We emphasize, that orientation of the field parallel to the surface has been essential for obtaining this result.

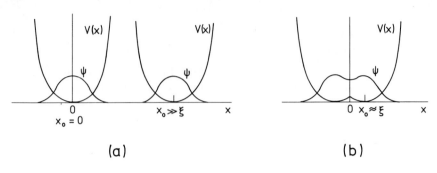

Fig.3.3. (a) Surface and interior nucleation of superconductivity at H_{c2}. (b) Surface nucleation at H_{c3}

Experimentally, the phenomenon of *surface superconductivity* in a magnetic field in the range $H_{c2} < H < H_{c3}$ oriented parallel to the sample surface has often been observed. We note that the nucleation field H_{c3} will limit the range of supercooling in type-I superconductors, since the superconducting phase in the surface sheath will spread into the sample interior. However, surface superconductivity can easily be suppressed by plating the surfaces with a normal metal. For a plated sample the boundary condition (3.64) is replaced by the less favorable condition

$$\frac{d\Psi}{dx} = \frac{\Psi}{b} \, , \tag{3.70}$$

where b is a parameter depending, among other things, upon the superconducting and normal material. The suppression of surface superconductivity arises, of course, from the pair-breaking process in the normal metal. If this pair-breaking is sufficiently strong, instead of (3.70) we have the more severe condition at the boundary

$$\frac{d\Psi}{dx} = \Psi = 0 \, . \tag{3.71}$$

3.7 Abrikosov Vortex State Near H_{c2}

In the following we discuss the famous Abrikosov solution of the GL equations near the upper critical field H_{c2}, predicting the vortex state. This theoretical result by ABRIKOSOV [1.3] essentially introduced a new phenomenon which is being referred to since as type-II superconductivity.

We consider a superconducting cylinder of type II placed in an external magnetic field H oriented along the z-direction and parallel to the axis of the cylinder. The diameter of the cylinder is small enough, such that demagnetization can be neglected. During reduction of the magnetic field, nucleation of the superconducting phase in the interior of the sample begins when H becomes equal to H_{c2}. When nucleation of the superconducting phase sets in, the order parameter $|\Psi|$ will be small, and the linearized GL equation (3.53) can be used for obtaining the quantity Ψ. If the field is decreased appreciably below H_{c2}, the order parameter $|\Psi|$ becomes larger, and the complete, nonlinear GL equations must be used.

In his theoretical analysis ABRIKOSOV treated the regime with H only slightly less than H_{c2}, where the solution Ψ of the complete GL equations must have strong similarity to a certain solution Ψ_L of the linearized equations. The solution Ψ_L satisfies the equation

$$\frac{1}{2m^*}\left(\frac{\hbar}{i}\underline{\nabla} - \frac{e^*}{c}\underline{A}_o\right)^2 \Psi_L = -\alpha\Psi_L \quad , \tag{3.72}$$

with

$$\text{curl } \underline{A}_o = (0, 0, H_{c2}) \quad . \tag{3.73}$$

Equation (3.72) has many degenerate eigenvalues corresponding to solutions describing nucleation in different parts of the sample. In Sect.3.6 we have seen that for $A_x = A_z = 0$ and $A_y = H_{c2}x$ the solutions have the form

$$\Psi_k = e^{iky} \exp\left[-\frac{1}{2}\left(\frac{x-x_o}{\xi(T)}\right)^2\right] \tag{3.74}$$

with

$$x_o = \frac{\hbar k c}{e^* H_{c2}}, \tag{3.75}$$

where k is an arbitrary parameter. The solutions describe a gaussian band of superconductivity of width $\xi(T)$ extending perpendicular to the x-axis at the location $x = x_o(k)$. A general solution Ψ_L must be a linear combination of the Ψ_k. We are interested in a solution which is periodic both in x- and y-direction. Periodicity in y-direction is achieved by setting

$$k = k_n = nq \tag{3.76}$$

yielding the period

$$\Delta y = \frac{2\pi}{q} \quad . \tag{3.77}$$

The general solution will have the form

$$\Psi_L = \sum_n C_n\, e^{inqy} \exp\left[-\frac{1}{2}\left(\frac{x-x_n}{\xi(T)}\right)^2\right] \tag{3.78}$$

with

$$x_n = \frac{n\hbar qc}{e^* H_{c2}} \quad . \tag{3.79}$$

The function Ψ_L of (3.78) is periodic in y-direction. Periodicity in x-direction can also be established, if the coefficients C_n are periodic functions of n, such that $C_{n+\nu} = C_n$, where ν is some integer. The particular choice of ν determines the type of periodic lattice structure ($\nu = 1$: square lattice; $\nu = 2$: triangular lattice). From (3.77,79) we note that the periodicity in x-direction is

$$\Delta x = \frac{\hbar c}{e^* H_{c2}} \cdot \frac{2\pi}{\Delta y} \quad , \tag{3.80}$$

yielding

$$\Delta x \cdot \Delta y \cdot H_{c2} = \varphi_0 \quad , \tag{3.81}$$

i.e., each unit cell of the periodic array contains one flux quantum.

From the form of Ψ_L in (3.78) we can draw some general conclusions independent of the choices for C_n and q. Inserting Ψ_L from (3.78) into (3.13b) we obtain for the current

$$J_{Lx} = -\frac{e\hbar}{2m}\frac{\partial}{\partial y}|\Psi_L|^2 \tag{3.82}$$

$$J_{Ly} = \frac{e\hbar}{2m}\frac{\partial}{\partial x}|\Psi_L|^2 \quad . \tag{3.83}$$

From the last two equations, with curl $\underline{h}_s = (4\pi/c)\underline{J}_L$, we find the local flux density $\underline{h}_s$ associated with the supercurrents

$$\underline{h}_s = -2\pi \frac{e\hbar}{mc} |\Psi_L|^2 . \tag{3.84}$$

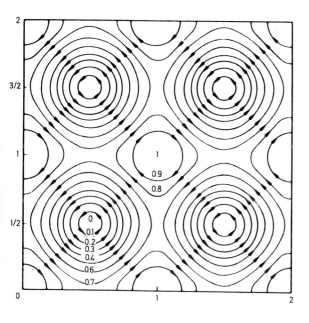

Fig.3.4. Contour diagram of $|\Psi_L|^2$ for the square vortex lattice. The axes are marked in units of $(2\pi)^{\frac{1}{2}}\xi(T)$ [1.3]

Fig.3.5. Contour diagram of $|\Psi_L|^2$ for the triangular vortex lattice. The vertical distance between vortex cores is $2\pi^{\frac{1}{2}}\xi(T)/3^{\frac{1}{4}}$ [3.13]

Equations (3.82-84) indicate that the lines of constant $|\Psi_L|^2$ coincide with the lines of constant local field h_s and with the streamlines of the current $\underline{J}_L$. Contour diagrams of the quantity $|\Psi_L|^2$ for the square and the triangular vortex lattice are shown in Figs.3.4 and 3.5, respectively.

Because of the nonlinearity of the complete GL equations the *normalization of the function* Ψ_L is rather important. The normalization will determine the strength of the supercurrent and therefore the flux density $\underline{B}$ and the free-energy density f. We assume that the free energy F remains stationary, if Ψ_L is replaced by the function $(1 + \varepsilon)\Psi_L$, where ε is a small quantity independent of the spatial coordinate $\underline{r}$. To first order in ε, the variation in the free energy is

$$\delta F = 2\varepsilon \int d\underline{r} \left[\alpha |\Psi_L|^2 + \beta |\Psi_L|^4 + \frac{1}{2m^*} \left| (\frac{\hbar}{i} \underline{\nabla} - \frac{e^*}{c} \underline{A}) \Psi_L \right|^2 \right] . \qquad (3.85)$$

Writing integrals such as $\int d\underline{r} |\Psi_L|^2$ in the form $\Omega \cdot \overline{|\Psi_L|^2}$, where Ω denotes the macroscopic volume, the condition $\delta F = 0$ yields

$$\overline{\alpha |\Psi_L|^2} + \overline{\beta |\Psi_L|^4} + \overline{\frac{1}{2m^*} \left| (\frac{\hbar}{i} \underline{\nabla} - \frac{e^*}{c} \underline{A}) \Psi_L \right|^2} = 0 . \qquad (3.86)$$

We set

$$\underline{A} = \underline{A}_0 + \underline{A}_1 , \qquad (3.87)$$

where $\underline{A}_0$ is the vector potential for $H = H_{c2}$. The correction $\underline{A}_1$ arises from the facts that the applied field is slightly less than H_{c2} and that the supercurrents also contribute to the field. Noting that Ψ_L must satisfy (3.72) and keeping only terms up to first order in $\underline{A}_1$, we obtain

$$\overline{\beta |\Psi_L|^4} - \frac{1}{c} \overline{\underline{A}_1 \cdot \underline{J}_L} = 0 , \qquad (3.88)$$

with

$$\underline{J}_L = \frac{e^*\hbar}{2m^*i} \left[\Psi_L^* \underline{\nabla}\Psi_L - \Psi_L \underline{\nabla}\Psi_L^* \right] - \frac{e^{*2}}{m^*c} \Psi_L^* \Psi_L \underline{A}_0 . \qquad (3.89)$$

We note that $\underline{J}_L$ is the current associated with the unperturbed solution. Integrating the second term in (3.88) in parts and setting curl $\underline{A}_1 = \underline{h}_1$ and curl $\underline{h}_s = (4\pi/c)\underline{J}_L$, one finds

$$\overline{\beta|\Psi_L|^4} - \frac{1}{4\pi}\overline{\underline{h}_1 \cdot \underline{h}_s} = 0 \quad . \tag{3.90}$$

Noting that $\underline{h}_1$ and $\underline{h}_s$ are everywhere parallel to the z-direction, we can write

$$h_1(\underline{r}) = H - H_{c2} + h_s(\underline{r}) \quad . \tag{3.91}$$

Inserting (3.84) and (3.91) into (3.90) we have

$$\overline{\beta|\Psi_L|^4} + \frac{e\hbar}{2mc}\overline{|\Psi_L|^2(H - H_{c2} - \frac{2\pi e\hbar}{mc}|\Psi_L|^2)} = 0 \quad . \tag{3.92}$$

Setting again $|\Psi_L| = \Psi_o f$ and using (3.2,17,23,26,60) we finally obtain

$$\overline{f^4}\left(1 - \frac{1}{2\kappa^2}\right) - \overline{f^2}\left(1 - \frac{H}{H_{c2}}\right) = 0 \quad . \tag{3.93}$$

Equation (3.93) represents a rather general result, which is independent of the detailed behavior of the function Ψ_L, i.e., independent of the type of the periodic lattice configuration of Ψ_L. For a particular lattice type, as determined by selecting the wave number q and the periodicity of the coefficients C_n, one can calculate the quantity

$$\beta_A \equiv \frac{\overline{f^4}}{(\overline{f^2})^2} \quad , \tag{3.94}$$

which is independent of the normalization of Ψ. The ratio β_A takes the value unity if Ψ is spatially constant and becomes increasingly large for functions which are more and more peaked locally. It is only the quantity β_A, which must be determined numerically for obtaining $\overline{f^2}$ and $\overline{f^4}$ from (3.93,94). The flux density $\underline{B}$ and the free-energy density can then be calculated immediately. (Although f denotes both the free-energy density and the normalized wave function Ψ/Ψ_o, confusion can be avoided with only a small amount of attention).

Using (3.84), the magnetic flux density is

$$B = H + h_s = H - \frac{2\pi e\hbar}{mc}|\Psi_L|^2 \quad . \tag{3.95}$$

Inserting (3.2,35,58) this can be written as

$$B = H - \frac{H_c}{\sqrt{2}\kappa} \overline{f^2} \quad . \tag{3.96}$$

Eliminating $\overline{f^2}$ with the help of (3.93,94), we find

$$B = H - \frac{H_{c2} - H}{(2\kappa^2-1)\beta_A} \quad . \tag{3.97}$$

This yields for the magnetization M

$$M = \frac{B-H}{4\pi} = - \frac{H_{c2} - H}{4\pi(2\kappa^2-1)\beta_A} \quad . \tag{3.98}$$

Using the identity for the Gibbs free-energy density g

$$\left(\frac{\partial g}{\partial H}\right)_T = - M \tag{3.99}$$

we can calculate g by integrating down from the normal state at H_{c2}, where $g_s(H_{c2}) = g_n(H_{c2})$. In this way we obtain

$$g_s(H) = g_n(H_{c2}) - \frac{(H_{c2} - H)^2}{8\pi(2\kappa^2-1)\beta_A} \quad . \tag{3.100}$$

Note, that the last result refers to the *average* density taken as the Gibbs free-energy G per volume Ω.

Equation (3.100) indicates that the configuration with the smallest value of β_A is thermodynamically most stable. (Note, that (3.100) applies to the regime $H < H_{c2}$ and $\kappa > 1/\sqrt{2}$). Numerical calculations show that the square lattice and the triangular lattice yield the values $\beta_A = 1.18$ and $\beta_A = 1.16$, respectively [3.13]. Thus the triangular lattice is the most stable form of all possible periodic solutions. This theoretical result is well supported experimentally, as will be discussed in Sect.4.

We see from (3.98) that the slope $|dM/dH|$ becomes very large as one approaches the value $\kappa = 1/\sqrt{2}$ from above and that it diverges at this value. It is such a discontinuous rise in $|M|$, which characterizes type-I superconductors, which is shown in Fig.2.1b.

It is apparent from our discussion, that Abrikosov's theory of the vortex state is based on the phenomenological concepts developed in the Ginzburg-Landau theory. However, as shown by GOR'KOV [1.5,6] for superconductors of arbitrary electron mean free path the Ginzburg-Landau-Abrikosov theory follows from the microscopic theory for the temperature region near T_c, provided that the order parameter is small, and that the spatial variation of the order parameter and of the magnetic field is slow. From GOR'KOV's reformulation of the theory, the order parameter $\Psi(\underline{r})$ can be identified with the gap parameter $\Delta(\underline{r})$ of the microscopic theory. This theoretical frame based on the work of Ginzburg, Landau, Abrikosov, and Gor'kov is often referred to as the *GLAG* theory. Subsequently, the restricted regime of validity of the GLAG theory has been extended by numerous authors in different directions. These developments will be summarized in Sect.4.4.

4. Magnetic Properties of Type-II Superconductors

4.1 Mixed State

As we have discussed in Sect.3.7, Abrikosov first predicted the existence of the vortex state in superconducting materials with $\kappa > 1/\sqrt{2}$. In type-II superconductors magnetic flux penetrates the material in the form of flux lines, each carrying a single flux quantum φ_0. Usually the flux lines are arranged in the form of a triangular flux-line lattice. This distribution of magnetic flux is referred to as the *mixed state*.

Type-II superconductivity was encountered first in experiments by SHUBNIKOV et al. [4.1]. These authors investigated the superconducting properties of alloys and discovered persistence of superconductivity up to unusually high magnetic fields. However, it was only after the appearance of Abrikosov's paper that their results could be understood in terms of type-II superconductivity. Because of Shubnikov's early experiments, often the mixed state is also referred to as the *Shubnikov phase*. In view of (3.27,28) we expect type-II superconductivity to occur preferentially in alloys, or more generally, in impure systems. On the other hand, pure metals usually are expected to display type-I superconductivity.

The possibility in type-II superconductors to form single-quantum vortex lines and to assume the mixed state causes a more gradual transition in an applied magnetic field than that displayed by superconductors of type-I. Figure 4.1 shows the magnetic flux density B (a) and the magnetization M (b) plotted versus the magnetic field H for a type-II superconductor. Up to the *lower critical field*, H_{c1}, magnetic flux is completely expelled and the Meissner effect is established. Flux penetration in form of vortex lines sets in at H_{c1}. In the field range $H_{c1} < H < H_{c2}$ the material resides in the mixed state. The field H_{c2}, at which superconductivity disappears in the bulk, is called the *upper critical field*. If one calculates the *thermodynamic critical field* H_c, defined according to (2.1) from the difference in the free-energy densities of the normal and superconducting phase in zero field, one finds $H_{c1} < H_c$ and $H_c < H_{c2}$. As indicated in Fig.4.1b, the area under the M(H) curve is identical to the area obtained by assuming

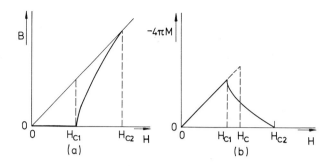

Fig.4.1a and b. Magnetic flux density (a) and magnetization (b) versus magnetic field for a type-II superconductor with a demagnetization coefficient $D = 0$

perfect diamagnetism up to H_c with a discontinuous disappearance of M at H_c. The continuous disappearance of the magnetization at the upper critical field H_{c2} establishes the transition at H_{c2} as a *second-order phase transition*.

Experimentally, Abrikosov's prediction of the vortex state has found ample verification [4.2,3]. Perhaps the most direct proof of the triangular flux-line lattice of the mixed state has been obtained from experiments by ESSMANN and TRÄUBLE [4.4] utilizing a Bitter method in conjunction with electron microscopy. In such an experiment the magnetic flux structure is decorated using small magnetic particles and subsequently observed with some optical technique. Figure 4.2 shows a typical example of a triangular flux-line lattice as studied by ESSMANN and TRÄUBLE. The microscopic vortex structure in the mixed state and the magnetic field distribution h(r) of individual vortex lines has been investigated from neutron diffraction through the interaction of the magnetic moment of the neutron with the magnetic field gradients in the mixed state [4.5-9]. Measurements of the line shape in nuclear magnetic resonance experiments with type-II superconductors have been used to determine the local magnetic field at the maximum, minimum, and saddle point of the distribution h(r) in a vortex structure [4.10-12]. The various experiments and their technical aspects will be discussed in more detail in Sects.6 and 8.

Following the experimental verification of the triangular vortex structure as the dominant feature of the mixed state, in recent years the detailed properties of flux-line lattices have been investigated. The structure of an individual vortex line has been analyzed. Further, anisotropy effects arising from the interaction between the vortex lattice and the

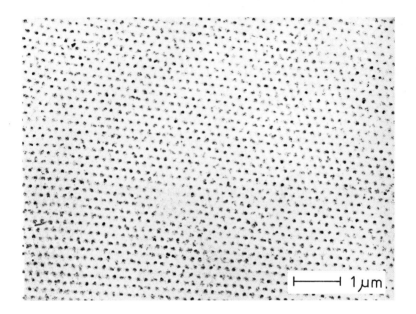

Fig.4.2. Triangular vortex lattice in a Nb disk (thickness = 0.5 mm; diameter = 4 mm) in a perpendicular magnetic field of 985 G at 1.2 K. The experimental observation utilizes a high-resolution Bitter method. (Courtesy of U. Essmann)

crystal lattice, and various kinds of lattice defects in the vortex lattice have been discovered. These developments will be discussed in Sect.4.5.

Theoretically, a rather simple model can be constructed for describing the behavior of vortex lines in the limit $\kappa \gg 1$. This model assumes the vortex to consist of a normal core imbedded in the superconducting phase, where the London equations can be applied (London model). It is valid in good approximation at fields in the range $H_{c1} < H \ll H_{c2}$. An extension of the London model has recently been proposed by CLEM [4.13,14] providing a more realistic description of the vortex core and incorporating a field-dependent penetration depth in order to account for arbitrary κ values and flux-line spacings. Although these models are quite useful in providing a qualitative understanding of many of the properties of type-II superconductors, one must go back to the Ginzburg-Landau theory and its extensions to obtain accurate quantitative results. As we have pointed out in Sect.3.7, a satisfactory theoretical description of the mixed state, although over a restricted regime, has been developed in terms of the GLAG theory. Subsequently, the GLAG theory has been extended in various ways. These theoretical developments will be described in Sect.4.2-4.

4.2 London Model

In the following we discuss the London model of vortex lines in type-II superconductors. This model assumes a normal vortex core of radius ξ imbedded in the superconducting phase. It is valid in the limit $\kappa \gg 1$, i.e, the diameter of the vortex core is assumed to be very small. With this restriction, the London model represents a good approximation at all temperatures, electron mean free paths, and magnetic fields in the range $H_{c1} < H \ll H_{c2}$, where the interaction between vortices is not too strong. The London model is often quite useful for a qualitative understanding of type-II superconductivity also at low κ values.

4.2.1 Isolated Vortex Line

We assume the magnetic field to be only slightly larger than H_{c1}, such that the vortex lines are far apart from each other and their interaction can be neglected. The structure of an isolated vortex line is shown schematically in Fig.4.3. The density n_s of the superconducting electrons (or the order parameter) is suppressed to zero in the center of the vortex and reaches its full value at a radius approximately equal to the coherence length ξ. The local magnetic field h near a vortex line attains its maximum in the center of the vortex and vanishes approximately exponentially beyond a radius of the order of the penetration depth λ. Crudely speaking, a vortex line can be taken as a tube of normal phase with radius ξ imbedded in the superconducting phase. We note that the average magnetic flux density is

$$B = n\varphi_0 , \qquad (4.1.)$$

where n is the vortex density (number of vortices per unit area). The density J_s of the circular supercurrents generating the field h(r) of the vortex line reaches its maximum near the radius λ and vanishes in the normal vortex core.

For $\kappa \gg 1$ the local magnetic field $\underline{h}(r)$ outside the normal vortex core can be described by the London equation

$$\underline{h}(r) + \lambda^2 \,\text{curl curl}\, \underline{h}(r) = 0 , \quad r > \xi , \qquad (4.2)$$

where λ is the penetration depth. However, to account for flux quantization a two-dimensional delta function $\delta_2(r)$ must be added to the right-hand side of (4.2),

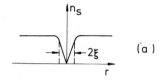
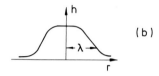
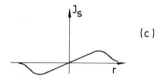

Fig.4.3a-c. Structure of an isolated vortex line. Density of the superconducting electrons (a), magnetic flux density (b), and supercurrent density (c) versus the distance from the vortex axis

$$\underline{h}(r) + \lambda^2 \, \text{curl curl} \, \underline{h}(r) = \underline{\varphi}_0 \, \delta_2(r) \, , \tag{4.3}$$

where the vector $\underline{\varphi}_0$ is parallel to the direction of the vortex line. The term on the right hand can be understood, if (4.3) is integrated over an area S with radius $r \gg \lambda$ surrounding the vortex line

$$\int \underline{h} \, dS + \lambda^2 \oint \text{curl} \, \underline{h} \, ds = \underline{\varphi}_0 \, . \tag{4.4}$$

Here the line integral is taken along the circumference of the area S. Because curl $\underline{h} = 0$ for $r \gg \lambda$ we have

$$\int \underline{h} \, dS = \underline{\varphi}_0 \, . \tag{4.5}$$

Because of Maxwell's equation div $\underline{h} = 0$ we obtain from (4.3)

$$\lambda^2 \nabla^2 \underline{h}(r) - \underline{h}(r) = - \underline{\varphi}_0 \delta_2(r) \, . \tag{4.6}$$

Equation (4.6) has the exact solution

$$\underline{h}(r) = \frac{\underline{\varphi}_0}{2\pi\lambda^2} K_0\left(\frac{r}{\lambda}\right) \, , \tag{4.7}$$

which yields further

$$J_s = \frac{\varphi_0 c}{8\pi^2 \lambda^3} K_1(\frac{r}{\lambda}) \quad . \tag{4.8}$$

Here K_0 and K_1 are the Hankel functions of imaginary argument of order zero and of order one, respectively [4.15]. With the asymptotic approximations for K_0 we have

$$h(r) \approx \frac{\varphi_0}{2\pi\lambda^2} \ln(\frac{\lambda}{r}) \quad , \quad \xi < r \ll \lambda \tag{4.9}$$

and

$$h(r) \approx \frac{\varphi_0}{2\pi\lambda^2} \left(\frac{\pi\lambda}{2r}\right)^{\frac{1}{2}} \exp(-r/\lambda) \quad , \quad r \gg \lambda \quad . \tag{4.10}$$

Next we can calculate the energy E_1 per unit length of vortex line (or the line tension). Neglecting the contribution from the small normal core, we only need to consider the magnetic field energy and the kinetic energy of the currents

$$E_1 = \int \left[\frac{h^2}{8\pi} + \frac{1}{2} m v_s^2 n_s\right] dS = \frac{1}{8\pi} \int \left[h^2 + \lambda^2 (\text{curl } h)^2\right] dS \quad . \tag{4.11}$$

In (4.11) the vortex core is excluded from the integrals. Noting the cylindrical symmetry of $h(r)$ and integrating the second term on the right of (4.11) in parts, we obtain

$$E_1 = \frac{1}{8\pi} \int (\underline{h} + \lambda^2 \text{ curl curl } \underline{h}) \cdot \underline{h} \, dS + \frac{\lambda^2}{8\pi} \oint (\underline{h} \times \text{curl } \underline{h}) \cdot d\underline{s} \tag{4.12}$$

and by insertion of (4.3)

$$E_1 = \frac{1}{8\pi} \int h \varphi_0 \delta_2(\underline{r}) dS + \frac{\lambda^2}{8\pi} \oint (\underline{h} \times \text{curl } \underline{h}) \cdot d\underline{s} \quad . \tag{4.13}$$

Here the line integral is taken along the inner and outer boundary of the integration area S. Since the core is excluded from the integration, the first term contributes nothing. The second term vanishes at the outer perimeter and yields from the core boundary

$$E_1 = \frac{\lambda^2}{8\pi} \left[2\pi r h(r) |\text{curl } h(r)|\right]_\xi = \frac{\lambda^2}{8\pi} \left[2\pi r h(r) \frac{dh}{dr}\right]_\xi \quad . \tag{4.14}$$

Inserting (4.9) we find

$$E_1 \approx \left(\frac{\varphi_0}{4\pi\lambda}\right)^2 \ln(\lambda/\xi) \quad . \tag{4.15}$$

From (4.15) we see that the energy of a vortex line depends only logarithmically upon the coherence length ξ. Therefore, no serious error should be introduced from our crude treatment of the vortex core. A second important feature of (4.15) is the quadratic dependence of E_1 upon the flux φ_0 per vortex line. This result clearly establishes a single flux quantum as the amount of flux per vortex line which is energetically most favorable. (A vortex line containing two flux quanta requires the energy $4E_1$ compared to the energy $2E_1$ associated with two vortex lines containing each a single flux quantum).

So far we have neglected the contribution of the normal core to the energy of the vortex line. Since superconductivity is destroyed in the core, this contribution will be approximately $(H_c^2/8\pi)\pi\xi^2$ per unit length. Equation (4.15) can be transformed using relations (3.59,60), thereby yielding

$$E_1 \approx \frac{H_c^2}{8\pi} 4\pi\xi^2 \ln(\lambda/\xi) \quad . \tag{4.16}$$

We see that the contribution of the vortex core is smaller than this value by a factor of about $4 \ln(\lambda/\xi)$.

4.2.2 Lower Critical Field H_{c1}

From the energy E_1 of the vortex line we can calculate immediately the lower critical field H_{c1} in the following way. Starting from the Gibbs free-energy density g

$$g = f - \frac{BH}{4\pi} \quad , \tag{4.17}$$

we can write for very low vortex densities n

$$g = n E_1 - \frac{BH}{4\pi} = B\left(\frac{E_1}{\varphi_0} - \frac{H}{4\pi}\right) \quad . \tag{4.18}$$

For $H < 4\pi E_1/\varphi_0$, g increases with B, and the minimum of g occurs for B = 0, i.e., for complete flux expulsion. On the other hand, for $H > 4\pi E_1/\varphi_0$, g decreases with increasing B resulting in a value B > 0, i.e., some flux penetration. We see that the lower critical field H_{c1}, at which flux penetration sets in, is given by

$$H_{c1} = 4\pi E_1/\varphi_0 \approx \frac{\varphi_0}{4\pi\lambda^2} \ln(\lambda/\xi) \quad . \tag{4.19}$$

Again, using (3.59,60), this can be cast in the form

$$H_{c1} \approx \frac{H_c}{\sqrt{2}\kappa} \ln \kappa \quad , \tag{4.20}$$

which corresponds to relation (3.59) between H_{c2} and H_c.

4.2.3 Interaction Between Vortex Lines

Next we consider the interaction energy between two vortex lines oriented *parallel* to each other. From the quadratic dependence of the energy E_1 in (4.15) upon the flux φ_0 per vortex line we see immediately that the vortex-line interaction must be *repulsive*, since the single-quantum flux line represents the energetically most favorable configuration. (For two *antiparallel* vortex lines the interaction is *attractive*, of course, since the free energy will be reduced when the two vortices coincide and annihilate each other). If the location of the center of both vortex lines is indicated by $\underline{r}_1$ and $\underline{r}_2$, the magnetic field is by superposition

$$\underline{h}(\underline{r}) = \underline{h}_1(|\underline{r} - \underline{r}_1|) + \underline{h}_2(|\underline{r} - \underline{r}_2|) \quad , \tag{4.21}$$

where $\underline{h}_1(|\underline{r} - \underline{r}_1|)$ and $\underline{h}_2(|\underline{r} - \underline{r}_2|)$ are obtained from (4.7). (We note that we are dealing with the case $\kappa \gg 1$, where the medium is linear and superposition can be applied). Again, the total energy E_{1+2} can be written as

$$E_{1+2} = \frac{1}{8\pi} \int [\underline{h}^2 + \lambda^2 (\text{curl } \underline{h})^2] dS \quad , \tag{4.22}$$

where $\underline{h}$ must be substituted from (4.21). Following the same procedure as before, in addition to the energies of both individual vortex lines one obtains the term representing the interaction energy E_{12}

$$E_{12} = \frac{\lambda^2}{8\pi} \oint (\underline{h}_1 \times \text{curl } \underline{h}_2) \cdot d\underline{s}_2 + \frac{\lambda^2}{8\pi} \oint (\underline{h}_2 \times \text{curl } \underline{h}_1) \cdot d\underline{s}_1 \quad . \tag{4.23}$$

Noting the symmetry and that

$$\text{curl } \underline{h}_2 = \frac{\varphi_0}{2\pi\lambda^2} \frac{1}{|r - r_2|} \quad \text{for} \quad |r - r_2| \ll \lambda , \tag{4.24}$$

we find

$$E_{12} = \frac{\varphi_0}{4\pi} h_{12} , \tag{4.25}$$

where h_{12} is given from (4.7)

$$h_{12} = \frac{\varphi_0}{2\pi\lambda^2} K_0(|r_1 - r_2|/\lambda) . \tag{4.26}$$

We see that the interaction energy decreases as

$$(|r_1 - r_2|)^{-\frac{1}{2}} \cdot \exp(-|r_1 - r_2|/\lambda) \quad \text{for large distances} \quad |r_1 - r_2|.$$

The *force* due to the interaction is obtained as follows. Taking the force f_{2x} on line 2 in x-direction, we have

$$f_{2x} = -\frac{\partial E_{12}}{\partial x_2} = -\frac{\varphi_0}{4\pi} \frac{\partial h_{12}}{\partial x_2} . \tag{4.27}$$

Noting Maxwell's equation curl $\underline{h} = 4\pi \underline{J}/c$ we find

$$f_{2x} = \frac{\varphi_0}{c} J_{1y}(\underline{r}_2) . \tag{4.28}$$

In general, the force per unit length on line 2 arising from the supercurrents of line 1 is

$$\underline{f}_2 = \frac{1}{c} \underline{J}_1(\underline{r}_2) \times \underline{\varphi}_0 . \tag{4.29}$$

This represents a special case of the *Lorentz force*

$$\underline{f}_L = \frac{1}{c} \underline{J}_s \times \underline{\varphi}_0 \tag{4.30}$$

per unit length of flux line arising from the supercurrent density $\underline{J}_s$. Here, in addition to the contribution from all other vortices, $\underline{J}_s$ includes an applied transport current. We see that a vortex array can be stationary

4.2.4 Magnetization Near H_{c1}

For fields slightly higher than H_{c1}, the Gibbs free-energy density is from (4.18,19,25,26)

$$g = \frac{B}{4\pi}\left[H_{c1} - H + \frac{1}{2}z\frac{\varphi_0}{2\pi\lambda^2}K_0(a/\lambda)\right] \quad . \tag{4.31}$$

Here the last term represents the interaction energy

$$g_{int} = \frac{n\varphi_0^2}{4\pi 2\pi\lambda^2}K_0(a/\lambda)\frac{1}{2}z \quad , \tag{4.32}$$

where a is the lattice parameter of the flux-line lattice, and z is the number of nearest neighbors of a vortex line. (z = 6 for the triangular lattice). Expression (4.32) is valid in the field regime where $a > \lambda$ such that contributions beyond the nearest-neighbor interaction can be neglected.

For a given magnetic field $H > H_{c1}$ the flux density B will be obtained from the condition

$$\frac{\partial g}{\partial B} = \frac{1}{4\pi}(H_{c1} - H) + \frac{\partial}{\partial B}g_{int} = 0 \quad , \tag{4.33}$$

yielding the function B(H). For distances $a > \lambda$ we can use the approximation

$$g_{int} = \frac{B\varphi_0 z}{16\pi^2\lambda^2}\left(\frac{\pi\lambda}{2a}\right)^{\frac{1}{2}} \cdot e^{-a/\lambda} \quad . \tag{4.34}$$

Here, the lattice parameter a is, of course, a function of B. In calculating the derivative $(\partial/\partial B)g_{int}$, the exponential variation with a/λ will dominate, and (4.33) can be approximated by

$$H_{c1} - H + \frac{z\varphi_0}{8\pi\lambda^2}\left(\frac{\pi a}{2\lambda}\right)^{\frac{1}{2}} \cdot e^{-a/\lambda} = 0 \quad . \tag{4.35}$$

Taking the logarithm we obtain the implicit form

$$\frac{a}{\lambda} = \ln\left[\frac{z\varphi_0}{8\pi\lambda^2(H-H_{c1})}\right] + \frac{1}{2}\ln\left(\frac{\pi a}{2\lambda}\right) \quad . \tag{4.36}$$

In first order this expression yields

$$\frac{a}{\lambda} = \ln\left[\frac{z\varphi_0}{8\pi\lambda^2(H-H_{c1})}\right] + \frac{1}{2}\ln\left\{\frac{\pi}{2}\ln\left[\frac{z\varphi_0}{8\pi\lambda^2(H-H_{c1})}\right]\right\} \quad . \tag{4.37}$$

For the triangular lattice we have $z = 6$ and

$$B = n\varphi_0 = \frac{2}{\sqrt{3}}\frac{\varphi_0}{a^2} \quad . \tag{4.38}$$

Inserting this into (4.37) and keeping only the leading term we find

$$B = \frac{2\varphi_0}{\sqrt{3}\lambda^2}\left\{\ln\left[\frac{3\varphi_0}{4\pi\lambda^2(H-H_{c1})}\right]\right\}^{-2} \quad . \tag{4.39}$$

From (4.39) we see that the slopes $\partial B/\partial H$ and $\partial M/\partial H$ become infinite at H_{c1}. This rapid entry of magnetic flux at H_{c1} is physically plausible in view of the variation of the repulsion between flux lines proportional to $\exp(-a/\lambda)$, thereby vanishing rapidly at large flux-line spacings.

An interesting exception from the properties of the vortex state discussed so far has been observed over a narrow κ range near the critical value $\kappa = 1/\sqrt{2}$. This phenomenon is associated with an attractive vortex interaction. We shall return to this situation in Sect.4.7.

At intermediate flux densities, corresponding to the range $H_{c1} \ll H \ll H_{c2}$, for $\kappa \gg 1$ the London model can still be used when the electromagnetic regions of different vortices overlap, as long as the vortex cores remain well separated. Interaction between many neighbors must now be included, of course. At flux densities near the upper critical field H_{c2}, where the order parameter is small, the functions $B(H)$ and $M(H)$ can be obtained from the Abrikosov theory, as was shown in Sect.3.7 (3.97,98).

4.3 Clem Model

The London model described in the previous section provides us with a useful phenomenological theory of type-II superconductors and simple expressions for the local magnetic flux density and supercurrent density near a vortex line (4.7,8). However, this simplicity is obtained at the expense of the unrealistic behavior that both expressions diverge on the axis of the vortex line. This divergence results from the fact that the depression of the order parameter to zero on the axis is not built into the model. In our calculations using (4.7,8) we have avoided this problem by excluding the vortex core $r < \xi$ in such integrals as contained in (4.11).

An improvement of the original London model with a more realistic description of the behavior in the vicinity of the vortex core has been proposed recently by CLEM [4.13]. In his model, inside the vortex core the magnitude of the order parameter is obtained from a variational trial function. Well outside the vortex core the behavior according to the London model is retained.

In treating first the core of an *individual vortex line*, CLEM assumes a variational model for the magnitude of the normalized order parameter $\Psi(r)$, which is of the form

$$\Psi(r) = f(r)e^{-i\varphi} \quad . \tag{4.40}$$

Here r is the radial coordinate. The function f(r) is written as

$$f(r) = r/R \quad , \tag{4.41}$$

with

$$R = (r^2 + \xi_v^2)^{\frac{1}{2}} \quad , \tag{4.42}$$

where ξ_v is a variational core-radius parameter. From the second GL equation one finds for the magnetic flux density

$$h = \frac{\varphi_0}{2\pi\lambda\xi_v} \frac{K_0(R/\lambda)}{K_1(\xi_v/\lambda)} \tag{4.43}$$

and for the supercurrent density

$$J_s = \frac{c\varphi_0}{8\pi^2\lambda^2\xi_v} \frac{r}{R} \frac{K_1(R/\lambda)}{K_1(\xi_v/\lambda)} \quad . \tag{4.44}$$

Here $K_0(x)$ and $K_1(x)$ are the same modified Bessel functions introduced in (4.7,8), respectively, and λ is the penetration depth.

Figure 4.4 shows plots of the normalized flux density $h(r)/h(0)$, supercurrent density $J_s(r)/J_s(\xi_v)$, flux $\varphi(r)/\varphi_0$ and the order parameter magnitude $f = r/(r^2 + \xi_v^2)^{1/2}$ versus r/ξ_v calculated with this model for $\lambda/\xi_v = 4$. The behavior of these quantities is similar to that found by numerically solving the GL equations [4.16].

For the energy E_1 per unit length of vortex line the Clem model yields

$$E_1 = \frac{\varphi_0}{4\pi} \sqrt{2} H_c \left[\frac{\kappa}{8} (\xi_v/\lambda)^2 + (8\kappa)^{-1} + \frac{\lambda}{2\kappa\xi_v} \frac{K_0(\xi_v/\lambda)}{K_1(\xi_v/\lambda)} \right] \quad . \tag{4.45}$$

Finally, from minimizing this energy a relation determining the variational parameter ξ_v is obtained. In the κ range $1/\sqrt{2} \leq \kappa \leq 10$ CLEM found that this energy minimum is, at most, only a few percent higher than the value obtained from an exact numerical solution of the GL equations.

The Fourier transform of the magnetic flux density is according to this model

$$\bar{h}(q) = \int d^2r\, h(r) \exp(-i\underline{q}\cdot\underline{r}) = \frac{\varphi_0 K_1(Q\xi_v)}{Q\lambda K_1(\xi_v/\lambda)} \quad , \tag{4.46}$$

where

$$Q = (q^2 + \lambda^{-2})^{1/2} \quad . \tag{4.47}$$

Note that $\bar{h}(0) = \varphi_0$. The Fourier transform $\bar{h}(q)$ can be measured by neutron diffraction and is useful in analyzing such experiments.

An extension of the model for an individual vortex line to a *flux-line lattice* is possible by linear superposition of the magnetic field contributions of individual flux lines [4.14]. It can be shown that linear superposition can be used for obtaining the local flux density for arbitrary κ values and flux-line spacings, provided the correct spatially-dependent magnitude of the order parameter is used in the calculation. Overlapping between vortices can be accounted for approximately by introducing a *field-dependent penetration depth* λ_{eff}, which reduces to the weak-field pene-

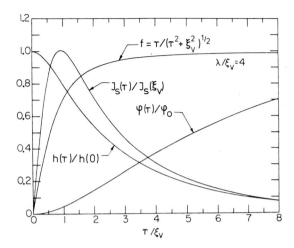

Fig.4.4. Clem model: normalized flux density $h(r)/h(0)$, supercurrent density $J_s(r)/J_s(\xi_v)$, flux $\varphi(r)/\varphi_0$, and the order parameter magnitude $f = r/(r^2+\xi_v^2)^{1/2}$ versus r/ξ_v, where r is the radial coordinate and ξ_v is the variational core radius parameter, calculated from (4.43,44) for $\lambda/\xi_v = 4$ [4.13]

tration depth λ at low average flux density B, increases with increasing B, and diverges at the upper critical field H_{c2}

$$\lambda_{eff} = \frac{\lambda}{(\overline{f^2(B)})^{1/2}} \quad (4.48)$$

Here $\overline{f^2(B)}$ is the spatial average of f^2. This mean-field approximation is incorporated into the variational model by replacing λ with the field-dependent quantity λ_{eff} and using the same procedure as before. Linear superposition of the individual vortex contributions leads to the periodic field distribution

$$h(\underline{r}) = B \sum_q F(q) e^{i\underline{q}\cdot\underline{r}} \quad (4.49)$$

where the sum extends over all reciprocal lattice vectors $\underline{q}$, and

$$F(\underline{q}) = \bar{h}_0(\underline{q})/\varphi_0 \quad (4.50)$$

is the form factor, whose magnitude is measured in neutron diffraction experiments. The quantity $\bar{h}_0(\underline{q})$ is the Fourier transform of the field contribution of a vortex line centered at the origin. The form factor is found to be given by

$$F(\underline{q}) = \frac{K_1(Q\xi_v)}{Q\lambda_{eff}K_1(\xi_v/\lambda_{eff})} \tag{4.51}$$

where

$$Q = (q^2 + \lambda_{eff}^{-2})^{\frac{1}{2}} \quad .$$

The attractive feature of this model consists of the fact that it is *analytic* and requires rather little computational effort and that it rests only on a *single fitting parameter* λ. The usefulness of the model can be seen from Figure 4.5, where the form factor F_{hk} for niobium, calculated from the model, is plotted versus the reciprocal lattice vector q_{hk} (solid curves). The theoretical results are compared with experimental data from neutron diffraction experiments [4.7]. By fitting (4.51) to the experimental data at the point indicated in Fig.4.5, the value for the remaining free parameter $\lambda = 263$ Å has been derived. The Clem model is seen to agree satisfactorily with experiment. The dashed curve in Fig.4.5 refers to the Fourier transform calculated with a field-independent penetration depth λ.

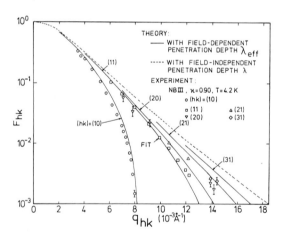

Fig.4.5. Form factor F_{hk} for niobium versus the reciprocal lattice vector q_{hk}. The solid curves are calculated from (4.51) and fitted to the experimental data at the point indicated. The source of the experimental data points is given in the text [4.14]

4.4 Theory of the Static Vortex Structure

In Sects.4.2,3 we have described two rather simple models which are useful for a qualitative or semiquantitative understanding of the properties of vortex structures. Both models are analytic and do not involve lengthy numerical calculations, which makes them particularly attractive. However, for a rigorous theoretical treatment one must return to the Ginzburg-Landau theory and its various extensions.

In Sect.3.7 we have outlined the Abrikosov theory of the vortex state which is based on a particular class of solutions of the GL equations. As shown by Gor'kov, this theory follows from the microscopic theory for the temperature range close to T_c. Over this restricted regime the GLAG theory provides a satisfactory description of type-II superconductivity. Subsequently, the range of validity of the GLAG theory has been extended through various generalizations which are applicable to different regimes of temperature, magnetic field, and mean free path of the electrons. An account of these developments has been given by FETTER and HOHENBERG [1.7] and by WERTHAMER [4.17]. A comparison of the various theoretical approximations can also be found in a recent paper by BRANDT [4.18]. Therefore, in the following we present only a brief summary of these directions in which the GLAG theory has been extended.

Since the original formulation of the GLAG theory, there have been four main paths taken to derive, from the Gor'kov equations, results that are valid outside the limited temperature range near T_c, to which the GLAG theory is restricted. Along the first path, one relaxes the requirement that the order parameter be small, but expands in powers of the spatial variation of the order parameter about the equilibrium BCS value. This procedure, developed by Eilenberger, Maki, Tewordt, Tsuzuki, Werthamer, and others during the period 1963-1966 [4.17], leads to a deeper understanding of the validity of the GL equations but does not produce a theory generally valid for the mixed state except close to T_c.

Along the second path, one relaxes the requirement that the spatial variations be slow, but expands in powers of the order parameter. This method applies to the mixed state at magnetic fields close to the upper critical field H_{c2}, where a second-order transition to the normal state occurs. The resulting theories, developed by de Gennes, Eilenberger, Helfand, Maki, Tsuzuki, Werthamer, and others during the period 1964-1967 [4.17,1.7], extend the GLAG theory to all temperatures and mean free paths, but are limited to fields near $H_{c2}(T)$.

Along the third path, one expands all terms to higher order in powers of $(T_c - T)$, the deviation from the transition temperature. This procedure, developed primarily by NEUMANN and TEWORDT [4.16,19a,b,20], extends the theory to a wider range of temperatures near T_c and, within this limitation, is appropriate for arbitrary magnetic fields and electron mean free paths.

Along the fourth path, one relaxes all requirements in order to obtain theoretical results valid for all temperatures, magnetic fields, and mean free paths. This approach was initiated by the work of EILENBERGER [4.21], who transformed the Gor'kov equations into a set of much simpler, transport-like integro-differential equations. In their appropriate ranges of validity, all previous theoretical results could be shown to follow from Eilenberger's equations. However, the generality of these equations is gained at the expense that they must be solved numerically with the aid of high-speed computers. As shown by USADEL [4.22,23], for the case of very impure superconductors (dirty limit), the Eilenberger equations can be transformed into even simpler diffusion-like equations. More recently, BRANDT [4.18,24] has reformulated the Gor'kov equations to construct a theory that permits the calculation of truly two-dimensional periodic solutions with rather modest numerical effort. So far, this treatment is restricted to clean superconductors.

During the last few years numerous results have been obtained from numerical solutions of the Eilenberger equations (or the Usadel equations for the dirty limit). In these calculations the structure of vortex lines (variation of the magnetic field and of the order parameter with the radial coordinate) has been determined for arbitrary temperatures and electron mean free paths, for different κ values, and both for isolated vortices and the vortex lattice within the circular cell approximation. Further, the electronic density of states and thermodynamic properties, including heat capacity and magnetization, have been derived for arbitrary temperatures, magnetic fields, and electron mean free paths [4.25-29]. Brandt's method permits the calculation of the Gibbs free-energy and the spatial variations of the local magnetic field and order parameter in clean superconductors not only for arbitrary temperatures and κ-values, but, since the circular cell approximation is avoided, also for arbitrary fields and lattice symmetry [4.18,24,30].

In context with the structure of the vortex core the question of *bound states and excitations of the core* represents an important subject [1.7]. This question was first treated by CAROLI and co-workers [4.31,32], and subsequently by BARDEEN et al. [4.33]. More recently, this subject has been discussed by KRAMER and co-workers [4.34,35].

4.5 Flux-Line Lattices

The dominant feature of the mixed state is the existence of the triangular vortex lattice, which is satisfactorily described by the GLAG theory and its various extensions. Here the theoretical description is based on the assumption of an isotropic superconductor, and particular features resulting from anisotropic properties of the crystal lattice are ignored. However, because of the interaction between the vortex lattice and the crystal lattice of the superconductor, the orientation of the (single-crystalline) vortex lattice can be correlated with the orientation of the crystal lattice. Departures from the ideal triangular lattice structure can occur sometimes in the form of deviations from the triangular symmetry, square lattices, and various lattice defects. Again, deviations from the triangular lattice structure are associated with the interaction between the vortex lattice and the crystal lattice of the superconductor. In this section we discuss these special features of flux-line lattices.

4.5.1 Correlation Between the Vortex Lattice and the Crystal Lattice

In Sect.2.4 we have discussed the correlation between the domain orientation and the crystallographic direction in monocrystalline type-I superconductors, resulting in highly anisotropic domain patterns. A similar correlation also exists for flux-line lattices in monocrystalline type-II superconductors. The early experiments, demonstrating this effect, utilized the Bitter method [4.36,37]. Subsequently, the influence of the crystal lattice on the morphology of flux-line lattices in monocrystalline type-II superconductors has also been studied by neutron diffraction [4.6,9,38,39]. The experimental results have been reviewed recently by OBST [4.40] and SCHELTEN [4.41].

These experiments on the flux-line-lattice morphology were performed on monocrystalline, low-κ type-II superconductors (Nb, V, Tc, and Pb alloys). In these low-κ materials, a strong correlation between the orientation of the flux-line lattice and the crystal lattice can be observed such that the symmetry of the flux-line lattice fits the symmetry of the crystal axis parallel to the applied magnetic field. If the magnetic field is parallel to a fourfold symmetry axis of the crystal, the vortices form a square lattice. If the field is parallel to a threefold symmetry axis, a hexagonal vortex lattice is observed. Orientation of the magnetic field along a twofold symmetry axis results in a distorted triangular vortex lattice, in which two sides of the elementary triangular cell have the same length and the third is either smaller or larger.

With increasing κ-value, the influence of the anisotropy of the crystal lattice decreases. The experiments indicate a gradual transition from the highly anisotropic flux-line lattice to the triangular lattice predicted by the GL theory for an isotropic superconductor.

Figure 4.6 shows an example of a square vortex lattice observed in a Pb-1.6% Tl sample using the Bitter method. The external field is parallel a fourfold [001] symmetry axis. The variation of the morphology of the vortex lattice with the orientation of the external field is summarized for a few examples in Table 4.1.

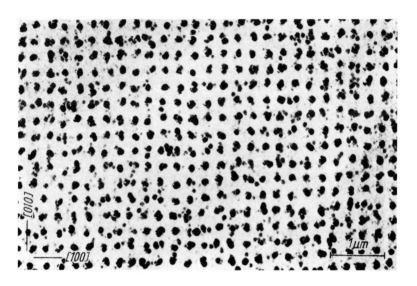

Fig.4.6. Square vortex lattice in a Pb-Tl single crystal with the applied field in [001] direction. The vortex lattice is oriented parallel to [100] crystal directions. κ ≈ 0.72, T = 1.2 K, H_e: 0 → H_{c2} → 365 G [4.37]

Strong anisotropy of the vortex lattice can also be observed in the "intermediate-mixed state", where regions of the Meissner phase and of the Shubnikov phase coexist in the material. We shall discuss this situation in Sect.4.7.

For explaining the deviations of monocrystalline vortex lattices from the simple triangular lattice structure, several theories have been proposed, which are based on various anisotropic properties of the crystal lattice. OBST [4.37] considered the *anisotropy of the energy gap*. In calculating the interaction energy between two vortex lines in a low-κ type-II superconductor, he assumed the same functional relations that are valid for κ ≫ 1. Under

Table 4.1. Morphology of the vortex lattice in single-crystalline type-II superconductors for different orientations of the applied magnetic field [4.41]

Field direction	[001]	[011]	[111]
Pb-1,6% Tl	[010], [100] square	65°/65°, 50° triangle, [100]	60° [112], 60° [121], 60° [211] triangle
Pb-Bi	Pb-2% Bi, [010], [100] square	Pb-4% Bi, 62°/62°, 56°, [011] triangle	
Nb	T=1.5K B=B₀: [110], [110] square, 30° rotated; T>1.5K B>B₀: α=53.2°, β=63.4°, [100], [010]	61.5°/61.5°, 57°, [011] triangle	60°/60°, 60°, [110], [101], [011] triangle

	[112]	[205]	⊥[111]
Nb	57°/57°, 66°, [110] triangle	55°/55°, 70° triangle	[111], ∢=60° triangle
Tc	⊥[0001]: 75°, 52.5°/52.5°, [0001] triangle		

this assumption, he showed that the interaction energy is anisotropic and has a maximum and a minimum in those directions where the energy gap reaches its minimum and maximum, respectively. A theory, based on the *anisotropy of the elastic properties* of the crystal and the magnetoelastic interactions, has been proposed by ULLMAIER et al. [4.42] and qualitatively explains some of the experimental results. Another theoretical treatment is due to TAKANAKA [4.43-46] who calculated the free energy of the mixed state including an *anisotropic Fermi surface*. This treatment starts from the GL equations and contains higher-order terms with respect to the spatial derivatives of the magnetic field. It is restricted to pure superconductors and temperatures near T_c. Takanaka's theory seems to explain some, but not all, of the experimental results.

A more general theoretical treatment of the mixed state in cubic superconductors has been given by TEICHLER [4.47-50]. Starting from Eliashberg's theory of the superconducting electron-phonon system, this theory takes into account the *microscopic anisotropies of the electronic band structure, the phonon system, and the phonon-induced electron-electron coupling*. Apparently, all different anisotropy parameters contribute to the macroscopic anisotropy phenomena. TEICHLER's theory seems to provide a reasonable explanation of the experimental results. Further, it generates useful physical insight into the significance of the different microscopic contributions to the anisotropy.

4.5.2 Defects in the Vortex Lattice

In an ordered structure like the vortex lattice, deviations from the perfectly ordered state and the appearance of various kinds of lattice defects can be expected. Such defects in the vortex lattice have, indeed, been observed as soon as the decoration method for the observation of vortex-line arrangements [4.4] became available. It did not take long to discover in vortex lattices the analogs of the different typical lattice defects we know from other regular structures like crystal lattices [4.51-53]. However, because of the two-dimensional nature of the vortex lattice, usually the defects are observed only within a plane perpendicular to the vortex lines and represent *line defects* extended in one dimension.

If locally a vortex line is missing, or if an excess line exists, we are dealing with the analogs of point defects. Figure 4.7a shows the configuration of a *single line vacancy*, analogous to the monovacancy, as observed with the Bitter decoration method. Figure 4.7b indicates schematically the vortex-line configuration associated with such a vacancy. The *interstitial line* is shown in Fig.4.8 (dumb bell configuration) and Fig.4.9 (crowdion configuration). Around an interstitial line the lattice distortions are much larger than for a line vacancy. Therefore, interstitial lines are observed less frequently. Besides the simple defect configurations shown in Figs.4.7-9, more complicated arrangements such as multiple line vacancies (the analogs of multivacancies) and multiple interstitial lines can also be observed. Generally, the line defects are observed only if the flux-line density is not too high.

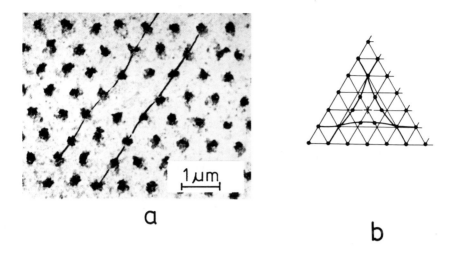

Fig.4.7a and b. Single line vacancy. (a) Bitter pattern, (b) schematic [4.51]

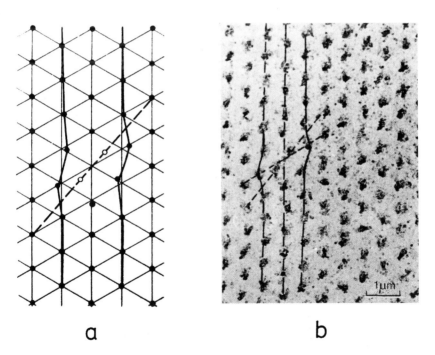

Fig.4.8a and b. Interstitial line in dumb bell configuration, (a) schematic, (b) Bitter pattern [4.51]

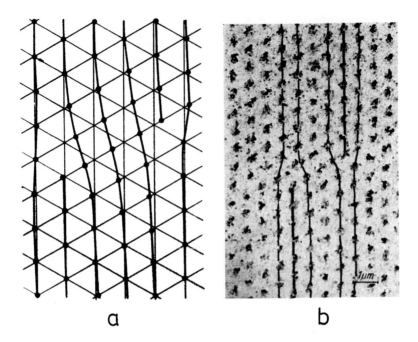

Fig.4.9a and b. Interstitital line in crowdion configuration, (a) schematic, (b) Bitter pattern [4.51]

In addition to the analogs of point defects, *dislocations* are frequently observed in flux-line lattices. Actually, in vortex lattices dislocations appear to be the most common defect. Figure 4.10 shows an example of an edge dislocation observed with the Bitter decoration method. The detail can be seen from the schematic model. An edge dislocation can be constructed in the following way. Imagine the perfect lattice being cut open along the y-axis for y > 0. Then we introduce two extra half-planes (each at the location b/2) into the cut section and let the lattice relax. At the dislocation core (indicated by the symbol ⊥) we have locally a square lattice (points 1-4 in Fig.4.10a). By looking at the schematic model parallel to the directions $\underline{a}$ and $\underline{b}$, we recognize one extra half-plane (y > 0) along these directions. The half-planes terminate at points 1 and 2, respectively, at the dislocation core. Edge dislocations play a significant role in the formation of density gradients in the vortex lattice, as shown schematically in Fig.4.11 for a square lattice.

Besides the simple edge dislocation, such as shown in Fig.4.10, other dislocation configurations can be observed in vortex lattices. Here, we mention only stacking faults consisting of two partial dislocations. Further, reactions between dislocations and dislocation movement take place, as one would expect. *Plastic deformations* of the vortex lattice can be understood in terms of flux-line dislocations, similar to the situation in metallic crystals.

At rather low vortex densities, the ordered structure of the flux-line lattice shows a gradual transition to a disordered, liquid-like vortex arrangement.

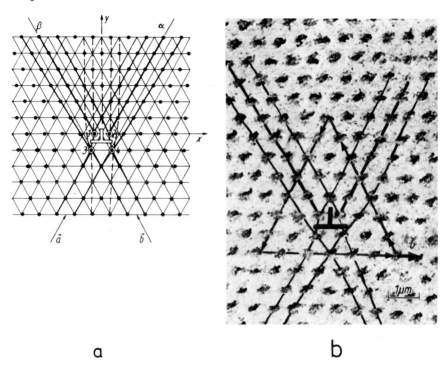

Fig.4.10a and b. Edge dislocation (a) schematic, the two additional half-planes (parallel y) terminate at the lattice points 1 and 2 (most clearly seen if viewed at small angle from top), (b) Bitter pattern [4.51]

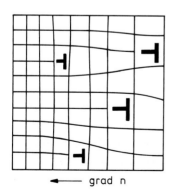

Fig.4.11. Edge dislocations in the formation of density gradients in the vortex lattice

A theoretical investigation of the properties of flux-line dislocations has been performed by LABUSCH [4.54] already prior to their experimental observation. He showed that the energy of an edge dislocation per unit length is approximately equal to the vortex-line energy $E_1 = H_{c1} \varphi_0/4\pi$. Calculations of the energy and lattice distortion associated with various defects in the vortex lattice can readily be performed for low vortex densities, i.e., near H_{c1}, using the London model. Here, the total energy including the interaction between vortex lines can be obtained for the particular defect configuration using procedures similar to those outlined in Sect.4.2. Starting with a reasonable ansatz for the lattice distortion, the configuration yielding the minimum defect energy is found by numerically varying the lattice displacements. Calculations of this type have been carried out by BRANDT [4.55,56] for vacancies and stacking faults in the vortex lattice, and by HILL et al. [4.57] for vacancies, interstitials, and edge dislocations.

On the other hand, for high vortex densities, i.e., near H_{c2}, the properties of defects in the vortex lattice (like the energy, the lattice distortion, the order parameter, and the local magnetic field) can be obtained by extending Abrikosov's treatment based on the linearized GL equations (see Sect.3.7) to defects in the vortex lattice. Again, the lattice displacements are found from a variational procedure. Such an approach has been developed by BRANDT [4.58,59] and has been applied to vacancies in the vortex lattice. This method has recently been extended to obtain the elastic energy, order parameter, and microscopic field of an arbitrarily distorted lattice in the entire field range $H_{c1} < H < H_{c2}$ [4.60-63].

4.6 Surface Effects

As we have seen in Sect.3.6, the presence of a surface facilitates the nucleation of superconductivity resulting in the phenomenon of surface superconductivity in the field range $H_{c2} < H < H_{c3}$.

In the following we restrict our attention to the field range $H_{c1} < H < H_{c2}$ and consider the *energy of a vortex line* near the planar surface of a type-II superconductor. The axial direction of the vortex line and the magnetic field are assumed parallel to the planar surface. We shall see that, depending on the magnetic field, an *energy barrier* can exist near the surface, strongly influencing vortex motion in the direction perpendicular to the surface. Finally, in this section we shall address ourselves to the *vortex nucleation* at the surface of type-II superconductors.

4.6.1 Energy Barrier Near a Surface

In the calculation of the vortex-line energy from the London model in Sect.4.2, the vortex line was assumed to be located inside the bulk of a superconducting sample. If the vortex line is placed near the surface of the sample, the plane of the surface being parallel to the vortex line, additional considerations are necessary [4.64].

We assume the superconductor to occupy the half-space $x > 0$, its surface coinciding with the y-z plane at $x = 0$. The magnetic field and the vortex line are parallel to the z-direction. Further, we assume the validity of the London model ($\kappa \gg 1$). We start from the analog of (4.3) for the field $\underline{h}(\underline{r})$,

$$\underline{h}(\underline{r}) + \lambda^2 \text{ curl curl } \underline{h}(\underline{r}) = \varphi_0 \delta_2(\underline{r} - \underline{r}_\ell) , \qquad (4.52)$$

where $\underline{r}_\ell$ is the line coordinate. In the following we take $\underline{r}_\ell = (x_\ell, 0)$. At the surface we have the boundary condition

$$\underline{h} = \underline{H} \quad \text{and} \quad (\text{curl } \underline{h})_x = 0 \text{ (zero normal current)} , \qquad (4.53)$$

where $\underline{H}$ is the applied magnetic field. The solution will be the sum

$$\underline{h} = \underline{h}_1 + \underline{h}_2 , \qquad (4.54)$$

where $\underline{h}_1 = \underline{H} \exp(-x/\lambda)$ describes the field penetration in the absence of a vortex line. The component $\underline{h}_2$ results from the vortex line and can be obtained from the method of images in the following way. To the vortex line at $(x_\ell, 0)$ we add the image line of opposite sign at $(-x_\ell, 0)$. The field $\underline{h}_2$ is the sum of the field of the vortex line and of its image. We note that $\underline{h}_2$ automatically vanishes at the surface, thus satisfying the boundary condition (4.53).

From (4.11) the Gibbs free energy is

$$G = \int d\underline{r} \left[\frac{h^2 + \lambda^2 (\text{curl } h)^2}{8\pi} - \frac{\underline{H} \cdot \underline{h}}{4\pi} \right] , \qquad (4.55)$$

where the integral is taken over the sample volume, $x > 0$, except for the core region of the vortex line, which is excluded. The integral in (4.55) can be transformed into a surface integral using a procedure similar to that leading to (4.13). In this way we obtain

$$G = \frac{\lambda^2}{4\pi} \int_{\text{(core and plane)}} d\underline{\sigma} \cdot (\frac{1}{2}\underline{h} - \underline{H}) \times \text{curl } \underline{h} \quad . \tag{4.56}$$

Here the surface integral $\int d\underline{\sigma}$ is taken over the surface of the vortex core and the surface of the specimen. Evaluating this integral [3.2] results in the following expression for the Gibbs free energy per length L of the vortex line

$$\frac{G}{L} = \frac{\varphi_0}{4\pi} \left[H \exp(-x_\ell/\lambda) - \frac{1}{2} \frac{\varphi_0}{2\pi\lambda^2} K_0(2x_\ell/\lambda) - (H - H_{c1}) \right] \quad . \tag{4.57}$$

Here the first term contains the interaction between the vortex line and the external field. It has the same form as (4.25) and represents a repulsive force. The second term describes the attractive interaction between the flux line and its image line. The magnitude of this contribution differs from (4.25) by the factor 1/2. However, the factor of 2 is recovered in calculating the force from the derivative with respect to x_ℓ. The third term represents the energy of the line inside the superconductor far away from the surface.

The variation of the vortex-line energy with the coordinate x_ℓ, as contained in (4.57), is shown schematically in Fig.4.12 for different values of the applied field. For $H < H_{c1}$, inside the superconductor the flux line is thermodynamically unstable, as indicated also in (4.18). In the range $H_{c1} < H < H_{en}$, an energy barrier develops near the surface, essentially due to the combined action of the first (repulsive) and the second (attractive) term in (4.57). For increasing field H, this barrier shrinks and finally disappears at the field $H = H_{en}$. For $H > H_{en}$, the flux line attains its lowest energy far inside the superconductor. We note that our discussion refers to the situation for a *single flux line*, where no other flux lines are residing already in the superconductor. The field H_{en} represents a *critical entry field* and is obtained from the condition that the slope $(\partial G/\partial x_\ell)_{x_\ell = \xi}$ becomes zero at $H = H_{en}$. In this way one finds

$$H_{en} = \frac{\varphi_0}{4\pi\lambda\xi} \quad . \tag{4.58}$$

Comparing this value with (4.19) we see that H_{en} is larger than H_{c1} by the factor $(\frac{\lambda}{\xi})/\ell n(\lambda/\xi)$.

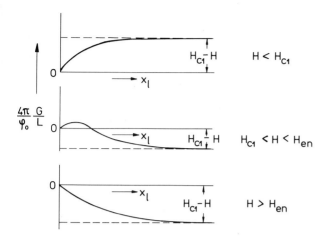

Fig.4.12. Vortex-line energy as function of the distance x_ℓ from the surface for different values of the applied magnetic field, according to (4.57)

As a consequence of the energy barrier, in an applied magnetic field, flux penetration into a superconductor sets in only when the field $H = H_{en} > H_{c1}$ is reached. Of course, this result and (4.58) refer to the case where the demagnetization coefficient D of the sample is zero. For finite D, the value of H must be modified accordingly. A treatment of the case with finite demagnetization coefficient, as represented by a flat superconducting strip, shows [4.65] that the critical entry field depends on the cross-sectional dimensions of the specimen. In such a geometry the magnetic energy of the vortex line *outside* the superconductor contributes significantly to the Gibbs free-energy barrier near the sample edges.

In addition to the entry of magnetic flux in an applied magnetic field, the Gibbs free-energy barrier near the surface also affects the penetration of magnetic flux associated with an *applied transport current*, say, for zero applied magnetic field. This siutation becomes important for a current-carrying superconducting strip or wire. Here, the energy barrier leads to an enhancement of the critical current, at which dissipation sets in, and thereby to a modification of Silsbee's rule [4.65]. Again, for a flat superconducting strip, the cross-sectional dimensions and the magnetic energy of the vortex outside the strip play an important role in the energy barrier. On the other hand, a current-carrying wire with circular cross-section in zero applied magnetic field represents another case, to which the one-dimensional treatment given above for an infinite superconducting half-space and the result in (4.58) can be literally applied. (The straight vortex line and its image must be replaced by circular lines in azimuthal direction). We shall return to the current-induced resistive state in Sect.14.

Finally, we note that our conclusions on the energy barrier near the surface carry over immediately also to type-I superconductors. Here, the single-quantum vortex line must be replaced by a flux tube containing the flux $\varphi = n\varphi_0$ where n is an integer. The magnitude of the energy barrier is found to be proportional to the flux φ per flux tube [4.65] and, therefore, becomes quite appreciable for large values of n. Such a result can be anticipated already from an inspection of (4.57), where the flux quantum φ_0 appears on the right hand as a common factor.

4.6.2 Vortex Nucleation at the Surface

Experimental studies of the microwave surface impedance in the superheated Meissner state have indicated an additional absorption mechanism which is incompatible with the expected behavior for a surface of a superconductor in a parallel magnetic field [4.66,67]. These results can be interpreted in terms of the formation of an array of lines of zero order parameter at the sample surface parallel to the field. These "nascent vortices" act as nucleation sites for vortex lines. Since the cores of the nascent vortices are located at the sample surface and are not enclosed by a superconducting path, fluxoid quantization need not be established.

A theoretical study of the stability of the superheated Meissner state at the surface of a superconductor in a parallel magnetic field leads to rather similar conclusions [4.68]. Superheating is limited due to the development of an instability at the surface. The instability can be derived from two-dimensional Ginzburg-Landau solutions including periodic solutions resembling the nascent vortices postulated by WALTON and ROSENBLUM. A three-dimensional view of the half-period of such a two-dimensional periodic solution (obtained numerically) is shown in Fig.4.13. Here the superconductor occupies the half-space $x > 0$, and the periodicity exists along the y-direction.

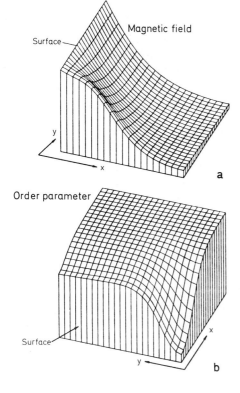

Fig.4.13a and b. Three-dimensional view of the magnetic field (a) and the order parameter (b) for a half-period of a two-dimensional periodic Ginzburg-Landau solution near the surface. The superconductor occupies the half-space $x > 0$ and the periodicity exists along the y-direction [4.68]

4.7 Attractive Vortex Interaction

Generally, two vortices of the same sign repel each other at all distances. Within the London model, this repulsive interaction can be seen in the quadratic dependence of the vortex-line energy upon the flux φ_0 per vortex line, as indicated in (4.15). However, in low-κ type-II superconductors an exception of this repulsive behavior between vortex lines takes place, and the interaction becomes *attractive* over a narrow κ-range. This attractive vortex interaction results in a new type of magnetic flux structure consisting of a mixture of flux-free domains (Meissner phase) and domains containing a vortex lattice (Shubnikov phase) with constant lattice parameter. This new configuration is often referred to as the *intermediate-mixed state*. It is accompanied by a first-order phase transition at H_{c1}.

4.7.1 Experiments

The intermediate-mixed state has first been revealed by the decoration technique in experiments by TRÄUBLE and ESSMANN [4.69], SARMA [4.70], and KRÄGELOH [4.71]. As shown primarily in decoration experiments by OBST [4.37, 40] using monocrystalline type-II superconductors, also in the intermediate-mixed state the orientation of the vortex lattice in the Shubnikov phase is correlated with the crystallographic direction. It appears that the existence of the flux-free domains does not affect fundamentally the correlation between the vortex lattice and the crystal lattice. An interesting feature of the intermediate-mixed state in monocrystalline samples is the arrangement of the flux-free domains within the Shubnikov phase in the form of a *macroscopic superlattice* with long-range order. In general, this superlattice is correlated to the crystal lattice in a characteristic way [4.40]. However, for superlattices with fourfold symmetry, absence of any correlation between the two lattices has also been observed. Depending on the material, the "lattice points" of the superlattice observed at the surface of the specimen can consist of complicated agglomerates of small Meissner and Shubnikov domains. It appears that these complicated structures are due to Landau branching near the surface.

Figure 4.14 shows an example of the intermediate-mixed state observed by the decoration method in a Pb-Tl alloy. In the monocrystalline sample the superlattice imbedded in the Shubnikov phase can clearly be seen. Figure 4.15 shows the superlattices for the same monocrystalline alloy for different crystal orientations. We note the mixture of Meissner and Shubnikov domains (due to Landau branching) constituting the lattice points of the superlattice. It is evident from results such as shown in Figs.4.14,15 that the intermediate-mixed state in monocrystalline samples displays a rich variety of features depending on material, temperature, magnetic field, and crystal orientation.

The attractive vortex interaction deduced from the coexistence of the Shubnikov and the Meissner phase in low-κ type-II superconductors results in an upper limit a_0 of the vortex-lattice parameter, which cannot be exceeded in these materials. As a consequence, at H_{c1} the magnetic flux density increases abruptly to a value B_0 corresponding to the lattice parameter a_0, and we have a phase transition of first order. Experimentally, this effect can be seen as a discontinuous change in the magnetization curve at H_{c1}, as shown schematically in Fig.4.16. Of course, this discontinuous behavior at H_{c1} only occurs when the demagnetization coefficient D

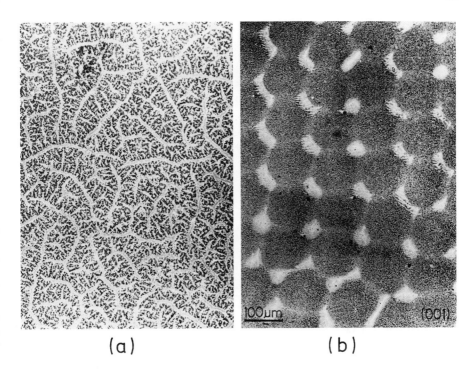

Fig.4.14a and b. Intermediate-mixed state of a Pb-Tl alloy ($\kappa \approx 0.72$, $T = 1.2$ K, $H_e: 0 \to H_{c2} \to 210$ G). (a) polycrystal, (b) single crystal with the applied field in [001] direction; the superlattice of flux-free domains is oriented parallel to [100] crystal directions; the Shubnikov phase is dark (courtesy of B. Obst)

of the sample is zero. For values $D > 0$, flux penetration sets in below H_{c1}, and the M(H) curves always display a finite slope. Magnetization measurements on low-κ type-II superconductors demonstrating this behavior have been performed by KUMPF [4.72], ASTON et al. [4.73], FINNEMORE et al. [4.74], and WOLLAN et al. [4.75]. As pointed out by ESSMANN and SCHMUCKER [4.76], a finite slope near H_{c1} of the magnetization curve for low-κ type-II superconductors can also result from flux pinning, even when $D \approx 0$.

An experimental study of the κ-regime, in which the interaction between vortices is attractive, has been performed by AUER and ULLMAIER [4.77] using magnetization measurements. For obtaining sufficient accuracy, such measurements require samples with high reversibility, i.e., vanishing influence of flux pinning, and a careful control of the Ginzburg-Landau parameter. Both requirements could be achieved by doping polycrystalline wires of high-purity tantalum and niobium with nitrogen. Figure 4.17 shows a

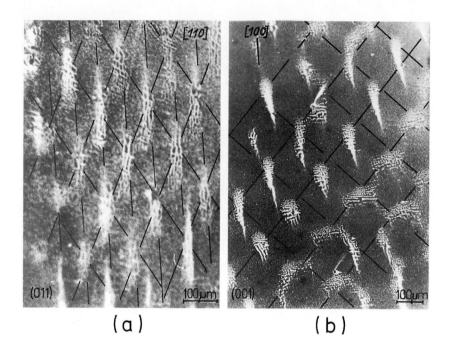

Fig.4.15a and b. Superlattices in the same monocrystalline Pb-Tl alloy as in Fig.4.14b for different crystal orientations ($\kappa \approx 0.72$, $T = 1.2$ K). (a) (011) oriented disk, H_e: $0 \to H_{c2} \to 70$ G; (b) (001) oriented disk, H_e: $0 \to H_{c2} \to 210$ G. The Shubnikov phase is dark (courtesy of B. Obst)

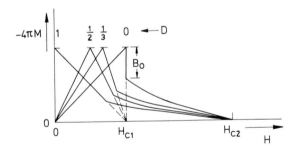

Fig.4.16. Magnetization versus applied magnetic field for low-κ type-II superconductors with an attractive vortex interaction. The different curves refer to different values of the demagnetization coefficient D

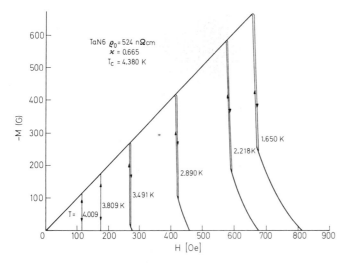

Fig.4.17. Highly reversible magnetization curves for a tantalum sample doped with nitrogen [4.77]

series of magnetization curves for a TaN sample (κ = 0.665) at different temperatures. We see that the material exhibits type-I behavior for the high-temperature curves and type-II behavior for the lower-temperature curves with a discontinuity at H_{c1}. From data such as shown in Fig.4.17 AUER and ULLMAIER were able to construct a phase diagram of the magnetic behavior for the TaN system, indicating the κ-range with an attractive vortex interaction (and a first-order phase transition at H_{c1}) as a function of temperature. The phase diagram obtained in this way is shown in Fig.4.18. We see that vortex attraction occurs in a narrow κ-range near $\kappa = 1/\sqrt{2}$, this range becoming larger with decreasing temperature.

Because neutron diffraction has proved to be a useful tool for studying the microscopic properties of vortex lattices, such experiments can also be expected to demonstrate the attractive vortex interaction in materials of the proper κ-range. Neutron diffraction in pure niobium has, indeed, verified that the lattice parameter of the vortex lattice remains constant, independent of the magnetic field in the range from $H = + H_{c1}$ down to $H = - H_{c1}$ [4.6]. These measurements were performed by decreasing the external field below H_{c1}. Because of surface and bulk pinning effects, flux remained in the sample, with a constant lattice parameter of the vortex lattice. Such behavior is consistent with the establishment of the intermediate-mixed state, where the volume fraction of the Shubnikov phase de-

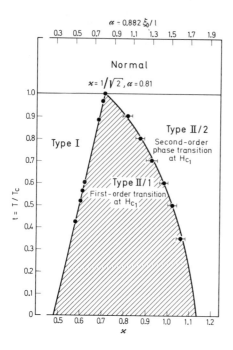

Fig.4.18. Phase diagram of the magnetic behavior for the TaN system. The Ginzburg-Landau parameter κ (lower abscissa) and the impurity parameter α (upper abscissa) are proportional to the concentration of dissolved nitrogen [4.77]

creases with decreasing field and the lattice parameter of the Shubnikov domains remains constant.

4.7.2 Theory

Motivated by the experimental results, various theoretical attempts have been undertaken to explain the attractive vortex interaction in the narrow κ-range near $\kappa = 1/\sqrt{2}$. Clearly, the GLAG theory is unable to account for such an attraction between vortices, and further extensions of this theory must be invoked.

In a theoretical study of the detailed electronic and magnetic structure of single vortices in type-II superconductors, EILENBERGER and BÜTTNER [4.78] found for $\kappa < 1.7$ that the gap parameter and the vector potential display damped oscillatory behavior as a function of the distance $\underline{r}$ from the center of the vortex. Their results were based on numerical solutions of the Eilenberger equations in the form of asymptotic expansions for large distances $\underline{r}$. Similarly, magnetic field reversal of an isolated flux line for a material with small κ was found by DICHTEL [4.79] and by BRANDT [4.80] for large distances $\underline{r}$ from a nonlocal approximation.

Although the existence of the Eilenberger-Büttner oscillations seems to suggest an explanation for an attractive vortex interaction, this calculation has been criticized by CLEARY [4.81]. Subsequently, the situation has been clarified, mainly through the work of JACOBS and LEUNG [4.82-89]. From a generalization of the Neumann-Tewordt extension of the GLAG theory, JACOBS has shown, that the Eilenberger-Büttner oscillations are unrelated to the question of an attractive vortex interaction and a first-order transition near H_{c1}. Apparently, a term in the vector potential, representing the effect of the surface and neglected in the Eilenberger-Büttner calculations, removes the oscillatory behavior of the vector potential. However, as shown by JACOBS, for widely separated vortices in type-II materials with $\kappa \approx 1/\sqrt{2}$ and not too small values of the electron mean free path, the vortex interaction is attractive. A similar conclusion, also based on a generalized Neumann-Tewordt free-energy expression, has been reached by HUBERT [4.90]. The Ginzburg-Landau functional and its extension by nonlinear terms has also been used by KRAMER [4.91] and by GROSSMANN and WISSEL [4.92] for treating the question of an attractive vortex interaction and a first-order phase transition near H_{c1}.

Finally, the conditions for vortex attraction were calculated in detail from Eilenberger's reformulation of the Gor'kov equations [4.86,87,89,93]. From these calculations it is predicted that attractive vortex interaction takes place in a narrow κ-range near $\kappa = 1/\sqrt{2}$. This κ-range increases with decreasing temperature, similar to the experimental observations by AUER and ULLMAIER [4.77]. A similar result is obtained by the two-dimensional calculations of the periodic lattice based on a reformulation of the Gor'kov free-energy functional [4.18,24,30]. In summary, the attractive vortex interaction and the nature of the first-order phase transition near H_{c1} in type-II superconductors now appears to be understood from microscopic theory, at least semiquantitatively, mainly through the work of JACOBS, KRAMER, LEUNG, and BRANDT.

5. Thin Films

In Sect.2.3 we have discussed the Landau domain theory which gives a reasonable description of the dimensions of the normal and superconducting domains in the intermediate state of type-I superconductors. According to the result of Landau's nonbranching model, indicated in (2.40), with decreasing sample thickness the domain width of the intermediate-state structure decreases and the (positive) wall energy contributes more and more to the free energy of the superconductor. Because of this, the critical field perpendicular to a planar superconducting film decreases with decreasing film thickness and would become zero when the film thickness is about equal to the wall energy parameter δ [2.3]. TINKHAM [5.1] was the first to point out that films of type-I superconductors with a thickness less than the coherence length assume an Abrikosov vortex state in a perpendicular magnetic field, similar to the mixed state in type-II superconductors.

In this theoretical concept TINKHAM has been motivated by the experimental result that very thin films of type-I superconductors show a second-order phase transition in a perpendicular magnetic field. We consider a superconducting film near its transition temperature T_c, where the order parameter is vanishingly small. Hence the penetration depth λ will be rather large (2.7), and the applied field will completely penetrate through the film. Because of the large value of λ, nonlocal effects are negligible, and the supercurrent density $\underline{J}_s$ satisfies the London equation (2.14). The kinetic energy

$$f_{kin} = \frac{1}{2} m^* v_s^2 |\psi|^2 = \frac{1}{2} \frac{4\pi}{c^2} \lambda_L^2 J_s^2 \quad , \tag{5.1}$$

associated with the supercurrent, must be included in the Ginzburg-Landau free-energy expression, see (3.38). Considering a circular loop of radius r centered on a normal spot and using (3.46,50) for the fluxoid quantization, we have

$$J_s(r) = \frac{(n\varphi_0 - r^2\pi H)c}{2r\pi \cdot 4\pi\lambda_L^2} \quad , \tag{5.2}$$

yielding

$$f_{kin} = \left(\frac{n\varphi_0 - r^2\pi H}{2r\pi c}\right)^2 \frac{c^2}{8\pi\lambda_L^2} \quad . \tag{5.3}$$

Here, n is the number of flux quanta contained in the loop area. In addition, since the order parameter will depend on the coordinate r, we must include the Ginzburg-Landau term in $|\nabla\psi|^2$. This term is

$$f_\nabla = \frac{\hbar^2}{2m^*}\left|\frac{d\psi}{dr}\right|^2 = \frac{\hbar^2}{32mn_s}\left(\frac{dn_s}{dr}\right)^2 \tag{5.4}$$

with $|\psi|^2$ given by (3.11).

From (5.3) we see, that for n = 0 the energy f_{kin} increases toward ∞ as $n_s r^2 H^2$. In this case, f_{kin} would soon exceed the condensation energy, and the critical field would depend on specimen size and be rather small in general. On the other hand, if magnetic flux penetrates the film in the form of close-packed vortices (n > 0), the kinetic energy will be reduced, and superconductivity can be maintained in the presence of finite perpendicular fields. The situation is somewhat similar to that of the experiment by LITTLE and PARKS [3.8,9], where the kinetic energy associated with the screening supercurrents in a hollow cylinder, placed in a magnetic field oriented parallel to the cylinder axis, varies with the amount of flux which has penetrated into the interior of the cylinder. As shown by TINKHAM, the configuration, which is energetically most favorable, consists of the formation of an array of single flux quanta within the film. In this case, the critical field, at which superconductivity completely disappears in the film, is identical to the upper critical field in type-II superconductors given by (3.59).

Experimentally, the Abrikosov vortex structure in very thin films of type-I superconductors placed in a perpendicular magnetic field has been observed in different ways. Recently, the most direct observations have been performed by RODEWALD and co-workers [5.2-4], and DOLAN and co-workers [5.5-8] using the Bitter method in the version introduced by ESSMANN and TRÄUBLE [4.4]. These experiments were carried out in thin films of Pb, Sn, and In (film thickness $\leq$ 1500 Å) and clearly established the single-quantum nature of the individual flux spots. In well prepared films, rather defect-free, triangular vortex lattices could be seen. We note that, as the film thickness increases, the single-quantum vortices grow into the multi-quanta flux tubes discussed in Sect.2.4. Under favorable conditions these flux tubes can also form a triangular lattice structure [2.21] as seen in Fig.2.12.

Following TINKHAM's prediction of the Abrikosov vortex state in sufficiently thin films of type-I materials, the fluxoid structure in superconducting films in the presence of a perpendicular magnetic field has been theoretically studied from the Ginzburg-Landau theory in more detail by GUYON et al. [5.9], PEARL [5.10,11], MAKI [5.12] and LASHER [5.13]. FETTER and HOHENBERG [5.14] have performed a stability analysis of the vortex lattice in thin superconducting films from the Landau superfluid hydrodynamics. The main result of these calculations is the derivation of a *critical film thickness*, d_c, below which the Abrikosov vortex lattice is energetically stable and the phase transition in a perpendicular field is of second order. For films with thickness $d < d_c$ the perpendicular critical field decreases with increasing film thickness, whereas for $d > d_c$ the critical field increases with increasing film thickness and approaches the bulk value at large values of d.

In the regime $d > d_c$, where the film shows type-I behavior, we can estimate the perpendicular critical field $H_{c\perp}$ from the wall energy $\alpha = \delta \cdot H_c^2/8\pi$ in the following way. Just below the transition to the normal state we consider an isolated superconducting domain with circular cross section of radius r. The contribution of the wall energy to the Gibbs free-energy density of the superconducting domain is

$$\frac{2r\pi\delta}{r^2\pi} \frac{H_c^2}{8\pi} = \frac{2\delta}{r} \frac{H_c^2}{8\pi} \quad . \tag{5.5}$$

By subtracting this from the condensation energy $H_c^2/8\pi$, we obtain the transition field

$$H_{c\perp} = H_c \cdot \left(1 - \frac{2\cdot\delta}{r}\right)^{\frac{1}{2}} \quad . \tag{5.6}$$

According to the Landau domain theory, (2.40), we expect $r \approx (\delta \cdot d)^{\frac{1}{2}}$, yielding

$$H_{c\perp} \approx H_c(1 - 2\sqrt{\delta/d})^{1/2} \quad . \tag{5.7}$$

A result qualitatively similar to (5.7) has been obtained by DAVIES [5.15]. More generally, we can write

$$H_{c\perp} = H_c(1 - \sqrt{C \cdot \delta/d})^{1/2} \quad , \tag{5.8}$$

where the constant C depends on the model and will be of order 1.

In the regime $d < d_c$, we have according to TINKHAM [5.1]

$$H_{c\perp}(d) = \sqrt{2}\kappa(d)H_c \quad . \tag{5.9}$$

Here, the Ginzburg-Landau parameter $\kappa(d)$ depends on film thickness d through the influence of d upon the electron mean free path ℓ. Using (3.28) and taking

$$\frac{1}{\ell} = \frac{1}{\ell_b} + \frac{A}{d} \quad , \tag{5.10}$$

we have

$$H_{c\perp} = 0{,}715\sqrt{2}\,\lambda_L(0)\left[\frac{1}{\ell_b} + \frac{A}{d}\right]H_c \quad . \tag{5.11}$$

Here, ℓ_b is the electron mean free path of the bulk material, and A is a constant. We see from (5.11) that for very thin films $H_{c\perp}$ is proportional to d^{-1}.

Comparing expressions (5.8,9), we can obtain an estimate of the critical thickness d_c where the two transition fields are equal. In this way we find

$$d_c \approx \frac{C\,\delta}{(1-2\kappa^2)^2} \quad . \tag{5.12}$$

If $(C \cdot \delta/d)^{1/2} \ll 1$, we can expand the right hand in (5.8) and obtain

$$d_c \approx \frac{C\,\delta}{4(1-\sqrt{2}\kappa)^2} \quad . \tag{5.13}$$

The transition from type-I to type-II behavior in superconducting films in a perpendicular magnetic field has been studied in decoration experiments using the high-resolution Bitter method [2.25,5.3,5-8] and in magnetic measurements yielding the perpendicular critical field $H_{c\perp}$ [5.9,16-21]. These experiments have been performed by varying the film thickness, and the critical film thickness, d_c, was determined. The transition from type-I to type-II behavior has also been studied in In films from a combination of measurements of the tunneling density of states and the magnetic reversibility [5.22]. A compilation of the values of the critical film thickness obtained in the different experiments is given in Table 5.1.

Table 5.1. Experimental values of the critical film thickness d_c, at which the transition from type-I to type-II behavior takes place, for some type-I superconductors

Material	T (K)		d_c (Å)	
Pb	4.2		10000	a
	1.2	~	1000	b
	4.2	<	1000	c
	2.1	<	1000	c
Sn	1.6		2500	d
	3.7		9700	e
	2.9		5600	e
	1.9		5000	e
	1.4		3600	e
	2.1		1000	c
In	1.2		1000	b
	2.1	<	800	c
	0.4	~	2700	f
	3.2		6000	g
	2.7		6000	g
	1.7		5800	g
	1.4		5900	g
Al	1.12		19900	g
	0.95		18900	g
	1.07		100000	h
	0.60		70000	h

References to Table 5.1
[a] G.D. Cody, R.E. Miller: Phys. Rev. *173*, 481 (1968)
[b] U. Kunze, B. Lischke, W. Rodewald: Phys. Status Solidi (b) *62*, 377 (1974)
[c] G.J. Dolan: J. Low Temp. Phys. *15*, 133 (1974)
[d] J.P. Burger, G. Deutscher, E. Guyon, A. Martinet: Phys. Rev. *137*, A 853 (1965)
[e] R.E. Miller, G.D. Cody: Phys. Rev. *173*, 494 (1968)
[f] K.E. Gray: J. Low Temp. Phys. *15*, 335 (1974)
[g] B.L. Brandt, R.D. Parks, R.D. Chaudhari: J. Low Temp. Phys. *4*, 41 (1971)
[h] M.D. Maloney, F. de la Cruz, M. Cardona: Phys. Rev. B *5*, 3558 (1972)

From the values in Table 5.1 we note that the results from the different authors differ appreciably. In general, the direct observations of the magnetic flux structure with the Bitter method yields values for the critical film thickness considerably below those determined from measurements of the perpendicular critical magnetic field.

6. Experimental Techniques

In previous sections we have referred frequently to experimental observations of magnetic flux structures without describing the details of the various methods. In the following we deal with the most important experimental techniques utilized for investigating flux structures in superconductors. In this section we are restricting ourselves to those methods which concentrate primarily on the *static* behavior. Experiments particularly designed for studying the *dynamic* behavior of magnetic flux structures will be described in Sect.8. The major portion of these techniques relies on the strong inhomogeneity of the magnetic field protruding from the *sample surface*, as shown in Fig.2.3. A summary of the developments in the methods for the direct observation of magnetic flux structures in superconductors has been given some time ago by KIRCHNER [6.1].

6.1 Bitter Method

About 45 years ago, for the first time BITTER [6.2] utilized small ferromagnetic particles for decorating the magnetic domain structure in ferromagnetic materials. When the ferromagnetic particles are sprinkled on a material displaying at its surface an inhomogeneous distribution of the magnetic flux density, the particles are attracted to the regions with the largest value of the local magnetic field. In this way the boundaries of ferromagnetic domains can be decorated and be made visible. A similar decoration effect can also be accomplished with small diamagnetic particles. The only difference consists in the fact, that such particles will be repelled by the regions of large values for the local magnetic field, thereby providing again a decoration of the domain structure.

The Bitter method has been applied with great success to the intermediate and the mixed state in superconductors. The spatial resolution depends, of course, on the diameter of the particles used for the decoration. The early reliable observations of the intermediate-state structure have been performed by SHARVIN and co-workers [6.3,2.29-31] using nickel particles of about 1 μm diameter. Because of their perfectly diamagnetic behavior, superconducting particles can also be used for decorating magnetic flux structures at low temperatures. SCHAWLOW and co-workers [2.26,27,6.4] have studied the intermediate state using niobium particles of about 60 μm diameter. Further work has been done by FABER [2.28] using tin particles and by HAENSSLER and RINDERER [2.33,11] using niobium particles. Bitter-pattern experiments on the intermediate state in type-I superconductors are reasonably straightforward. They do not require excessibly small particle sizes for the decoration process. Different domain configurations (attained, say, for different magnetic fields) can be observed during the same experimental run simply by shaking the specimen following a change in the domain pattern and thereby allowing the decorating powder to rearrange itself according to the new domain configuration. According to (2.40), for a type-I superconductor with thickness $d = 10^{-2}$ cm and wall-energy parameter $\delta = 500$ Å, the characteristic length of the domain structure is about $2 \cdot 10^{-3}$ cm. Observations of the intermediate state from the Bitter patterns usually can be made using standard light optics.

The observation of flux-line lattices in type-II superconductors requires rather high spatial resolution. According to (4.38), the lattice parameter of the vortex lattice at a field of 500 G is about 2000 Å. Hence, the Bitter method can only successfully be applied to the mixed state, if decorating particles with a diameter as small as about 100 Å can be produced and if the spatial resolution of the observations can be extended to the same level. These two goals have been achieved for the first time by TRÄUBLE and ESSMANN [2.34]. Their Bitter-pattern experiment provided the first direct experimental verification of the existence of the Abrikosov vortex lattice [4.4]. The principal part of their method is often referred to as the *Träuble-Essmann technique*.

The method is based on the fact that small ferromagnetic particles with diameters down to about 80 Å can be generated by evaporating ferromagnetic materials in a rare-gas athmosphere (argon or helium in the pressure range 0.2-10 torr). Presumably, particles of such small size are single-domain magnets. In a magnetic field gradient they experience a rather strong force. Therefore, they are well suited for directional diffusion in an inhomogenous magnetic field. Preferential deposition of these particles at the specimen surface in form of agglomerates separated from each other, requires the condition

$$(\underline{m} \cdot \nabla \underline{H}) > 3 k_B T \quad , \tag{6.1}$$

where $\underline{m}$ is the magnetic moment of a particle. Hence, low temperatures are in favor of the particle agglomeration. The sample is mounted on top of a supporting rod such that good thermal contact is established. The lower end of the rod is immersed in a bath of liquid helium. The evaporation source is arranged above the specimen. For temperature stabilization, a shield is placed in the direct path between sample and evaporation source. Apparently, the evaporation of iron yields better spatial resolution than the evaporation of nickel, presumably due to smaller particle size in the first case. For obtaining optimum resolution, TRÄUBLE and ESSMANN found that the pressure of the helium gas during evaporation must be below 0.8 torr.

Following the evaporation of the ferromagnetic particles, the sample is slowly brought to room temperature, carefully avoiding condensation of matter on the specimen surface. Subsequently, a carbon replica of the domain structure is made, by depositing a carbon film over the sample surface and the ferromagnetic particle decoration. Finally, the carbon replica is examined in an electron transmission microscope. A resolution as high as about 100 Å can be obtained.

Examples of magnetic flux structures, obtained with the Träuble-Essmann technique, are given in Figs.4.2,6-11. Clearly, the spatial resolution of this method is impressive and cannot be surpassed presently by any other technique, which operates in a similarly direct way. However, an important drawback lies in the fact, that the method is restricted to static situations and excludes the possibility for observing dynamic phenomena.

6.2 Magneto Optics

Faraday rotation of the polarization vector of a beam of polarized light has proven to be a useful phenomenon for investigating the domain structure in ferromagnetic materials. Similarly, this principle can be successfully applied for studying magnetic domains in superconductors. The essential arrangement is shown schematically in Fig.6.1. A superconducting plate, SC, residing in the intermediate state, is covered on top with a sheet of some magneto-optical material, MO. Between the two, a highly reflecting film, RE, is attached to the surface of the superconductor. An external magnetic field may be applied perpendicular to the plate SC. Polarized light is directed upon this arrangement from the top. It passes through the magneto-

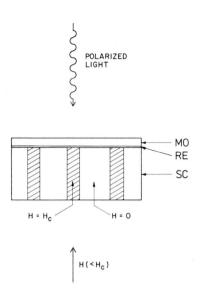

Fig.6.1. Principle of the magneto-optical flux detection (SC: Superconductor; MO: magneto-optical material; RE: reflecting metal film)

optical sheet before and after its reflection at the film RE. During both passages through MO, the light experiences Faraday rotation, the rotation angle depending on the local magnetic field. By passing the outgoing light through an analyzer, a magneto-optical image of the magnetic field distribution near the surface of the specimen can be obtained.

This use of the Faraday effect for the observation of magnetic flux structures in superconductors was first demonstrated by ALERS [6.5] and subsequently further developed by DeSORBO [6.6,2.32]. In contrast to the Bitter method, the magneto-optical technique allows the continuous observation of dynamic phenomena. Hence, motion picture studies, including high-speed photography, become possible. From Fig.6.1 we see that the spatial resolution of the magneto-optical method cannot exceed roughly the thickness of the magneto-optical plate. Therefore, it becomes imperative, to reduce this thickness to a value as small as possible.

The early experiments performed by ALERS, DeSORBO, and others, utilized sheets of magneto-optical material (cerous nitrate-glycerol mixture or cerium phosphate glass) with 0.3-3 mm thickness. The spatial resolution of these experiments was rather low, and the search for magneto-optical materials providing sufficient Faraday rotation at much smaller thickness became an important task. Here, a highly successful development has been initiated by KIRCHNER [6.7,2.20], when he noticed that films of the rare-earth compounds EuO, EuS, or EuSe of only 1000-2000 Å thickness provided what was needed. KIRCHNER's work essentially started *high-resolution magneto-optics* for the observation of magnetic flux structures in superconductors.

KIRCHNER's technique utilizes the same scheme as shown in Fig.6.1. In the above materials a high Verdet constant (Faraday rotation per unit thickness and unit magnetic field) also implies strong optical absorption, so the two must be considered together. In Table 6.1 we have listed the Verdet constant, V, at 4.2 K and the absorption coefficient, β, near room temperature for three europium compounds. Also given are values of the ratio V/β, which can be thought of as a magneto-optical figure of merit. Typical absorption curves for EuS and EuSe are shown in Fig.6.2. From Table 6.1 we see that EuS at a wave length of 5780 Å (Hg-yellow line) would give the largest specific Faraday rotation per unit absorption.

Table 6.1. Verdet constant, V, at 4.2 K and absorption coefficient, β, for three europium compounds

Material	λ [Å]	V [deg/KOe·µm]	β [µm^{-1}]	V/β [deg/KOe]
EuO	5461	1.0 [a]	9.0	0.11
EuO	5780	1.6 [a]	9.5	0.17
EuS	5461	4.8 [b]	12 [b]	0.40
EuS	5780	9.1 [b]	12 [b]	0.76
EuSe	5461	9.6 [b]	10 [b]	0.96
EuSe	5780	5.2 [b]	8 [b]	0.65

[a] K.Y. Ahn, J.C. Suits: IEEE Trans. MAG-*3*, 453 (1967)
[b] J. Schoenes: Z. Physik B *20*, 345 (1975)

For our magneto-optical application it is important to note that the europium compounds EuO, EuS, and EuSe are magnetic semiconductors displaying magnetic ordering at low temperatures [6.8]. EuO and EuS are ferromagnetic with a Curie temperature of 66.8 and 16.3 K, respectively. EuSe shows metamagnetic behavior with antiferromagnetic and ferromagnetic ordering below 4.6 and 2.8 K, respectively. In order to avoid large-scale ferromagnetic ordering in the magneto-optic material (which would result in a strong distortion of the magneto-optic image of the flux structure because of the influence of the ferromagnetic domains on the superconductor)

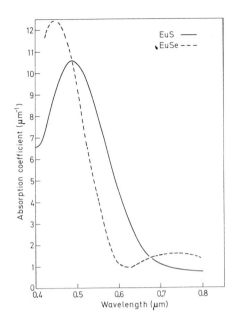

Fig.6.2. Absorption characteristics of EuS (solid curve) and EuSe (dashed curve) in the regime of visible light

ferromagnetism must be suppressed in the EuO, EuS, or EuSe film. This suppression can be accomplished through the admixture of the proper amount of the compound EuF_2 which remains paramagnetic also at low temperatures [2.20]. Subsequently, high-resolution magneto-optical studies of magnetic structures in superconductors were performed predominantly with films of EuS-EuF_2 mixtures. The composite films are usually prepared by vapor deposition. Here special care is needed since the melting point of EuS (about 2500° C) is approximately 1000° C above that of EuF_2. During the evaporation, a uniform mixture ob both compounds may quickly deteriorate due to preferential deposition of the low-melting-point component. Systematic studies of the influence of the amount of the EuF_2 admixture upon the quality of the magneto-optic image have been performed by KIRCHNER [2.20] and LAENG and RINDERER [6.9]. Empirically, the optimum EuF_2 admixture has been found to vary between 10 and 60 weight %, depending on the details of the evaporation procedure.

Usually, the film RE between specimen and magneto-optic material consists of a highly reflecting, thin aluminum film. The light intensity, I_1, reflected out of the magneto-optic layer, MO, of thickness t and absorption coefficient β is

$$I_1 = I_o e^{-2\beta t} \quad , \tag{6.2}$$

where I_o is the intensity of the incident light. The Faraday rotation θ per unit thickness is

$$\frac{\theta}{t} = V H \, , \tag{6.3}$$

V being the Verdet constant and H the magnetic field within the layer MO. The light intensity I_2 behind the analyzer is for the crossed position

$$I_2 = I_o \, e^{-2\beta t} \sin^2(2VHt) \, . \tag{6.4}$$

For most practical applications the Faraday rotation is very small, and the sin in (6.4) can be replaced by its argument. In Fig.6.3 we show the transmittance I_2/I_o calculated from (6.4) versus the thickness t of the magneto-optical film for EuS at 4.2 K and a wavelength of 5780 Å with the magnetic field as parameter. We note that the transmittance passes through a maximum at a thickness of about 800 Å. The thickness at which the maximum occurs does not depend strongly on H.

So far we have neglected any light which is reflected at the upper surface of the layer MO without passing through the magneto-optical film. If we operate the system with the analyzer in the crossed position, assuming that for the unrotated light complete extinction is achieved, we do not have to concern ourselves with the light reflected at the upper surface of MO. However, experience indicates that optimum contrast can often be achieved if the analyzer is set slightly off the complete extinction position. In this mode of operation the portion of the light beam reflected at the upper surface of MO will decrease the magneto-optical contrast since it is not undergoing Faraday rotation. This effect can be minimized through *destructive interference* between the beam reflected at the upper surface of MO and the beam passing through MO and reflected at its lower surface. This destructive interference should occur only for those regions where no Faraday rotation takes place, i.e., where H is zero, in order to maximize the magneto-optical contrast. Complete destructive interference requires that the two beams reflected at the top and the bottom of the layer MO have the same amplitude and the opposite phase at the top of MO. With respect to the *phase*, this interference condition can readily be met by making the thickness of MO an odd multiple of λ/4, λ being the wavelength of the light. Since the refractive index of the layer MO is *smaller* than that of the underlying metal, a 180° phase reversal takes place both during the reflection at the upper and lower surface of MO. The phase condition can simply

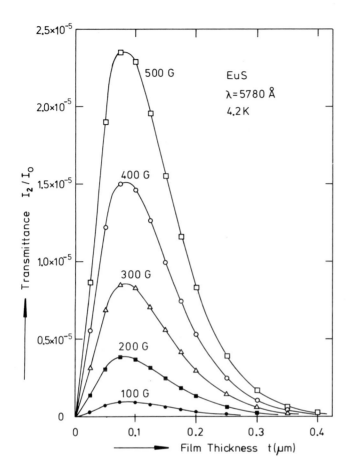

Fig.6.3. Transmittance I_2/I_0 calculated from (6.4) versus the thickness of the magneto-optical film for EuS and different magnetic fields (T = 4.2 K, λ = 5780 Å)

be met by increasing the thickness of the layer MO until a local minimum in reflectivity is obtained. With respect to the *amplitude*, the interference condition can be approximately satisfied using the fact that only a small portion of the incoming light beam is reflected at the top of MO. The other portion of the light is attenuated during its path through MO by absorption, the bottom of MO being highly reflective through the underlying metal. Experience shows that the amplitude condition is reasonably fulfilled for the second reflectivity minimum. We see that under optimum conditions MO plays the role of an *antireflective coating* of the underlying metal. The advantage of arranging MO such that it operates as an antireflective coating has already been demonstrated by KIRCHNER [6.10].

Practically, one meets the condition for antireflectivity by monitoring the light intensity reflected from MO continuously during the vacuum deposition and terminating the deposition of MO at a local reflectivity minimum. Typically, the thickness of the magneto-optical film turns out to be about 2000 Å. At such a small value, this thickness does not limit the spatial resolution any longer, and the resolution will be determined by the wavelength of the light used in the experiment.

The principle scheme shown in Fig.6.1 has been utilized in high-resolution experiments performed with different versions of this arrangement. A rather attractive version is such that the superconducting specimen is in direct contact with the liquid helium bath. The configuration of Fig.6.1 may be turned upside down and be located at the lower end of an optical cryostat. The magneto-optical film, the reflecting aluminum film, and the superconducting specimen may be vapour deposited in this sequence on a substrate, which constitutes the low-temperature window of the cryostat. Observation of the magneto-optical contrast can be made with a polarizing microscope, its objective being pointed upwards toward the room-temperature window. An apparatus of this type is shown in Fig.6.4. In such an arrangement, the two windows at the bottom of the cryostat must be positioned in maximum proximity in order ot utilize a high-resolution objective requiring a relatively short working distance.

We note that in the latter arrangement the important reflection occurs at the MO/glass interface and not at a MO/vacuum interface as we had assumed in our previous discussion. When the refractive index of glass is substituted for vacuum, the reflected wave will be much smaller and consequently less influential on the quality of the magneto-optical contrast.

The high-resolution magneto-optical technique provides a spatial resolution of 0.5 μm, and a magnetic field sensitivity of about 20 Oe. Examples of magnetic flux structures obtained with this method are given in Figures 2.8-10,14,15,17. Clearly, the spatial resolution of this technique is considerably less than that of the Träuble-Essmann method. However, the possibility for continuous observation of dynamic phenomena represents an important advantage of the magneto-optic technique. Special aspects of this technique arising in high-time resolution experiments will be discussed in Sect.8.1.

Fig.6.4. Magneto-optical apparatus. Below the optical cryostat the polarizing microscope (turned upside down) can be seen

6.3 Micro Field Probes

The magnetic field distribution near the surface of a superconductor can be measured using a localized magnetic field probe. For obtaining high spatial resolution, the active area of the field probe should be as small as possible, and the arrangement should allow the well-controlled translation of the probe along the sample surface. The high *magnetoresistance of bismuth* at low temperatures has been utilized rather early for localized field measurements. A review of this work has been given by SHOENBERG [2.5]. Experiments resolving the intermediate-state structure were performed for the first time by MESHKOVSKY and SHALNIKOV [6.11,12] using a probe of the dimensions 5 μm × 10 μm × 300 μm. *Hall probes* have also been employed by various authors. BROOM and RHODERICK [6.13] used such a probe with a spatial resolution of 0.25 mm. A Hall probe with an active area of 0.6 mm × 1.8 mm has been used by WEBER and RIEGLER [6.14].

Rather detailed experiments with a *high-resolution Hall probe* have been performed by GOREN and TINKHAM [2.18]. Their Hall probe consisted of a bismuth film of 0.6 μm thickness with a field-sensitive area as small as 4 μm × 4 μm. Translation of the probe along the specimen surface could be controlled within 0.5 μm, the total possible length of travel in all directions being 1 mm. During its operation, the probe was about 5 μm away from the sample. The Hall probe was run at dc, and a field sensitivity of 0.1 Oe was achieved.

For obtaining patterns of magnetic flux structures using localized field probes, the variation of the probe signal with the location at the specimen surface must be transformed into a two-dimensional plot. Although such a transformation may be automatized in various ways, it seems to represent a rather complicated procedure compared with the methods discussed in Section 6.1,2. However, using Hall probes it is possible to achieve extremely high magnetic field sensitivity. Utilizing lock-in techniques for signal detection, this field sensitivity may be increased even further.

6.4 Neutron Diffraction

So far we have only discussed various methods operating *outside* the superconductor near the sample surface, where the magnetic field protrudes from the sample interior. However, it seems desirable to study the magnetic flux structures also within the sample interior itself. An important technique providing this possibility is available in form of neutron diffraction experiments. Presently, neutron diffraction represents the only method yielding detailed information on the microscopic properties of magnetic flux structures within the interior of a superconductor. The method relies on the interaction between the magnetic moment of the neutron and the magnetic field modulation of the vortex structure.

In neutron scattering experiments, the neutron beam diffracted by the vortex lattice is studied. Usually, the incoming neutrons are directed perpendicular to the direction of the vortex lines. According to the Bragg equation

$$2a_{hk} \sin \frac{1}{2} \vartheta_{hk} = \lambda_n , \qquad (6.5)$$

maxima of the scattered neutron intensity occur at certain scattering angles ϑ_{hk}. Here, a_{hk} is the flux-line lattice spacing between the (hk) lattice planes. λ_n is the wavelength of the neutron beam. Reasonable intensity for neutron beams can be obtained up to about 10 Å wavelength. Since the flux-line lattice spacing is typically 1000 Å, we see from (6.5) that rather small scattering angles must be resolved. With drastic collimation of the incident beam, it has been possible to measure scattering angles as small as 12'. The principal arrangement of such a small-angle scattering apparatus, including the major dimensions, is shown in Fig.6.5. A detailed description has been given by SCHELTEN [6.15]. The typical sample volume is about 1 cm^3, and the scattering data are average values over a large number of flux lines. This number of flux lines is about 10^{10}, corresponding to a total length of investigated flux lines of 100 000 km.

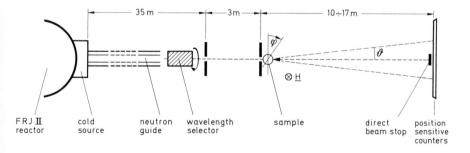

Fig.6.5. Apparatus for small-angle neutron scattering. H is the applied magnetic field, ϑ is the scattering angle, and φ is the rotation angle of the sample [6.16]

From neutron diffraction studies, the form factor $F(q_{hk})$, i.e., the normalized Fourier transform of the two-dimensional field distribution $h(\underline{r})$ of a single flux line, can be obtained. The form factor is

$$F(q_{hk}) = \frac{\int d^2r\, h(\underline{r})\, \exp(i\underline{q}_{hk}\underline{r})}{\int d^2\underline{r}\, h(\underline{r})} \quad ; \quad F(0) = 1 \quad , \tag{6.6}$$

where

$$\int d^2r\, h(\underline{r}) = \varphi_0 = 2.07 \times 10^{-7}\,\text{G cm}^2 \quad . \tag{6.7}$$

Here, q_{hk} is the reciprocal lattice vector. For a monocrystalline vortex lattice, the form factor is directly related to the integrated neutron reflectivity R_{hk} through the equation

$$R_{hk} = \frac{I_{hk}}{\Phi_n} = 2\pi\left(\frac{\gamma}{4}\right)^2 \frac{\lambda_n^2}{q_{hk}}\frac{\Omega}{F_c^2}|F(q_{hk})|^2 \quad , \tag{6.8}$$

where I_{hk} is the total intensity of the (hk)-reflection, Φ_n the unscattered neutron flux behind the sample, γ the ratio of the magnetic moment of the neutron and the nuclear magneton ($\gamma = 1.91$), and Ω the irradiated sample volume. F_c is the area of a two-dimensional unit cell of the vortex lattice. For a triangular lattice we have

$$F_c = \frac{\sqrt{3}}{2} a^2 \quad , \tag{6.9}$$

where a is the vortex-lattice parameter. Experimentally, q_{hk} is found from the scattering angle ϑ_{hk} using (6.5),

$$q_{hk} = \frac{2\pi}{\lambda_n} \vartheta_{hk} \quad . \tag{6.10}$$

Following the pioneering experiments by CRIBIER et al. [4.5], many investigations on the properties of vortex lattices in type-II superconductors have been performed by neutron diffraction. The dominant part of this work has been carried out in Saclay and Grenoble, France, and in Jülich, Germany. From these experiments, the form factors and the microscopic magnetic field distribution of the vortex structure has been obtained. Figure 6.6 shows a

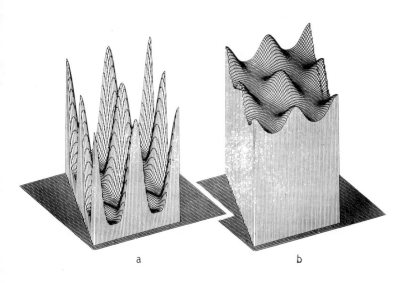

Fig.6.6a and b. Three-dimensional view of the microscopic field distribution in a niobium single crystal. (a) for a flux density B = 560 G = B_o and (b) for B = 2200 G = 0.7 H_{c2}. In (a) the nearest-neighbor distance between vortex lines is 2060 Å; the maximum field at the vortex axes is 2270 G. The corresponding numbers in (b) are 1040 Å and 2550 G, respectively (courtesy of H. Ullmaier)

three-dimensional plot of the microscopic field distribution for Nb at two flux densities, generated in this way. From such plots the maximum, minimum, and saddle point fields and their dependence on the flux density and the impurity concentration can be determined. Detailed studies of this kind were performed with pure and impure Nb and with Nb-Ta and Pb-Bi alloys [4.6-8, 6.16].

Besides the microscopic field distribution h($\underline{r}$) of a single vortex line, the correlation between the shape and orientation of the unit cell of the vortex lattice and the crystallographic structure of the superconducting metal represents an important subject of neutron diffraction experiments (see Sect.4.5.1). Further, the degree of misorientation of the vortex single crystallites in the symmetry plane of the vortex lattice and the quality of the vortex crystal (mosaic spread) can be studied from the shape of the rocking curves. Detailed experiments of this kind have been performed in single crystals of Nb, Tc, and PbBi [6.17,18,4.9]. The influence of various factors on the quality of the vortex lattice in monocrystalline Nb, as obtained from the shapes of the rocking curve, is summarized in Fig.6.7. The data are taken from the work of THOREL and KAHN [6.18]. From the width of the rocking curve, they found that with increasing magnetic field the vortex crystal becomes more perfect, as expected from the enhanced vortex-vortex interaction. An interesting correlation was found between the vortex crystal quality and the flux-flow behavior under the application of a direct electric current. At the onset of flux motion, a broadening of the rocking curve was observed, which was evident throughout the nonlinear regime of the flux-flow voltage. The rocking curve sharpened again for higher currents where the voltage-current relation became linear. These results indicate the motion of a rather perfect vortex crystal through the superconductor in the linear flux-flow regime. The orientation of the moving flux-line lattice relative to the crystallographic axes of the niobium crystal remained unchanged following a change in current direction by 90°. For the studied range of flux-flow velocities up to about 1 cm/sec no measurable effect on the magnetic field profile around a vortex core could be detected. A drastic sharpening of the rocking curve was observed following the application of a low-frequency ($\sim$ 40 Hz) alternating current or of a number of current pulses of sufficient amplitude. The frequency dependence of the effectiveness of these methods for growing vortex crystals is closely related to the skin depth for the penetration of the oscillatory field.

Although, so far, we have discussed only stationary properties of vortex lattices, as determined by neutron diffraction, this method is by no means restricted to the stationary case. Neutron diffraction experiments, providing information on the *dynamic properties* of vortex lattices, will be described in Sect.8.3.

6.5 Magnetization

Magnetization measurements represent another important tool for studying the magnetic behavior in superconductors. However, in contrast to the methods described previously, magnetization experiments only provide *information integrated over the sample volume* and do not resolve the microscopic nature of the magnetic flux structure within the material. From the variation of the magnetization with temperature and the external magnetic field, the macroscopic parameters of a superconductor (T_c, H_c, H_{c1}, H_{c2}, κ) can be determined.

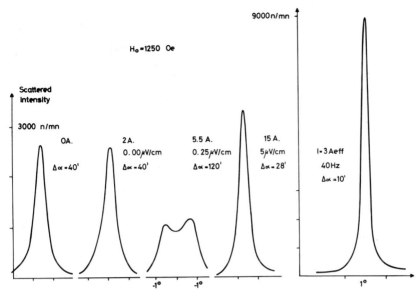

Fig.6.7. Different shapes of the rocking curve for a vortex crystal in monocrystalline Nb with a magnetic field of 1250 Oe. In each case, the integrated scattered intensity, i.e., the area under the rocking curves, remains constant. (a): The first curve is obtained on a vortex crystal generated by increasing the field above H_{c2} and decreasing it to 1250 Oe. The full-width-half-maximum of the rocking curve, $\Delta\alpha$, is a measurement of the mosaicity of this vortex crystal. (b): The second curve is done with an applied direct current of 2A in the zero voltage region of the I-V characteristics (no change in shape). (c): The third curve is for a transport current of 5.5 A (above the critical current = 4.5 A) in the nonlinear region of the I-V characteristics described by the flux creep model (strong broadening of the peak and details of the shape not very reproducible). (d): The fourth curve is for an applied current of 15 A, in the flux-flow region. The width of 28' is the limiting value when the current is increased to 30 A. (e): The last curve is for an applied ac current of 40 Hz and 3 A effective magnitude, the peak value being lower than the dc critical current (rocking curve is very sharp, with $\Delta\alpha$ = 10', representing the highest quality obtained for this sample; note change of scales) (courtesy of P. Thorel)

The magnetization M is defined as

$$M = \frac{1}{4\pi}(B - H_e) \quad , \tag{6.11}$$

where B is the magnetic flux density àveraged over the sample cross section, and H_e the external magnetic field. All methods for determining the magnetization are based on the induction law, relating the voltage V, induced in a coil of N turns, with the rate of change of magnetic flux, $d\varphi/dt$,

$$V = -N\frac{d\varphi}{dt} \quad . \tag{6.12}$$

The change of magnetic flux, $d\varphi/dt$, can be accomplished in different ways.

In the *ballistic method*, illustrated in Fig.6.8, the sample is moved quickly from the center of the upper pick-up coil A to the center of the lower coil B. Both coils A and B are identical, except that they are wound in opposite direction. An external magnetic field is provided by the solenoid S. During transfer of the sample from the upper to the lower coil, the total flux in both coils changes by the amount

$$\Delta\varphi = 8\pi \, M A_s \, , \qquad (6.13)$$

where A_s is the cross-sectional area of the sample. This flux change causes the voltage pulse $V(t)$, which can be integrated over time with a ballistic galvanometer. The magnetization is given by

$$M(H_e) = -\frac{\int V(t)dt}{8\pi \, N \, A_s} \, . \qquad (6.14)$$

By repeating the measurement at different fields H_e, magnetization curves $M(H_e)$ can be constructed. We note that the opposite direction of the windings in coils A and B results in a cancellation of any signals arising from fluctuations in the external magnetic field.

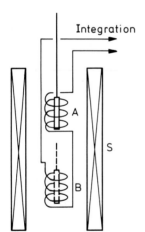

Fig.6.8. Ballistic method for measurement of the magnetization

Instead of the point-by-point measurement of the magnetization curve, just described, the function $M(H_e)$ can be recorded directly in an *electronic integrating method*. This method is illustrated in Fig.6.9. Again we have

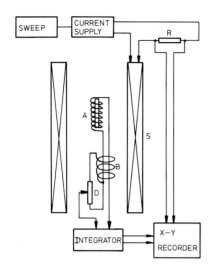

Fig.6.9. Electronic integration method for measuring the magnetization curve $M(H_e)$

two pick-up coils A and B connected in series opposition. The sample is now permanently located in the upper coil. The external magnetic field H_e, provided by the solenoid S, is increased linearly with time from field zero up to its maximum value. The voltage induced in coil A is electronically integrated. Since we are interested only in the flux change taking place within the sample volume, we must compensate for the flux changes in the remaining volume of coil A. This compensation can be performed through the coil B wound oppositely to A together with the voltage devider D. Before inserting the sample into coil A, the voltage devider is adjusted such that the total voltage is zero. In this way, after insertion of the sample, the remaining signal only arises from the flux changes within the sample volume. For the magnetization we have

$$M[H_e(t)] = -\frac{1}{4\pi \, NA_s} \int_{t'=0}^{t'=t} V(t')dt' \quad . \tag{6.15}$$

By plotting the signal of the integrator versus a voltage proportional to H_e using a x-y recorder, the $M(H_e)$ curves can directly be obtained. We note, that voltage fluctuations arising from thermoelectric effects are included in the integration procedure and may affect the results.

For samples with a small volume, yielding only a rather small signal, the *vibrating-specimen magnetometer* provides a possibility to improve the experimental accuracy. In the past, various types of this instrument have been developed. The version described by FONER [6.19] is shown schematically in Fig.6.10. The sample Sa is caused to vibrate in vertical direction using

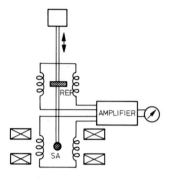

Fig.6.10. Vibrating-specimen magnetometer

a drive similar to a dynamic speaker. The dipole field of the sample induces an ac signal in a pair of coils placed on both sides of the specimen. This ac signal is compared with the signal produced by a reference sample REF located between a second pair of coils. Through the comparison between both signals, erroneous contributions arising from variations in frequency and amplitude of the vibration, and from amplifier noise can be eliminated.

We note that for obtaining the (reversible) magnetization curves characterizing the *thermodynamic equilibrium*, irreversibilities arising from flux pinning in the bulk or at the surface of the superconductor must be avoided. Flux pinning causes the formation of appreciable magnetic flux density gradients in the material and results in strongly hysteretic behavior. We shall return to the subject of flux pinning in Sect.11.

6.6 Miscellaneous

In the following we briefly describe several experimental approaches taken for investigating magnetic flux structures in superconductors, in addition to those discussed already in this section. However, we do not present a complete listing of all experiments performed in the past in this direction.

Nuclear magnetic resonance (NMR) has been utilized to measure the magnetic field distribution of the vortex structure through the broadening of the resonance line by the field inhomogeneity. Such experiments have been performed by PINCUS et al. [6.20], FITE and REDFIELD [6.21], DELRIEU and WINTER [6.22], REDFIELD [4.10], KUNG [4.11], ROSSIER and MacLAUGHLIN [6.23], and DELRIEU [4.12]. NMR measurements are particularly suitable for determining directly the local field value at the maximum, minimum, and saddle point of the field distribution within the vortex lattice. These field values can be obtained without any assumptions regarding the sign of the form factors, in contrast to neutron diffraction experiments. A disadvantage of the NMR measurements consists in the fact, that they can be performed only on powders or thin wires and foils.

An interesting *atomic-beam technique* for studying the periodic field modulation near the surface of a type-II superconductor containing a vortex lattice, has been reported by BROWN and KING [6.24]. Here, in a standard atomic-beam magnetic-resonance apparatus, the transition probability between two of the hyperfine states in a state-selected beam is measured as a function of the atomic velocity, after the beam has passed through the periodic magnetic field. If a is the wavelength of the field modulation and v the velocity of the atoms of the beam, the transition probability will increase whenever the frequency v/a is equal to the frequency of a transition between two of the hyperfine states. Motion of the vortex lattice parallel to the beam velocity (through the application of a transport current to the superconductor), causes a Doppler shift of the peaks in the transition probability when plotted versus the beam velocity. With the apparatus, periodic flux structures with wave lengths between 100 and 10^5 Å could be resolved.

Spin precession of polarized muons can be used as a microscopic probe of local magnetic fields. Positive muons can be considered as radioactive protons with known magnetic moment ($e\hbar/2m_\mu c$) and lifetime (2.2 μsek). The magnetic field on the muon is determined from the precession frequency of the anisotropic angular distribution of the decay positrons. Muon spin-precession experiments for studying the local magnetic field distribution in the mixed state of type-II superconductors have been performed by FIORY and co-workers [6.25].

In the past, various *electron-optical methods* have been tried or evaluated regarding their suitability for investigating magnetic flux structures at low temperatures. Here we have in mind experiments in which the electron beam is utilized directly for probing the magnetic field modulation of the superconductor (and not for observing the powder pattern as in the Träuble-Essmann technique). Transmission electron microscopy for studying the intermediate or mixed state has been attempted by various groups [6.26]. Similarly, scanning electron microscopy can be utilized for obtaining magnetic contrast [6.27]. Here, the contrast may result from the tilting of the dissipation profile in the presence of the field modulation *within* the superconductor [6.28-31]. In addition, the deflection of the secondary electrons by the strayfields *outside* the superconductor may provide the contrast for imaging the magnetic field modulation [6.32-35]. Further references on electron-optical methods may be found in the review by HAWKES and VALDRÉ [6.36]. So far it appears that these direct electron-optical methods have not shown satisfactory results and did not yield a spatial and magnetic resolution comparable to the methods described in Sects.6.1-4.

7. Lorentz Force and Flux Motion

7.1 Motion of Magnetic Flux Structures

In the previous sections we have dealt with the static properties of magnetic flux structures. An important development of recent years has been the investigation of the dynamic behavior of magnetic flux structures and the discovery of the intimate connection between flux motion and the transport properties of superconductors. Usually, the experimental geometry consists of a superconducting film or plate with a magnetic field applied perpendicular to the broad surfaces. Motion of the magnetic flux structure can be induced by the Lorentz force of an electric current or the thermal force of a temperature gradient. The dynamics of magnetic flux structures becomes important in various technological applications of superconductivity.

About 25 years ago, SHOENBERG [2.5] noted that the intermediate-state structure in a type-I superconductor would become unstable in the presence of an electrical current and that domain motion in a direction perpendicular to the current and the magnetic field is expected to occur. He appears to have been the first to suggest the phenomenon of current-induced flux motion. GORTER [7.1] first proposed flux flow as a resistive mechanism in a type-I superconductor. Later, the concept of flux motion was extended by GORTER [7.2,3], and ANDERSON [7.4] to vortex lines in type-II superconductors. The first experimental evidence for the appearance of flux-flow resistance in the mixed state was obtained by KIM et al. [7.5-7]. Subsequently, the Hall, Peltier, Ettinghausen, and Nernst effect induced in superconductors by the motion of magnetic flux structures as also observed. Further details regarding these developments may be found in recent reviews [7.8-12].

7.2 Lorentz Force

We now describe phenomenologically the motion of vortex lines or flux tubes under the influence of an electrical current. We consider a flat superconductor with its broad surfaces located in the x-y plane and a magnetic field applied along the z-direction. The superconducting plate is assumed to contain a lattice of vortex lines (mixed state) or some arrangement of flux tubes, each consisting of many flux quanta (intermediate state).

If an electric current with sufficient density $\underline{J}_x$ is applied in x-direction, the magnetic flux structure moves with velocity $\underline{v}_\varphi$ in a direction given by the Hall angle θ between the y-axis and $\underline{v}_\varphi$ (see Fig.7.1). The forces regulating the flux motion satisfy the following equation

$$(\underline{J}_x \times \underline{\varphi})/c - fn_s e(\underline{v}_\varphi \times \underline{\varphi})/c - \eta \underline{v}_\varphi - \underline{f}_p = 0 \quad . \tag{7.1}$$

The Lorentz force $(\underline{J}_x \times \underline{\varphi})/c$ is compensated by the Magnus force $fn_s e(\underline{v}_\varphi \times \underline{\varphi})/c$, the damping force $\eta \underline{v}_\varphi$, and the pinning force $\underline{f}_p$. Here φ is the flux contained in a vortex line or flux tube. In (7.1) the forces are given per *unit length* of vortex line or flux tube.

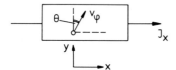

Fig.7.1. Sample geometry for current-induced flux motion

The Lorentz force can be interpreted in terms of the Maxwell stresses as in classical magneto-hydrodynamics. Using the Maxwell equation (2.10) and the vector identity

$$\frac{1}{2} \text{grad} \, (\underline{H} \cdot \underline{H}) = (\underline{H} \cdot \text{grad})\underline{H} + \underline{H} \times \text{curl} \, \underline{H} \tag{7.2}$$

we have

$$\frac{n}{c} (\underline{J} \times \underline{\varphi}) = \frac{\mu}{c} (\underline{J} \times \underline{H}) = -\frac{\mu}{8\pi} \text{grad}(H^2) + \frac{\mu}{4\pi} (\underline{H} \cdot \text{grad})\underline{H} \quad . \tag{7.3}$$

Here we have introduced the density of vortex lines or flux tubes

$$n = B/\varphi \tag{7.4}$$

in order to obtain the force per unit volume. From (7.3) we see that the Lorentz force per unit volume of the superconductor arises from the magnetic hydrostatic pressure $\mu H^2/8\pi$ and a term describing an additional tension along the lines of force. The combination of both terms yields a tension $\mu H^2/8\pi$ along the lines of force together with an equal pressure transverse to them. For illustration we consider the two limiting cases: an assembly of *straight* vortex lines with a density gradient and an assembly of *curved* vortex lines with constant density. In the first case the Lorentz force is directed down the density gradient and in the second case toward the center of curvature. The first case is realized in a geometry with little demagnetization, where the field distortion outside the superconductor arising from the vortex-line density gradient is insignificant. The second case is favored in a geometry with large demagnetization effects, like a thin plate in a perpendicular magnetic field.

It has been suggested by De Gennes [3.2,7.13] that the Magnus force plays an important role in the motion of vortex lines in superconductors similar to the situation for quantized vortices in superfluid helium. He derived for the Magnus force the expression $n_s e(\underline{v}_{-\varphi} \times \underline{\varphi})/c$. Neglecting dissipation and pinning, the Magnus force results in vortex motion perpendicular to the driving force, i.e., parallel to the current density $\underline{J}$. The Magnus force is the analog of the hydrodynamic lift force as expressed in the Kutta-Joukowski relation. These ideas have been generalized by VAN VIJFEIJKEN [7.14-16] through the introduction of the constant f in the Magnus term, which indicates the fraction of the Magnus force that is active. Except for extremely pure superconductors, f is much smaller than 1.

The damping force is described by a scalar damping factor η, containing all dissipative processes (like eddy-current damping within the normal vortex core). The pinning force $\underline{f}_p$ is directed opposite to the flux-flow velocity $\underline{v}_\varphi$. Flux pinning generally results from a local depression in the Gibbs free energy of the vortex line. Such a potential well can be caused by metallurgical inhomogeneities in the material or by geometrical effects, like the sample edges, grooves, indentations, voids, etc. For flux motion to occur, the Lorentz force must exceed the pinning force. Flux pinning will be discussed in more detail in Sect.11.

Besides describing the dynamics of flux structures under the influence of an applied electric transport current, (7.1) also regulates flux motion taking place in a time-dependent applied magnetic field for zero applied transport current. The shielding currents generated in the superconductor by the time-dependent magnetic field then produce the Lorentz force causing the propagation of the field change through the material. Of course, if the pinning force in the material exceeds the Lorentz force of the shielding currents, the magnetic field change cannot propagate and thermodynamic equilibrium will not be attained. In this case the magnetization curve M(H) is irreversible.

So far, our discussion extends equally well to single-quantum vortex lines in type-II and to multi-quanta flux tubes in type-I superconductors. In a microscopic description we must treat the detailed distribution of currents and magnetic fields in the vicinity of the flux lines. These theoretical aspects of the time-dependent behavior contained in (7.1) will be discussed in Sect.10.

7.3 Flux Flow Resistance

The steady-state motion of the magnetic flux structure induced by an applied electric current according to (7.1) causes the time-averaged macroscopic electric field following Faraday's law

$$\underline{E} = -\,\mathrm{grad}\ V = -\,(\underline{v}_\varphi \times \underline{B})/c \quad , \tag{7.5}$$

where V is the voltage. For the geometry underlying (7.1) the component $v_{\varphi y}$ causes the (longitudinal) resistive voltage, whereas the component $v_{\varphi x}$ causes the (transverse) Hall voltage. Essentially, the resistive voltage arises through the generation of normal eddy currents during flux motion.

From (7.1) we see that current-induced flux motion takes place in a direction which depends on the magnitude of the parameters f and η. Taking the case f = 1 and neglecting for simplicity the pinning force, (7.1) can be written as

$$\frac{n_s e}{c} (\underline{v}_s - \underline{v}_\varphi) \times \underline{\varphi} - \eta \underline{v}_\varphi = 0 \quad . \tag{7.6}$$

It follows that $\underline{v}_\varphi$ must be perpendicular to the difference $\underline{v}_s - \underline{v}_\varphi$. As seen from Fig.7.2, this condition requires that the component $v_{\varphi y}$ cannot be larger than $v_s/2$, independent of the value of η. For a given value of v_s (or of the current density J_s) this would result in a maximum value of the longitudinal electric field E_x. If we let $\eta \to 0$, we see that $\underline{v}_\varphi \to \underline{v}_s$, i.e., the vortex line would move along with the supercurrent velocity. Then the

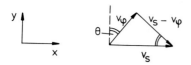

Fig.7.2. Vector diagram for $\underline{v}_s$, $\underline{v}_\varphi$, and $\underline{v}_s - \underline{v}_\varphi$ according to (7.6)

longitudinal electric field would be zero, and there would exist a transverse Hall field only. On the other hand, if we take the case f = 0, the flux-flow velocity $\underline{v}_\varphi$ is oriented perpendicular to $\underline{v}_s$. Then the Hall voltage would be zero and there would exist a longitudinal electric field only. We shall come back to this question of the Hall angle and its dependence on the parameters f and η in Sect.10.

The mechanism of flux flow for generating voltages in the mixed or intermediate state of a superconductor has been verified in a series of experiments [7.8]. Rather direct evidence for current-induced flux flow in the intermediate state has been obtained with the high-resolution magneto-optical method [2.22,7.17]. Similar observations on single-quantum vortex lines in type-II superconductors have not yet been possible because of the limitation in spatial resolution of the magneto-optical technique. However, current-induced flux flow in a type-II superconductor has been demonstrated convincingly with the Träuble-Essmann technique [2.37]. Recently, the properties of the vortex lattice during flux flow have also been investigated using neutron diffraction [7.18]. A series of experiments designed particularly to study the dynamics of magnetic flux structures in superconductors will be described in Sect.8.

Neglecting the Magnus force and inserting (7.5) into (7.1) we have

$$\partial V/\partial x = \varphi B(J_x - J_c)/\eta c^2 \quad . \tag{7.7}$$

Here J_c is the critical current density, defined by

$$J_c = f_p c/\varphi \quad . \tag{7.8}$$

According to (7.7) the longitudinal electric field increases linearly with the current density J_x, the slope being the flux-flow resistivity

$$\rho_f = \varphi B/\eta c^2 \quad , \tag{7.9}$$

which depends on the viscosity coefficient η. Following the original work of KIM et al. [7.5,7], the flux-flow resistivity has been measured in many materials. In Fig.7.3 we show typical data obtained with a niobium foil at different values of the perpendicular magnetic field. The voltage is seen to remain zero up to a certain current value. Above this critical current the voltage increases with current with increasing slope dV/dI until the linear flux-flow regime is attained indicated in (7.7). The slope of this linear regime is the flux-flow resistance. The variation of flux-flow resistivity in a type-II superconductor with magnetic field and temperature is shown schematically in Fig.7.4.

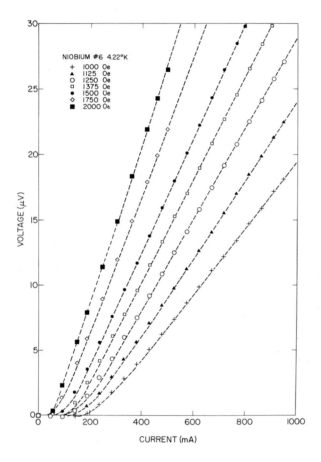

Fig.7.3. Flux-flow voltage versus current in a niobium foil for different applied magnetic fields [7.19]

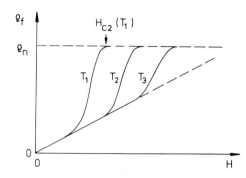

Fig.7.4. Schematic diagram of the flux-flow resistivity in a type-II superconductor versus magnetic field for different temperatures ($T_1 > T_2 > T_3$)

The critical current can be defined as the current value at which nonzero resistive voltage can first be detected. However, the results thus obtained will depend sensitively on the *voltage criterion* used in the experimental determination of the deviation from zero voltage. A procedure which is independent of any voltage criterion is obtained if the critical current is found by extrapolating the linear part of the V(I) curve linearly to zero voltage. Generally the critical current decreases with increasing magnetic field and temperature. However, in type-II superconductors the critical current often displays a peak just below H_{c2}. A discussion of this *peak effect* and of various models proposed for an explanation has been given by KRAMER [7.19a]. The curved region of the V(I) curve can be explained by nonuniform flux motion due to a spatial variation in the pinning force f_p.

For studying ideal flux-flow behavior it is desirable to reduce the influence of flux pinning as much as possible. This can be achieved by "shaking" the flux structure through the application of a small-amplitude alternating magnetic field or alternating current superimposed on the dc component [7.20]. Microwave surface resistance measurements represent an experimental method for investigating the dynamics of vortex structures which is practically free from the influence of flux pinning [7.21-24]. Here, the vortex lines are merely oscillating at the microwave frequency within the potential well of the pinning site and need not climb over an appreciable Gibbs free-energy barrier.

Extremely weak flux pinning has been found in a number of type-II alloys (like Nb with 15-20 at.% Mo). Aluminum films doped with the proper amount of oxygen appear to be an excellent material for studying ideal flux-flow behavior [7.25,26]. Complications arising in flux-flow studies from the entry and exit of vortices at the sample edges can be eliminated by replacing the usual rectangular geometry with a circular geometry (see Fig.7.5).

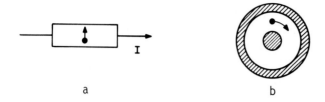

Fig.7.5. (a) Rectangular and (b) circular geometry. In the rectangular geometry the vortex lines are generated and annihilated at the sample edges during flux flow. In the circular geometry the applied current flows radially from the center to the circumference, and during flux flow the vortex lines perform a continuous circular motion around the center. (The crosshatched sections should consist of a superconductor with a critical current much larger than that of the material between these sections)

The early flux-flow resistivity experiments on type-II superconductors revealed the empirical relation in the low-temperature limit

$$\rho_f/\rho_n = H/H_{c2}(0) \;, \tag{7.10}$$

where ρ_n is the resistivity in the normal state. With (7.9) we obtain the relation

$$\eta = \varphi_0 H_{c2}(0)/\rho_n c^2 \;. \tag{7.11}$$

The quantity $H/H_{c2}(0)$ is approximately the fraction of the volume occupied by the vortex cores, and the result (7.10) indicates that the dissipation takes place predominantly in the vortex core. Deviations from the simple relation (7.10) have been found in type II superconductors with large κ values. At both low and high magnetic fields flux-flow resistivity higher than predicted by (7.10) is observed. At high fields the excess resistivity results from pair breaking caused by the paramagnetic effect [7.27,28]. This effect arises from a reduction of the normal-state energy due to the alignment of electron spins according to the normal-state paramagnetism. At low fields the excess resistivity as a function of temperature results in a resistivity minimum as observed first by AXT and JOINER [7.29] and subsequently by others [7.22,23,30]. A model accounting for this flux-flow resistivity minimum in high-κ type-II superconductors has been proposed by CLEM [7.31]. He considers the additional contribution to ρ_f arising from the entropy production associated with the generation of local temperature gradients by the moving flux lines. This contribution to ρ_f is appreciable only in impure superconductors and disappears in the pure limit.

Whereas the origin of the pinning force against the *onset* of flux motion is rather obvious from a consideration of local depressions in the Gibbs free energy of the vortex lines, the appearance of the frictional force f_p on the *moving* flux structure observed experimentally and contained in (7.1,7) is less clear. Experimentally, the V(I) characteristic is found to be generally shifted towards higher currents compared to a line passing through the origin. Further, it usually shows linear behavior following a curved region just above the onset of the resistive voltage. Hence, the frictional force f_p on the moving vortex lattice generally appears to be independent of the flux-flow velocity and approximately equal to the force that the pinning sites exert on the static vortex lattice. As a vortex line moves through the potential well of a pinning site, at first sight there appears to be no net change in free energy of the vortex line and no work done against the pinning forces. However, as shown by YAMAFUJI and IRIE [7.32], LOWELL [7.33], and GOOD and KRAMER [7.34], energy dissipation results from the elastic distortions of the vortex lines taking place as they pass through the location of a pinning site. This can lead to a frictional force independent of the flux-flow velocity and to the power dissipation $f_p v_\varphi$ per unit length of vortex line. An important assumption for this mechanism is the fact that the spatial distribution of the pinning sites is sufficiently random such that a uniform velocity v_φ of the vortex lattice can be defined. We note that the energy dissipation associated with this dynamic pinning force appears both for attractive and repulsive interaction between vortex line and pinning site.

The concept of flux-flow resistance, developed originally for type-II superconductors, holds equally well for type-I superconductors as shown directly in magneto-optical experiments [2.22,7.17]. Rather large normal domains or flux tubes of different shape also display current-induced flux motion. Highly regular domain motion under the influence of a transport current has been observed for the laminar Landau domain structure established in the Sharvin geometry [7.35-37]. However, in type-I materials, ohmic resistance in the normal laminae can cause dissipation in addition to flux flow. The ohmic resistance component in type-I superconductors generally becomes dominant for fields close to H_c, whereas the flux-flow component dominates at low fields. The accurate separation of the electric resistance into the ohmic and the flux-flow component for type-I superconductors appears to be somewhat difficult [7.38]. In addition, an electrical transport current usually causes a rearrangement of the magnetic flux structure in a type-I superconductor [7.17,39], leading to current hysteresis in the electrical resistance [7.40,41]. Following the application of a large transport current, the structure of the intermediate state generally consists of long normal domains oriented preferentially perpendicular to the current direction. Simultaneously, the electrical resistance is enhanced beyond its original value. This current hysteresis in type-I superconductors only occurs when the formation of such long normal laminae is possible. It disappears in regimes where the flux-tube configuration is energetically more favorable than the laminar structure, i.e., in type-I films below a distinct film thickness or at very low magnetic fields.

In the intermediate state also *current-induced flow of superconducting domains* can be observed near the critical field [7.42,43]. The superconducting domains move in the *same* direction as the normal domains at lower magnetic fields. Since the motion of a superconducting domain through the normal, flux-containing phase is equivalent to the motion of a corresponding flux tube in *opposite* direction (analogous to the situation of electrons and holes in semiconductors), the moving superconducting domains generate a "flux-flow voltage" opposite to the regular resistive voltage. Hence, in the presence of the superconducting domain motion near H_c the total resistive voltage is reduced below its value for the normal state, as one would expect.

Under certain conditions a combination of both dynamic and static superconducting laminae can be observed, in which trains of superconducting domains are flowing between long static domains. Whereas the static laminae are oriented nearly perpendicular to the current direction, the orientation of the dynamic laminae is nearly parallel to the current (and perpendicular to the Lorentz force).

The extension of this concept of the motion of superconducting domains near H_c to type-II superconductors seems interesting. Just below H_{c2} a type-II superconductor may be described by an ordered array of well separated islands with an order parameter of larger than zero imbedded in the normal phase (with order parameter zero).

7.4 Flux Penetration into a Superconductor

In addition to the current-induced, steady-state flux motion in the presence of a time-independent applied magnetic field, the speed of flux penetration into a superconductor in a *time-dependent* applied magnetic field is also regulated by the forces contained in (7.1).

For the one-dimensional case (like an infinite plate or a cylinder with circular cross section and infinite length oriented parallel to the field) flux penetration into a type-I superconductor has been treated theoretically by PIPPARD [7.44] and by LIFSHITZ [7.45]. At some time the external magnetic field H_e is switched on abruptly, the field being zero initially. Assuming that the process is isothermal and that the speed of flux propagation is such that the shielding eddy current just maintains the critical field at the moving phase boundary, they find a time τ for the field to completely penetrate into the superconductor given by

$$\frac{1}{\tau} = \frac{c^2}{\pi \sigma_n R_o^2} (\tilde{h} - 1) \quad . \tag{7.12}$$

Here R_o is the cylinder radius, σ_n the electrical conductivity of the superconductor in the normal state, and $\tilde{h} = H_e/H_c$ the reduced applied field. Except for the factor $(\tilde{h} - 1)$, the expression in (7.12) is identical with the result for the normal skin effect.

For finite demagnetization, i.e., for the two- and three-dimensional case, the mathematical treatment of flux penetration leads to considerable difficulties. These difficulties arise from the coupled differential equations which constitute Maxwell's equations for the different vector components. However, for the two- and three-dimensional case the relaxation time for magnetic flux penetration can be obtained in good approximation from an energy-balance argument due to FABER [7.46] independent of the detailed mechanism of the penetration. This argument states, that the excess of the Gibbs free energy ΔG associated with the unstable state assumed by the system immediately after the abrupt application of the magnetic field, is equal to the energy dissipated by the eddy currents generated during flux penetration. In this way (7.12) has been generalized by LAENG et al. [7.47] to the case of a circular disk of radius R_o and finite demagnetizing coefficient D suddenly subjected to an external field H_e within the range $(1 - D)H_c < H_e < H_c$. The time τ for attaining the equilibrium flux distribution through the specimen is found to be

$$\frac{1}{\tau} = \frac{9}{16} \frac{c^2}{\pi \sigma_n R_o^2 (1-D)} [\tilde{h} - (1 - D)] \quad . \tag{7.13}$$

We note that, aside from the numerical factor 9/16, for $D \to 0$ (7.13) reduces to (7.12).

So far we have dealt with the penetration of magnetic flux into a superconductor which originally had been in the flux-free Meissner state. In the following we consider the more general case where the superconductor initially resides in the intermediate or mixed state and where additional flux penetration takes place because of an *abrupt change in the external field* H_e by the amount ΔH_e. We assume a geometry where demagnetizing effects play a dominant role, the demagnetizing coefficient being close to 1. The following discussion refers to an *electrical conductor in general* and need not be restricted to the case of a superconductor. We consider a circular disk of radius R_0 and thickness d, with $R_0 \gg d$. An external magnetic field H_e is oriented perpendicular to the flat disk. This field is changed abruptly by the amount ΔH_e. We wish to calculate the time τ required for the change in magnetic flux to penetrate to the center of the disk. Immediately following the abrupt change in magnetic field, this additional flux will be completely expelled from the disk because of eddy-current screening, whereas above and below the disk the change in field is propagating through space with the speed of light. This results in a distortion of the additional field ΔH_e similar to the Meissner effect in a superconductor. Again we equate the excess Gibbs free energy of this unstable state with the energy dissipated by the eddy currents for calculating the relaxation time τ.

The Joule heat Q_J produced in the disk during flux penetration is

$$Q_J = \int_0^\tau dt \int d\Omega \, \sigma E^2(r,t) \quad , \tag{7.14}$$

where σ is the electrical conductivity and $E(r,t)$ the electric field at radius r and time t. Neglecting the Hall effect, the field $E(r,t)$ only contains a component in azimuthal direction. The second integral is taken over the sample volume Ω and is in principle a function of t. For the line integral along a circle of radius r we have

$$\oint ds \, E(r,t) = 2\pi r \, E(r,t) = -\frac{1}{c} \dot{\varphi}(r,t) \quad . \tag{7.15}$$

The time-derivative $\dot{\varphi}(r,t)$ of the flux threading the circle of radius r can be approximated by

$$\dot{\varphi}(r,t) = r^2 \pi \, \Delta H_e / \tau \quad , \tag{7.16}$$

thereby eliminating any time dependence of the electric field. Inserting (7.15,16) into (7.14) we find

$$Q_J = \tau\sigma d \int_0^{R_o} dr\, 2\pi r E^2(r) = \frac{\sigma\Omega}{8c^2\tau}\Delta H_e^2\, R_o^2 \quad . \tag{7.17}$$

In obtaining (7.17) we have neglected any dependence of $\dot{\varphi}$ and E on the coordinate in axial direction, which appears reasonable if the specimen is rather thin.

The excess Gibbs free energy ΔG per volume is

$$\frac{\Delta G}{\Omega} = -\int_0^{\Delta H_e} M(H)dH = \frac{1}{4\pi(1-D)}\int_0^{\Delta H_e} HdH = \frac{1}{8\pi}\frac{\Delta H_e^2}{1-D} \quad . \tag{7.18}$$

Equating ΔG with the dissipated energy of (7.17) we obtain

$$\frac{1}{\tau} = \frac{c^2}{\pi(1-D)}\frac{1}{\sigma R_o^2} \quad . \tag{7.19}$$

Of course, an expression of this type also describes the penetration depth of an oscillatory magnetic field into a circular disk with the conductivity σ. The result (7.19) (except for a factor of $\pi/2$) has also been empirically obtained from the maximum in the resistive loss per cycle of circular metal foils [5.11]. For $D \to 0$ (7.19) yields the regular skin effect in the one-dimensional case.

Equation (7.19) can directly be applied to a superconductor residing in the intermediate or mixed state. The relaxation time for the complete penetration of an abrupt change in magnetic field ΔH_e is expected to be given by (7.19) if σ^{-1} is replaced by the flux-flow resistivity ρ_f obtained for the field H_e, assuming $\Delta H_e \ll H_e$. In (7.19) σ is then, of course, a function of the field H_e. This expectation has been verified in an experimental study of the penetration of an oscillatory magnetic field component into thin-film superconductors and of its frequency dependence [7.48,49]. The relaxation time for magnetic flux penetration has also been investigated by ULLMAIER [7.57].

Finally, we note that the irreversible entry of magnetic flux into a superconductor can be strongly influenced by a barrier in the Gibbs free energy of a normal domain near the sample edges, as we have discussed in Sect.4.6.1.

7.5 Hall Effect

As a consequence of the Magnus force in (7.1) the flux-flow velocity contains a component in x-direction (parallel to the transport current) resulting in a transverse electric field, i.e., the Hall effect. A beautiful demonstration of the Hall effect was given by HAENSSLER and RINDERER [2.11] in a study of flux penetration into a circular disk of a type-I superconductor using the Bitter method for decoration of the magnetic flux. Upon the application of a perpendicular magnetic field, instead of penetrating straight to the center of the disk in radial direction, flux was observed to move to the center in spirals, the azimuthal component of flux motion being due to the Magnus force (Fig.7.6). The Hall effect has also directly been visualized magneto-optically from the angle of orientation of the laminar domain structure in a current carrying superconducting strip of type-I [7.39]. The domain orientation was found to deviate by the Hall angle from the direction perpendicular to the current.

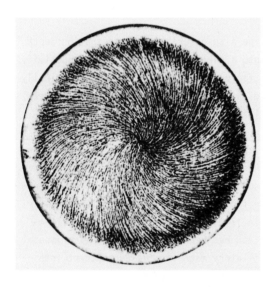

Fig.7.6. Spiral path of magnetic flux penetration into a superconducting In disk of 38 mm diameter and 3.98 mm thickness demonstrating the Hall effect. Purity of the In: 99.999 %; T = 1.755 K (Courtesy of L. Rinderer)

The Hall effect in the mixed state of intrinsic type-II superconductors and type-II superconducting alloys has been measured in many experiments. Perhaps the material mostly studied has been niobium [7.50-54]. The measurements included the dc Hall field, helicon resonances, and the surface resistance in circularly polarized fields. In niobium the Hall angle was found to depend strongly on the purity of the material. A similar behavior was observed in vanadium [7.54,55]. In Fig.7.7 the reduced Hall angle is plotted versus the reduced field for two vanadium specimens of different purity. In Fig.7.8 we show the low-field limit P(o) of the reduced Hall angle obtained for vanadium and niobium by various authors as a function of the ratio ξ_0/ℓ of the coherence length to the electron mean free path.

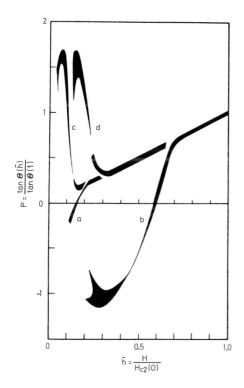

Fig.7.7. Reduced Hall angle $P = \tan\theta(\tilde{h})/\tan\theta(1)$ versus the reduced field $\tilde{h} = H/H_{c2}(0)$ for two vanadium samples of different purity. (a) 4.22 K and (b) 2.13 K for a specimen with the residual resistance ratio RRR = 14.5; (c) 4.22 K and (d) 3.76 K for a specimen with RRR = 7.2 [7.55]

Further references on measurements of the Hall effect in type-II alloys may be found in the papers by HAKE [7.58], BYRNAK and RASMUSSEN [7.59], WEIJSENFELD [7.60], and GILCHRIST and VALLIER [7.53]. These measurements were also extended to intermetallic compounds [7.61]. In Sect.10 we shall compare the experimentally observed Hall angles with theoretical models.

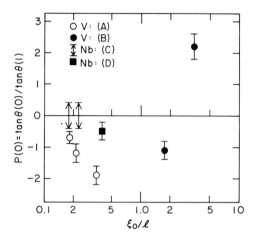

Fig.7.8. Low-field limit P(0) of the reduced Hall angle for vanadium and niobium versus the ratio ξ_0/ℓ of the coherence length and the electron mean free path. (A) and (C) [7.54], (B) [7.55], (D) [7.56] (from [7.54])

7.6 Ettinghausen and Peltier Effect

Motion of the magnetic flux structure across the superconductor, induced by the Lorentz force, is associated with the transport of entropy, since within the normal core of a vortex line or of a (multi-quantum) flux tube the entropy density is larger than in the surrounding superconducting phase. We consider again a flat superconductor placed perpendicularly in a magnetic field and use the same coordinates as in Fig.7.1. During flux flow vortex lines or larger flux tubes are generated at a sample edge and are annihilated at the opposite edge, resulting in the absorption of heat energy from the material at the first edge and its emission at the second edge. In this way a temperature gradient is established along the direction of flux flow, with a component perpendicular and parallel to the electric current direction, in analogy to the Ettinghausen and Peltier effect, respectively.

The heat current density $\underline{U}$ coupled with the motion of the magnetic flux structure is

$$\underline{U} = nTS_\varphi \underline{v}_\varphi \quad , \tag{7.20}$$

where n is the density of vortex lines or flux tubes and S_φ the transport entropy per unit length of vortex line or flux tube. In the stationary case the heat current density (7.20) is compensated by regular heat conduction, and we have for the component in y-direction

$$U_y = nTS_\varphi v_{\varphi y} = -K \frac{\partial T}{\partial y} \quad . \tag{7.21}$$

A similar equation holds for the component in x-direction. K is the heat conductivity. With (7.5) we have

$$\left|\frac{\partial T}{\partial y}\right| = c \frac{T}{K} \frac{S_\varphi}{\varphi} \left|\frac{\partial V}{\partial x}\right| \quad . \tag{7.22}$$

We see that the transverse temperature gradient of the Ettinghausen effect is proportional to the longitudinal electric field, the proportionality constant being determined by S_φ/φ, K, and T. A relation similar to (7.22) holds for the longitudinal temperature gradient of the Peltier effect, $|\partial T/\partial x|$ being directly connected with the Hall field $|\partial V/\partial y|$. Experimental data illustrating (7.22) are shown in Fig.7.9.

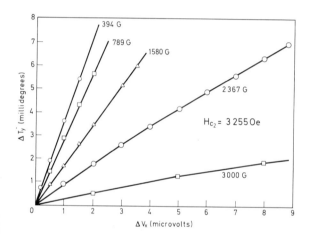

Fig.7.9. Transverse temperature difference caused by the Ettinghausen effect versus the longitudinal flux-flow voltage for the type-II alloy In + 40 at.% Pb at different magnetic fields [7.62]

The Ettinghausen and Peltier effect associated with the motion of the magnetic flux structure have been observed in various type-I and type-II superconductors [7.62-65]. Further references may be found in the reviews by CAMPBELL and EVETTS [7.10], and HUEBENER [7.11]. VIDAL [7.66] measured the entropy transported by the moving vortex structure in a type-II superconductor from the second-sound generation by an ac method. It is interesting to note that due to the nature of the current-induced flux motion the (transverse) Ettinghausen effect is usually 2-3 orders of magnitude larger than the (longitudinal) Peltier effect, in contrast to the behavior normally found in metals. Further, in the mixed or the intermediate state the Ettinghausen effect is 2-3 orders of magnitude larger than in the normal state. The ratio of the Peltier and the Ettinghausen heat current is, of course, given by the Hall angle θ, and we have

$$\tan \theta = \frac{V_{\varphi x}}{V_{\varphi y}} = \frac{U_x}{U_y} = \left|\frac{\partial T/\partial x}{\partial T/\partial y}\right| \quad . \tag{7.23}$$

From the Onsager reciprocity relations we can immediately conclude, that also the Nernst and Seebeck effect must exist in the mixed and the intermediate state of a superconductor. However, we postpone a discussion of these phenomena until Sect.9. The main quantity to be derived from all these thermomagnetic effects during flux flow is, of course, the transport entropy S_φ. We shall return to this quantity S_φ in a more detailed discussion in Sect.9.3 and 10.

7.7 Josephson Relation

Whereas (7.5) for the flux-flow voltage in superconductors is generally accepted, its direct interpretation in terms of Faraday's law has been the subject of many discussions [7.8,67]. Eddy-current damping of the flux motion is clearly a dissipative process and should result in a resistive voltage. On the other hand the question remains how an electric field can be sustained in a multiply connected superconductor. The answer is provided by the Josephson relation [7.68]

$$\frac{\partial \Phi}{\partial t} = 2 \text{ eV}/\hbar \quad . \tag{7.24}$$

Here

$$\Phi = \varphi_2 - \varphi_1 \tag{7.25}$$

is the phase difference of the complex order parameter Ψ for the two points in space (1) and (2), Ψ being written in the form of (3.9). Equation (7.24) connects the time derivative $\partial \Phi/\partial t$ with the chemical potential difference between the points (1) and (2) and may be interpreted in terms of the Einstein relation

$$\Delta E = \hbar\omega \quad , \qquad (7.26)$$

with $\Delta E = 2eV$. The factor 2 arises since we are dealing with a macroscopic quantum system based on Cooper pairs.

Treating the complex order parameter (3.9) as a quantum mechanical wave function we can obtain the Josephson relation in the followay way. We start with two superconductors (1) and (2) which are weakly coupled with each other via a thin electrically insulating barrier (see Fig.7.10). For simplicity we assume the two superconductors to be the same at both sides of the "Josephson junction". Denoting the wave function in each superconductor by Ψ_1 and Ψ_2, respectively, we have the relations

$$i\hbar \frac{\partial \Psi_1}{\partial t} = \mu_1 \Psi_1 + K^* \Psi_2 \qquad (7.27)$$

$$i\hbar \frac{\partial \Psi_2}{\partial t} = \mu_2 \Psi_2 + K^* \Psi_2 \quad . \qquad (7.28)$$

Here μ_1 and μ_2 is the particle energy on both sides of the junction, and K^* describes the coupling. If a potential difference V across the junction exists, we can set

$$\mu_1 = \frac{e^*V}{2} \quad \text{and} \quad \mu_2 = -\frac{e^*V}{2} \quad . \qquad (7.29)$$

Introducing the particle densities n^*_{s1} and n^*_{s2}, we have

$$\Psi_1 = \sqrt{n^*_{s1}}\, e^{i\varphi_1} \quad \text{and} \quad \Psi_2 = \sqrt{n^*_{s2}}\, e^{i\varphi_2} \quad . \qquad (7.30)$$

Inserting (7.30) into (7.27), one obtains

$$i\hbar\, e^{i\varphi_1} \frac{1}{2\sqrt{n^*_{s1}}} \dot{n}^*_{s1} - \hbar\sqrt{n^*_{s1}}\, \dot{\varphi}_1\, e^{i\varphi_1} = \frac{e^*V}{2} \sqrt{n^*_{s1}}\, e^{i\varphi_1} + K^* \sqrt{n^*_{s2}}\, e^{i\varphi_2} \qquad (7.31)$$

and a similar equation from (7.28). Finally, by equating the real and imaginary parts in each case, we find

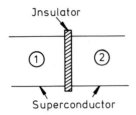

Fig.7.10. Josephson weak link between two superconductor

$$\dot{n}^*_{s1} = \frac{2K^*}{\hbar} (n^*_{s1} n^*_{s2})^{\frac{1}{2}} \sin(\varphi_2 - \varphi_1)$$

$$\dot{n}^*_{s2} = - \frac{2K^*}{\hbar} (n^*_{s1} n^*_{s2})^{\frac{1}{2}} \sin(\varphi_2 - \varphi_1)$$

(7.32)

$$\dot{\varphi}_1 = - \frac{e^*V}{2\hbar} - \frac{K^*}{\hbar} \left(\frac{n^*_{s2}}{n^*_{s1}}\right)^{\frac{1}{2}} \cos(\varphi_2 - \varphi_1)$$

$$\dot{\varphi}_2 = \frac{e^*V}{2\hbar} - \frac{K^*}{\hbar} \left(\frac{n^*_{s1}}{n^*_{s2}}\right)^{\frac{1}{2}} \cos(\varphi_2 - \varphi_1)$$

(7.33)

The current across the junction is

$$J = \dot{n}^*_{s1} = - \dot{n}^*_{s2} = \frac{2K^*}{\hbar} (n^*_{s1} n^*_{s2})^{\frac{1}{2}} \sin\varphi$$

(7.34)

using (7.25). This result can be written as

$$J = J_0 \sin \varphi$$

(7.35)

with $J_0 = 2K^* n^*_s/\hbar$. On the other hand, (7.33) yields the result given in (7.24), if we identify $e^* = 2e$. Both results (7.24) and (7.35) are obtained by equating $n^*_{s1} = n^*_{s2} = n^*_s$ and represent the fundamental equations of the Josephson effect.

From (7.24) we expect a *time-dependent* voltage between two points (1) and (2) in a superconductor if a vortex line traverses the region between both points. Hence the voltage associated with the uniform motion of an arrangement of many vortex lines consists of a superposition of individual voltage pulses which contribute significantly to the electrical noise power (see Sect.13). Calculations of the shape of the voltage pulses and its dependence on the geometrical details have been performed by CLEM [7.69].

Turning now to the *time-averaged* voltage generated by a moving arrangement of vortex lines, we replace the time variation of ∂Φ/∂t associated with the motion of each individual flux quantum by a phase jump of 2π as the vortex line crosses the straight line connecting points (1) and (2). One finds

$$\overline{\partial\Phi/\partial t} = 2\pi s\, v_\varphi n \quad , \tag{7.36}$$

where s is the distance between (1) and (2) and n the vortex-line density. With (7.24,3.51) we obtain

$$\overline{|E|} = \frac{\overline{V}}{s} = \frac{1}{c} v_\varphi B \quad , \tag{7.37}$$

i.e., the same result as in (7.5). For simplicity, in deriving (7.37) we assumed the orientation of v_φ, s, and of the vortex lines to be perpendicular to each other and omitted the vector notation. The close coupling between the flux-flow velocity and the electric field represents a useful feature for investigating vortex dynamics in superconductors.

7.8 Instabilities

Flux motion at very high velocities results in strong energy dissipation and can lead to thermal instabilities and to a temperature rise of the superconductor above the transition temperature. These thermal effects can be minimized by optimizing the heat transfer to the low temperature bath (large surface to volume ratio, film substrates with high heat conductivity, operation in superfluid helium). The influence of thermal effects can also be reduced through the utilization of pulse techniques. Although thermal instabilities represent an interesting and important subject, we do not pursue it here any further. Thermal effects in the current-induced resistive state of thin-film bridges will be discussed to some extent in Sect.14.4.

As the flux-flow velocity becomes larger and larger, something interesting can be expected to happen when the flux-flow speed approaches sound velocity, if for no other reason, because superconductivity is a phenomenon mediated by phonons. A moving flux structure passing through the sound barrier should generate "Čerenkov radiation" of phonons. This problem has been treated theoretically by SUGAHARA [7.70], who predicted the emission of elastic shock waves by flux lines accelerated to supersonic velocities. An experimental verification of these effects has not yet been reported.

THOMPSON and HU [7.71] recently extended their theoretical treatment of the vortex dynamics, based on the time-dependent Ginzburg-Landau equations for a gapless superconductor to nonlinear effects. For simplicity they restricted themselves to films with thickness much less than the penetration depth. They showed that above a maximum value of the transport current the flux-flow state is unstable. At the instability the resistive voltage is expected to jump discontinuously from the lower flux-flow value to the higher normal-state value. Under voltage-biased conditions THOMPSON and HU predicted an inhomogeneous state of the film with alternating superconducting regions at the maximum current and normal regions at the same current. This dynamic instability becomes most pronounced if the magnetic field applied perpendicular to the film is small. Although THOMPSON and HU assumed rather restricted conditions, their results are expected to be more generally valid. Very likely their treatment serves as a qualitative explanation of the instabilities observed in a series of experiments mentioned below. We note that these instabilities of vortex flow in an applied magnetic field are, of course, closely connected with the current-induced transitions in zero applied field. The theory by THOMPSON and HU is just the extension to the flux-flow state of the case of the critical current in a thin film or wire for zero applied field discussed in Sect.3.3. The current-induced resistive state in zero applied magnetic field will further be discussed in Sect.14.

Recently MEISSNER [7.72,73] observed an instability in the flux-flow behavior of thin type-I superconducting films at high flux-flow velocities, which manifests itself in a sharp kink upward in the V(I) curve and a rapid rise of resistance with current above an instability current. These experiments were performed using current pulses of 80 nsec duration. Thermal effects for explaining the results can be ruled out for several reasons. Subsequently, MEISSNER [7.74] extended his experiments to time-of-flight measurements of the flux-flow speed. Here, a legion of flux lines is generated by pulsing a drive coil on one side of a film carrying an appropriate electrical current. Following their current-induced motion across the film, the arrival of the flux lines on the other side of the film is detected with a pick-up coil. The time-of-flight experiments indicate that close to the instability current the flux-flow speed reaches the sound velocity in the metal. We note that the pulsed nucleation of the localized group of flux lines in a superconductor carrying an electrical current of high density

represents an interesting method for studying extreme flux-flow speeds
without encountering the thermal difficulties arising during rapid flux
motion of a vortex lattice filling the *whole* specimen. In this respect the
situation is similar to the current-induced resistive state in a constricted
thin-film superconductor in which a train of flux lines or flux tubes is
nucleated at the sample edge, as will be discussed in Sect.14.2. Qualitatively
the experimental results of MEISSNER agree with the theoretical prediction
by THOMPSON and HU [7.71]. However, the detailed nature of the observed
flux-flow instability is not yet completely understood. Further experiments
qualitatively supporting the theory by THOMPSON and HU have been reported
by MONCEAU [7.75] and GRAY [7.76].

7.9 Force-Free Configurations

So far in our discussion of current-induced flux motion we have assumed that
the current and the magnetic field are directed perpendicular to each other.
An interesting geometrical configuration consists of a superconducting wire
carrying a transport current in the presence of an applied *longitudinal
magnetic field*. Such systems display a longitudinal paramagnetic moment.
Apparently, the transport current flows along a helical path and thereby
generates an additional longitudinal component of magnetic flux within the
superconductor [7.77,78]. SEKULA et al. [7.79] observed that the critical
current in type-II superconducting wires in a longitudinal magnetic field
is much higher than in a transverse field. It was suggested by BERGERON
[7.80] that in a longitudinal field the transport current and the vortex
lines adopt a force-free configuration in which $\underline{J}$ and $\underline{B}$ are oriented parallel to each other and the Lorentz force vanishes.

Presently, the behavior of helical vortices in type-II superconductors
is not completely understood. The experimental situation and various theoretical models have been discussed in reviews by CAMPBELL and EVETTS [7.10],
and TIMMS and WALMSLEY [7.81]. Still unresolved is the question of the vortex arrangement in the critical state and above the critical current where
both a resistive voltage and a longitudinal paramagnetic moment can be observed. As shown by CLEM [7.82], a vortex line subjected to a sufficiently
large current density applied parallel to the vortex axis, is unstable
against the growth of helical perturbations. (An analogous instability is
known in magnetohydrodynamics). This instability may play an important role
in the complex motions of spiral vortices taking place in the dissipative
regime.

8. Special Experiments

In this section we discuss various experiments which are particularly useful for investigating the dynamics of magnetic flux structures in superconductors. Here flux motion is generally induced via the Lorentz force.

8.1 Magneto Optics

The magneto-optic method described in Sect.6.2 provides a continuous image of the magnetic flux structure at the surface of the superconductor and is a highly useful tool for studying flux-flow behavior including motion picture techniques. Ultimately the time resolution of the magneto-optic technique is limited by the relaxation time of the magneto-optic material. In nonmetallic ferromagnetic compounds the magnetization can be modulated at microwave frequencies indicating relaxation times less than 10^{-10} s [8.1]. Similar relaxation times can also be expected in the case of the magnetic semiconductors EuO, EuS, and EuSe mentioned in Sect.6.2.

Using a high-speed cine camera the propagation of flux jumps has been investigated in type-II superconductors with time intervals of about 100 µs between consecutive frames [8.2-5].

Flux motion with strong periodicity in time may be studied with rather high time resolution using stroboscopic illumination in combination with magneto-optic flux detection. The principle of this method is shown in Fig.8.1. The magnetic flux structure in a type-I superconductor is caused to undergo a periodic motion by superimposing an oscillatory component to the sample current (or to an external applied magnetic field). The same oscillator modulating the flux structure in the sample is used for driving an acousto-optical shutter, thereby generating short light pulses synchronously with the sample modulation. By varying the phase between the sample modulation and the light pulse the complete time evolution during one cycle can be investigated. Such a system has been successfully operated in a study

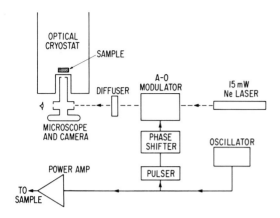

Fig.8.1. Stroboscopic method for magneto-optical flux detection (courtesy of D.E. Chimenti)

of the dynamics of the current-induced resistive state yielding a time resolution of better than 10^{-7} s [8.6].

8.2 Nuclear Magnetic Resonance

As shown by DELRIEU [8.7], a nuclear magnetic resonance method may be used to measure the speed of the flux-line lattice. If the flux-line lattice moves through the superconductor, the nuclear spins are exposed to an oscillating magnetic field, and the resonance frequency is frequency modulated. For high flux-line speeds this results in a narrowing of the linewidth, thus providing a measure of the flux-line speed. DELRIEU was able to observe such an effect.

8.3 Neutron Diffraction

A neutron wave travelling with speed v_n through a moving flux-line lattice encounters flux lines with their apparent positions displaced in the direction of the flux-line velocity v_φ. This displacement produces a shift of approximately v_φ/v_n in the Bragg scattering angle. For an experimental observation shifts in excess of 10^{-2} radians would be necessary. Therefore, flux-line speeds of, say, $v_\varphi = 1$ m/s and slow neutrons with $v_n = 100$ m/s or 40 Å wavelength are required. An experiment of this kind has been proposed by THOREL and KAHN [6.18]. The effect has been observed recently by SCHELTEN et al. [7.18] using a polycrystalline type-II NbTa superconductor. Their sample provided the high normal-state resistivity necessary for obtaining a large value of v_φ. At the same time it showed nearly ideal type-II behavior with very little flux pinning and rather narrow rocking curves. The authors obtained sufficient resolution for the scattering angle by analyzing the *small relative shifts of the steep wings of the rocking curves* attained upon reversal of the applied current. An example of their data is shown in Fig.8.2. The results confirmed the existence of current-induced motion of the vortex lattice and of the validity of (7.5) between the electric field $\underline{E}$ and the velocity $\underline{v}_\varphi$.

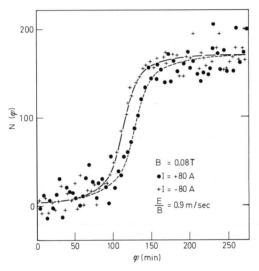

Fig.8.2. Shift $2\Delta\varphi$ of the wing of a rocking curve caused by reversing the transport current from + 80 A to - 80 A for a nearly reversible type-II NbTa superconductor. The observed shift $2\Delta\varphi$ corresponds to a velocity change from + 0.9 to - 0.9 m/s. $N(\varphi)$ is the diffracted intensity [7.18]

In principle, the Doppler shift in wavelength or speed of the neutron beam resulting from its interaction with the moving vortex lattice can also be utilized for measuring flux-flow speeds. As pointed out by THOREL and KAHN [6.18], by using a neutron spin-echo technique [8.8], flux-line speeds as low as 5 cm/s would be detectable with neutrons of 20 Å wave length.

As pointed out in Sect.6.4, the influence of current-induced flux flow on the *morphology of the vortex lattice* has also been investigated by neutron diffraction. Further, the vortex configuration in the *critical state* just before the onset of current-induced flux motion represents an interesting subject for neutron diffraction experiments. As a result of (7.3) the critical state is characterized by density gradients and curvature in the vortex lines. A spatial dependence of this behavior can be caused by variations in the pinning-force density. The spatial distribution of flux pinning (and the distribution of the transport current) has been studied in neutron diffraction experiments by SIMON and THOREL [8.9] and by KROEGER and SCHELTEN [8.10].

8.4 Sharvin Point Contact

As has been demonstrated by SHARVIN [8.11], the electrical resistance of a small contact consisting of a thin (about 25 μm diameter) wire of copper or some other metal welded to the surface of a superconductor changes when the material under the point contact passes through the superconducting phase transition. In this way the resistance of the Sharvin point contact is sensitive to the passage of normal domains through the area of the contact point and may be utilized for studying the dynamics of the intermediate state. The maxima and minima of the resistance correspond to the normal and superconducting state, respectively, of the material under the contact point. The effective contact area in such experiments has a diameter of the order of 1 μm and is appreciably smaller than the cross section of the wire [8.11, 12]. Measuring currents of about 1 mA are used. By attaching two or more contacts to the sample in close proximity, a "time-of-flight" experiment can be performed to determine the speed of the domain motion. This technique is well suited for investigating the dynamics of highly regular flux structures like the laminar Landau domains generated in an inclined magnetic field (Sharvin geometry).

Extremely regular motion of the Landau domain structure in the Sharvin geometry under the influence of a transport current has been observed by FARRELL [7.35,36]. In these experiments flux pinning effects could be minimized through the application of ancillary field coils at both sample ends. An example of these results is shown in Fig.8.3. Domain motion induced by the thermal force of a temperature gradient has also been studied with the point contact method [8.13]. All these experiments by FARRELL and co-workers have been performed with tin, and the measurements were subsequently extended to zinc [8.14].

From the shape and temperature dependence of the changes in resistance CHIEN and FARRELL [8.15] concluded that the resistance of the point contact is a measure of the local value of the order parameter right at the point and does not involve extended averaging over the current distribution in the

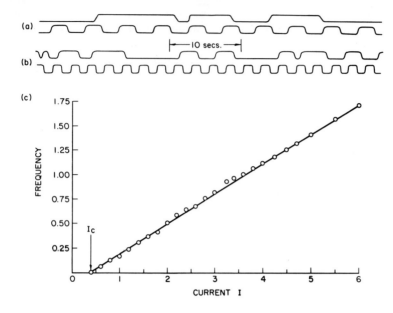

Fig.8.3. (a), (b): Resistance fluctuations of the Sharvin point contact caused by the passage of Landau domains under the contact before (upper trace) and after (lower trace) the ancillary field is applied. The vertical axis for each trace records resistance. The maxima and minima correspond to normal and superconducting material under the contact, respectively. Time increases to the right along the horizontal axis. (a) T = 3.34 K, H = 55 Oe, I = 2.0 A. (b) T = 1.60 K, H = 216 Oe, I = 3.0A. (c) Frequency (number of domains passing under the contact per second) as a function of transport current I (in amperes) with the ancillary field turned on; T = 1.82 K, H = 202 Oe [7.35]

neighborhood of the point. It appears that such a response of the point contact may be caused by the proximity effect. From these results it seems feasible that the Sharvin point contact may also yield details about the interface between normal and superconducting domains and may become useful for the study of type-II materials.

8.5 Magnetic Coupling

An important development in the study of flux motion in superconductors have been experiments on magnetically coupled sandwich structures. The first experiments of this kind were performed by GIAEVER [8.16,17]. The principle is shown in Fig.8.4. A primary and a secondary superconducting film are deposited on a substrate. Both films are electrically separated from each

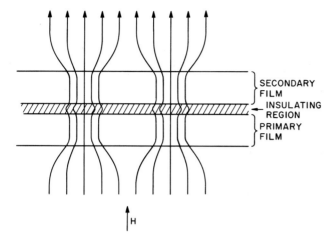

Fig.8.4. Magnetic coupling between two superconducting films separated from each other by a thin insulating layer

other by a thin insulating layer from, say, SiO_2. In a magnetic field oriented perpendicular to the sandwich, vortex lines or multi-quantum flux tubes exist in both superconducting films. If the insulating layer is sufficiently thin and in the absence of appreciable flux pinning in the superconducting films, the flux structures in both films will perfectly register with each other through magnetic coupling as indicated in Fig.8.4. This magnetic coupling effect is expected to occur if the thickness of the insulating layer becomes less than about the spacing of the vortex lines or flux tubes. If flux flow is induced in the primary film through the application of a primary current, flux motion will also take place in the secondary film because of the magnetic coupling between the two films. Therefore, in addition to the flux-flow voltage in the primary film, a voltage in the secondary film can be observed, although the transport current applied to the secondary film is zero. Because of this, the configuration in Fig.8.4 has been termed a dc transformer. GIAEVER observed these effects originally in type-II films of tin. Subsequently similar effects were also found in type-I superconductors [8.18,7.17].

When a transport current is applied only to the primary film, the ratio V_1/V_2, where V_1 and V_2 is the primary and secondary voltage, respectively, is generally referred to as the *coupling parameter*. If flux pinning can be neglected, magnetic coupling between the moving flux structures in both films is expected to take place with a coupling parameter near 1.0 as long as the viscous drag force in the secondary film is appreciably less than the coupling force. As soon as the viscous drag force exceeds the coupling force (at sufficiently large flux flow velocities), slippage between the flux structures in the primary and secondary film will occur, the coupling parameter will be less than 1.0 and will decrease with increasing primary current and flux flow speed. A model along these lines has been proposed by CLADIS et al. [8.19] to explain qualitatively their experimental observations. From a force balance as contained in (7.1) one can write

$$\eta_1 \dot{x}_1 = - A \sin\left[\frac{2\pi}{a} (x_1 - x_2)\right] + f_L \tag{8.1}$$

$$\eta_2 \dot{x}_2 = A \sin\left[\frac{2\pi}{a} (x_1 - x_2)\right] . \tag{8.2}$$

Here we have neglected the Magnus force and the pinning force. Current-induced flux flow is assumed to take place along the x-coordinate. The indices refer to the primary and secondary film, respectively. The first term on the right represents the force on the flux structure in one film arising from the magnetic coupling to the structure in the other film. The quantity A is a general constant. The length a is the lattice constant of the vortex lattice along the x-direction. The modelling of the magnetic coupling force with a sinusoidal function in (8.1,2) is a convenient simplification and does not represent an essential feature except for the periodicity with the lattice parameter a. In (8.1) the Lorentz force is denoted by f_L.

From (8.1,2) we obtain for the time-averaged quantities $\overline{\dot{x}_1}$ and $\overline{\dot{x}_2}$

$$\overline{\dot{x}_1} = \frac{f_L}{\eta_1 + \eta_2} + \frac{\eta_2}{\eta_1 + \eta_2} \overline{(\dot{x}_1 - \dot{x}_2)} \tag{8.3}$$

$$\overline{\dot{x}_2} = \frac{f_L}{\eta_1 + \eta_2} - \frac{\eta_1}{\eta_1 + \eta_2} \overline{(\dot{x}_1 - \dot{x}_2)} . \tag{8.4}$$

The following two limits are of special interest.

a) $\overline{(\dot{x}_1 - \dot{x}_2)} = 0$, or $\overline{\dot{x}_1} = \overline{\dot{x}_2}$:

We have complete coupling and

$$\overline{\dot{x}_1} = \overline{\dot{x}_2} = \frac{f_L}{\eta_1 + \eta_2} . \tag{8.5}$$

The total damping results from dissipation in *both* films. Because of the presence of the secondary film $\overline{\dot{x}}_1$ and $\overline{V}_1$ is reduced below the value obtained if the primary film would act alone.

b) $\overline{\dot{x}}_2 = 0$:

In this case we have

$$\overline{\dot{x}}_1 = \frac{f_L}{\eta_1} \quad . \tag{8.6}$$

The dissipation results only from the primary film.

The magnetic coupling behavior is summarized schematically in Fig.8.5. Up to the primary current I_1^* we have complete coupling. At $I_1 = I_1^*$ slippage between both flux structures sets in. For $I_1 > I_1^*$, with increasing I_1 the primary voltage is less and less affected by damping in the secondary film and rises rapidly, whereas the secondary voltage approaches zero. Even in this simple model the details depend of course on the magnitudes of η_1 and η_2, and on the thickness of both films.

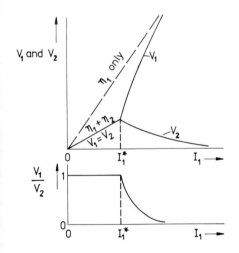

Fig.8.5. Magnetic coupling between two superconductors: primary voltage V_1, secondary voltage V_2, and the ratio V_1/V_2 versus the primary current I_1. Complete coupling exists up to $I_1 = I_1^*$

The simplified model we have outlined has been extended by CLEM [8.20,21] including the effects of bulk pinning and applied currents in both the primary and secondary film and incorporating a more detailed theory of the magnetic coupling force for quantitative comparison with experiment. Experiments performed on magnetically coupled, granular aluminum films [7.25,

8.22], showing extremely small flux pinning effects, are in excellent agreement with CLEM's theory. Because of the small depinning currents in the granular aluminum films, these experiments could be performed for the first time over a wide range of coupled vortex motion without complications from heating effects. The behavior of magnetically coupled thin-film superconductors has also been studied theoretically and experimentally by SHERRILL and co-workers [8.23-25]. Clearly, the dynamics of magnetically coupled superconducting films represents an interesting subject providing further insight into the forces on magnetic flux structures, dissipation in moving flux structures, and flux pinning.

8.6 Micro Field Probes

In addition to the Sharvin point contact, thin-film micro field probes attached to the surface of the superconductor can be utilized for investigating the dynamics of magnetic flux structures. This principle has been used in a study of the flux-tube nucleation rate in the current-induced resistive state of a thin-film type-I superconductor [8.26]. Here the passage of multi-quantum flux tubes underneath a microstrip consisting of granular aluminum causes temporary breakdown of superconductivity in the aluminum strip, which is detected via the resistive voltage change associated with a small bias current passing through the Al field probe. The experimental configuration is shown schematically in Fig.8.6. An indium film sample deposited on a substrate is weakened locally through a geometrical constriction, such that current-induced breakdown of superconductivity is well localized. In the current-induced resistive state, which will be discussed in more detail in Sect.14, flux tubes of opposite sign are nucleated at both film edges of the constriction and move rapidly toward the center of the In film where they annihilate each other. The microstrip of granular aluminum is about 4 μm wide and 1 mm long and is placed across the constricted region of the In sample such that it is traversed perpendicularly by the flux tubes in the indium. The films of indium and aluminum are electrically insulated from each other by a thin layer of SiO_2. Separate pairs of voltage and current leads are connected to the In and the Al film. In this way the Al film acts as a *magnetoresistive field probe* and allows counting of the rate with which the flux tubes are nucleated at the In film edge and subsequently move to the center. For the proper functioning of this device the width of the field probe must be similar to the diameter of the

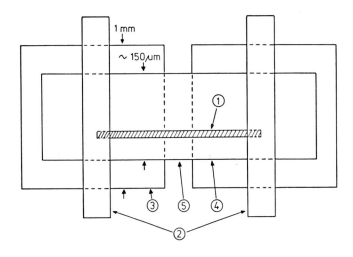

Fig.8.6. Magnetoresistive micro field probe for detecting flux tubes traversing a superconductor. Schematic diagram of composite sample: (1) micro field probe (granular Al); (2) Ag contacts; (3) In underlay; (4) In sample; (5) constriction. The numbers also indicate the order of film deposition. The insulating SiO_2 layer has been omitted for clarity [8.26]

flux tubes to be detected. Further, during their motion the flux tubes must be reasonably well separated in space. In addition, for detecting flux tubes nucleated at high rates, sufficiently fast switching of the aluminum field probe is required.

With an arrangement such as shown in Fig.8.6 flux-tube nucleation rates up to 1.5×10^6 s^{-1} have been measured between 1.5 and 2.0 K in the current-induced resistive state of In films with 2-3 µm thickness [8.26]. Further, the frequency bandwidth of the nucleation rate could be investigated. Figure 8.7 shows an example of the variation of the flux-tube nucleation rate with the resistive voltage across the indium constriction. The data display the expected linear behavior. The two straight lines extrapolate to the origin. The break in the curve at about 380 nV can be interpreted in terms of an abrupt change in the number of flux quanta per flux tube. (In the case shown in Fig.8.7 this number suddenly increases by about six as the sample voltage is increased through the region around 380 nV). Further details of these experiments will be discussed in Sect.14.2.3.

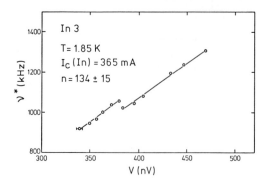

Fig.8.7. Flux-tube nucleation frequency ν^* measured with the micro field probe versus the resistive voltage across an indium constriction. The average number n of flux quanta per tube derived from the slopes is indicated. The frequency jump around 380 nV corresponds to a change $\Delta n = +6$ for increasing voltage [8.26]

8.7 Simulation Experiments

The static and dynamic properties of magnetic flux structures in superconductors can be investigated by various simulation experiments. ROSE-INNES and STANGHAM [8.27] simulated the vortex lines of a superconductor by magnetic needles inserted into boyant spheres and floating vertically on water. In these experiments the floating magnets form a triangular lattice similar to the flux-line lattice in a type-II superconductor. The usual lattice defects can also be reproduced by the floating magnet model. Flux pinning can be simulated by placing patterns of iron fillings on the bottom of the trough [8.28] or by arranging pinning magnets above the array of floating magnets [8.29-31]. From such experiments it can be shown that for an elastic flux lattice, flux pinning is much stronger than for a rigid lattice (see Sect.11). MEISSNER also simulated the Lorentz force by passing an electric current through the water to which some electrolyte had been added. His experiments show interesting details about simulated flux-tube motion, such as the degree of uniformity of such motion and the fact that at any one time only a fraction of all flux tubes is moving.

9. Thermal Force and Flux Motion

9.1 Thermal Force

In addition to the Lorentz force of an electric current, the thermal force of a temperature gradient across the superconductor may also act as a driving force on the magnetic flux structure. Again we assume a flat superconductor with its broad surfaces located in the x-y plane and a magnetic field applied in z-direction such that the plate resides in the mixed or the intermediate state. If a temperature gradient of sufficient magnitude is applied in the x-direction, the magnetic flux structure moves with velocity $\underline{v}_\varphi$ in a direction given by the Hall angle θ between the x-axis and $\underline{v}_\varphi$ (see Fig.9.1). The forces regulating the flux motion satisfy the equation

$$- S_\varphi \operatorname{grad} T - f n_s e (\underline{v}_\varphi \times \underline{\varphi})/c - n\underline{v}_\varphi - \underline{f}_p = 0 \quad . \tag{9.1}$$

Analog (7.1) the thermal force $-S_\varphi \operatorname{grad} T$ is compensated by the Magnus force $f n_s e(\underline{v}_\varphi \times \underline{\varphi})/c$, the damping force $n\underline{v}_\varphi$, and the pinning force $\underline{f}_p$. Again φ is the flux contained in a vortex line or flux tube. In (9.1) the forces are again given *per unit length* of vortex line or flux tube. It is important to note that all superconducting properties are in general temperature dependent. Hence, (9.1) only applies throughout the whole superconductor in the same numerical form if the temperature gradient is rather small. For appreciable values of grad T complications can occur (like piling up and sideways motion of vortex lines) because of the variation in the quantities S_φ, n and f_p along the direction of the temperature gradient [9.1].

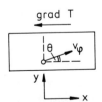

Fig.9.1. Sample geometry for flux motion induced by a temperature gradient

The thermal force $-S_\varphi \, \text{grad} \, T$ is fundamental to all thermal-diffusion phenomena, S_φ being referred to as the *transport entropy* [9.2]. In our case the thermal force is based on the fact that the entropy density in the vortex core, or within a flux tube, is higher than in the surrounding superconducting medium, and is directed from the hot end to the cold end of the sample. The thermal force is, of course, a direct consequence of the Onsager relations from irreversible thermodynamics. The quantity S_φ in the expression for the thermal force and for the heat current density of a moving vortex structure given in (7.20) is identical as will be shown in Sect.9.2. Again, (9.1) and the flux motion induced by the thermal force apply equally well to single-quantum vortex lines and to multi-quanta flux tubes.

9.2 Nernst and Seebeck Effect

From (9.1) we see that in general a temperature gradient in x-direction will cause flux motion in the x- and y-direction. The velocity components $v_{\varphi x}$ and $v_{\varphi y}$, together with (7.5), will then lead to a transverse and longitudinal electric field, in analogy to the Nernst and Seebeck effect, respectively. The voltages due to the Nernst and Seebeck effect are expected to increase proportional to the temperature gradient for small values of grad T. Neglecting the Magnus force and inserting (7.5) into (9.1) we have for the transverse Nernst voltage

$$\frac{\partial V}{\partial y} = (- S_\varphi) \frac{B}{\eta c} (\text{grad}_x T - \text{grad}_c T) \quad , \tag{9.2}$$

where the critical temperature gradient $\text{grad}_c T$ is defined by

$$\text{grad}_c T = - f_p / S_\varphi \quad . \tag{9.3}$$

We note the strong similarity between (7.7) and (9.2). Equation (9.2) is illustrated in Fig.9.2 which shows the Nernst voltage versus the longitudinal temperature difference in a tin film for different magnetic field applied perpendicular to the film.

From (9.1) we note that without the Magnus force the component $v_{\varphi y}$ would be zero and the Seebeck effect would vanish. Again, in the mixed and the intermediate state the (transverse) Nernst effect is 2-3 orders of magnitude larger than the (longitudinal) Seebeck effect, in contrast to the behavior

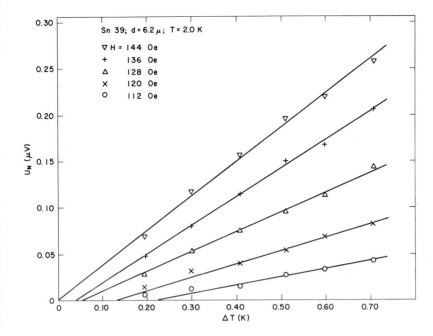

Fig.9.2. Nernst voltage versus the longitudinal temperature difference in a tin film of 6.2 μm thickness at 2.0 K and different magnetic fields [9.3]

generally found in metals. Further, in the mixed and the intermediate state the Nernst effect is 2-3 orders of magnitude larger than in the normal state and can be distinguished from the normal contribution without difficulty. For the Seebeck effect the normal contribution and the contribution from flux flow are expected to have similar magnitudes below the superconducting transition temperature. The ratio of the electric fields associated with the Nernst and Seebeck effect is, of course, given by the Hall angle θ. For the geometry of Fig.9.1 we then have

$$\tan\theta = \frac{V_{\varphi y}}{V_{\varphi x}} = \left|\frac{\partial V/\partial x}{\partial V/\partial y}\right| \quad . \tag{9.4}$$

The Nernst effect associated with thermally induced flux motion has been observed in a number of type-I and type-II superconductors [7.63,9.1,3-7]. Measurements of the Seebeck effect caused by thermally induced flux motion have not been reported. Besides its manifestation through a flux-flow voltage, flux motion induced by the thermal force has also been observed with the Sharvin point contact method [8.13]. These experiments were performed in a single crystal of tin, in which laminar Landau domains with high regularity were established in an inclined magnetic field. The temperature gradient was perpendicular to the orientation of the laminae, and domain motion down the temperature gradient could be observed. Magneto-optical studies of flux motion in a type-I superconductor induced by the thermal force have been reported by RINDERER [9.8].

As indicated by (7.7) and (9.2), flux flow induced by an electric current or a temperature gradient is rather similar, both cases displaying a linear flux-flow regime. In both cases the onset of flux motion is determined by the magnitude of the pinning force, which is related to the critical current density and the critical temperature gradient, see (7.8) and (9.3). An interesting discrepancy between the critical current density and the critical temperature gradient measured in the same specimens has been reported [9.9,10]. In comparing the critical current density and the critical temperature gradient it is important to note that the spatial distribution of the current density and thereby of the Lorentz force may strongly vary with the detailed state of flux pinning and flux motion. On the other hand, an influence of flux pinning and flux flow on the spatial dependence of the temperature gradient and thereby of the thermal force is practically absent.

The Nernst and Seebeck effect can be complicated by the fact that finite temperature differences are required to overcome flux pinning, resulting in an appreciable variation of the superconducting properties along the direction of the temperature gradient. On the other hand, the Ettinghausen and Peltier effect can be measured at exactly uniform temperature throughout the sample. The small temperature gradients arising from these effects can even be compensated with some heaters in a "potentiometric" procedure. Further, in the measurements of the latter effects the influence of flux pinning is eliminated, since the temperature gradient (or the power input to the compensating heater) is directly proportional to the flux-flow electric field, see (7.22). For these reasons the Ettinghausen and Peltier effect is more useful experimentally than the Nernst and Seebeck effect.

The Righi-Leduc effect, i.e., the thermal analog of the Hall effect, has also been observed in the mixed state of a pure type-II superconductor [9.11] and in type-I and type-II alloys [9.12-14]. Flux flow is not necessary for the observation of this effect in the mixed state, since a temperature gradient can exist in a superconductor with a static vortex structure (in contrast to an electric field). The experimental results were, indeed, obtained for a static vortex structure.

Finally, we show that the quantity S_φ in the expression for the thermal force is identical with the quantity S_φ in expression (7.20) for the heat-current density of a moving vortex structure. For the geometry of Fig.9.1 the Nernst coefficient ν is defined by [9.2]

$$\partial V/\partial y \equiv - \nu B \partial T/\partial x \quad . \tag{9.5}$$

With $\partial V/\partial y = v_{\varphi x} \cdot B/c$ we have

$$v_{\varphi x} = - c\nu \partial T/\partial x \quad . \tag{9.6}$$

For simplicity we neglect the Magnus force and the pinning force. The thermal force per unit length of vortex line, f_{th}, then satisfies the equation

$$f_{th} = \eta v_{\varphi x} \tag{9.7}$$

and with (9.6)

$$f_{th} = - \eta c \nu \partial T/\partial x \quad . \tag{9.8}$$

For the geometry of Fig.7.1 the Ettinghausen coefficient ε is defined by [9.2]

$$\varepsilon \equiv \frac{1}{BK} \frac{U_y}{J_x} \quad , \tag{9.9}$$

where U_y is the heat current density in y-direction, J_x the electrical current density in x-direction, and K is the heat conductivity. With (7.20) we obtain

$$\varepsilon = \frac{1}{BK} \frac{nTS_\varphi v_{\varphi y}}{J_x} \quad . \tag{9.10}$$

Neglecting again the Magnus force and the pinning force, the Lorentz force per unit length of vortex line, f_L, satisfies the equation

$$f_L = \frac{1}{c} J_x \cdot \varphi = \eta v_{\varphi y} \quad . \tag{9.11}$$

By insertion into (9.10) and with $B = n\varphi$ we obtain

$$\varepsilon = \frac{TS_\varphi}{c\eta K} \quad . \tag{9.12}$$

Using the Bridgman relation of irreversible thermodynamics [9.2]

$$T\nu = \varepsilon K \tag{9.13}$$

we find from (9.8) and (9.12)

$$f_{th} = -S_\varphi \, \partial T/\partial x \quad . \tag{9.14}$$

Here the quantity S_φ has originated from the expression given in (7.20).

9.3 Transport Entropy

The main quantity derived from thermomagnetic measurements during flux flow is, of course, the transport entropy S_φ. Experimentally S_φ/φ can directly be obtained from the Ettinghausen effect using (7.22). On the other hand, according to (9.2) measurements of the Nernst effect yield the proportionality coefficient $S_\varphi B/\eta$ between the transverse Nernst field and the longitudinal temperature gradient (see also Fig.9.2). From flux-flow resistance measurements the coefficient $\varphi B/\eta$ can be found according to (7.7). The ratio of both coefficients then yields the quantity S_φ/φ. We note that the experiments only yield the transport entropy *per unit flux* and do not provide information about the total transport entropy contained in the individual multi-quantum flux tubes present in type-I superconductors.

As a function of temperature, S_φ is found to increase with temperature at low temperatures, to pass through a maximum and to vanish at the critical temperature T_c. As a function of magnetic field, S_φ is found to decrease gradually from a maximum value at low fields reaching zero at the critical field H_c or H_{c2}. This behavior is illustrated in the three-dimensional plot of Fig.9.3.

For a type-I superconductor an expression for S_φ can be derived as follows. From the difference between the entropy density in the normal and the superconducting state,

$$S_n - S_s = -\frac{H_c(T)}{4\pi} \frac{\partial H_c(T)}{\partial T} \quad , \tag{9.15}$$

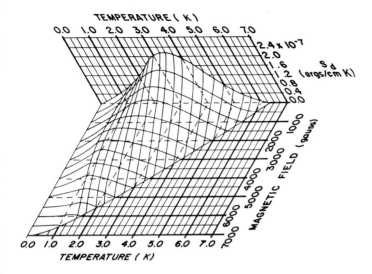

Fig.9.3. Three-dimensional plot of the transport entropy S_φ per vortex line of unit length versus B and T for the type-II alloy In + 40 at.% Pb [7.62]

we find for the excess entropy per unit length of a flux tube containing the flux φ

$$\frac{S_\varphi}{\varphi} = \frac{H_c(0)}{2\pi} \frac{T}{T_c^2} \quad , \tag{9.16}$$

using the approximation

$$H_c(T) \approx H_c(0)(1 - T^2/T_c^2) \quad . \tag{9.17}$$

Equation (9.16) is expected to be valid only if the flux tubes are well separated from each other and if they are sufficiently large such that the contribution from the wall energy can be neglected. At not too high temperatures and magnetic fields there is fair agreement between the experimental values [9.3,7.65,9.7] and expression (9.16). However, (9.16) fails to predict the decrease of S_φ to zero as one approaches H_c or T_c. With increasing magnetic field the normal domains in a type-I superconductor increase in size and become less localized until at H_c the whole specimen consists of a single normal domain. Therefore, S_φ vanishes as one approaches H_c. Similarly, as we approach T_c, the individual flux tubes become larger, less localized, and S_φ vanishes. CLEM [9.15] has suggested the existence of local temperature gradients due to flux-tube motion, leading to an expression for S_φ of the form given in (9.16) but multiplied with the factor

$$g = \frac{[1-H/H_c(T)][1-K_s/K_n]}{1-[H/H_c(T)][1-K_s/K_n]} \quad . \tag{9.18}$$

Here K_s and K_n are the heat conductivity in the superconducting and normal state, respectively. The expression for S_φ corrected in this way reduces to zero at H_c and T_c. However, its value at low magnetic fields and its field dependence is inconsistent with experiment [7.65]. A calculation similar to CLEM's has been performed by ANDREEV and DZHIKHAEV [9.16] resulting in the same correction factor given in (9.18). De LANGE and OTTER [9.17] considered tunneling of the core excitations between moving flux tubes for explaining the reduction of S_φ to zero as one approaches H_c. The same authors [9.18] proposed a model consisting of a proper treatment of the core contribution S_C plus a term S_∇ arising from the local temperature gradients according to CLEM. The transport entropy then becomes the sum

$$S_\varphi = S_C + S_\nabla \quad , \tag{9.19}$$

resulting in

$$S_\varphi = [2 - (K_s/K_n)] S_C \quad . \tag{9.20}$$

In analyzing thermomagnetic measurements performed in type-I superconductors it is important to take into account that resistive voltages may arise both from the flux-flow behavior and from the ohmic dissipation in stationary normal laminae (see Sect.7.3).

Similar to the derivation of (9.16), for a type-II superconductor the transport entropy can be obtained from thermodynamic arguments [9.19]. For small magnetic fields, such that the vortices are well separated, one finds for the transport entropy per unit length of vortex line

$$S_\varphi/\varphi_0 = -\frac{1}{4\pi} \frac{\partial H_{c1}}{\partial T} \quad . \tag{9.21}$$

Equation (9.21) may be used in conjunction with expression (4.19) or with an empirical function $H_{c1}(T)$ for obtaining S_φ. VAN VIJFEIJKEN [9.20] calculated S_φ from the excess entropy contained in the normal core. Relation (9.21) and Van Vijfeijken's model are in reasonable agreement with experiment well below T_c [7.62,9.21]. However, they fail to reproduce the reduction of S_φ to zero as one approaches T_c. This is not surprising, because near T_c the vortices become rather large, and their interaction cannot be neglected any more. The model by DE LANGE and OTTER [9.18] contained in (9.19), and combining the core contribution S_c with the contribution S_∇ arising from local temperature gradients, has also been extended to type-II superconductors, leading to the result

$$S_\varphi = [1 + (K_n - K_s)/(K_n + K_s)] S_c \quad . \tag{9.22}$$

At low temperatures where $K_s/K_n \ll 1$, (9.22) yields $S_\varphi = 2 S_c$. It appears that the magnitude of the contribution S_∇ depends sensitively on the electron mean free path, S_∇ being significant in dirty type-II superconductors and absent in pure type-II materials.

Recent theoretical work based on the time-dependent Ginzburg-Landau equations and a comparison with experiment will be discussed in Sect.10.2.

It is interesting to compare the transport entropy in type-II superconductors with the incremental entropy S_i per unit length of vortex determined from isothermal calorimetry. The quantity S_i is defined as

$$S_i(T,B) = \varphi_0 \left.\frac{\partial S}{\partial B}\right|_T = \frac{\varphi_0}{T} \left.\frac{\partial Q}{\partial B}\right|_T \quad , \tag{9.23}$$

where S is the entropy density of the superconductor and Q the heat input per volume required for an isothermal transition in a slowly varying magnetic field. As expected from theory, S_i is found to depend only on temperature and to be nearly independent of B up to H_{c2} [9.22-25]. We note that the quantity S_i is a second derivative of the thermodynamic potential and discontinuous at H_{c2} and T_c, similar to the specific heat. This discontinuous behavior of S_i at H_{c2} and T_c is in contrast to the continuous decrease to zero of the transport entropy as one approaches the second-order phase transition. We must look upon the transport entropy as a quantity which measures the *local difference in entropy density* associated with the magnetic flux structure relative to the dominant phase of the background. Therefore, near the second-order phase transition this difference must be taken relative to the *normal* phase and vanishes as one approaches H_{c2} and T_c. The temperature dependence of the quantities S_i, S_c, and S_φ is shown schematically in Fig.9.4.

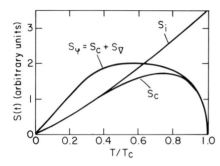

Fig.9.4. Temperature dependence of the transport entropy S_φ, of the contribution S_c of the vortex core to S_φ, and of the incremental entropy S_i of a vortex line [9.18]

It is well known that near the critical field the intermediate state consists of *isolated superconducting domains* which are imbedded in the normal phase. As we have seen in Sect.7.3, these superconducting inclusions also display domain motion in the presence of a transport current in the same direction as the normal domains at lower magnetic fields. Since in the superconducting phase the entropy density is smaller than in the normal phase, this current-induced flow of superconducting domains amounts to a heat current in the opposite direction. From these considerations we might expect a *sign reversal of the Ettinghausen and the Peltier effect* in the intermediate state of type-I superconductors as one approaches H_c. An experimental observation of such an effect has not yet been reported.

Because of the smaller value of the entropy density in the superconducting phase, the thermal force on a superconducting domain should be opposite to that for a normal domain (imbedded in the superconducting phase) for the same temperature gradient across the sample. Therefore, a superconducting domain should experience thermally induced domain motion from the cold to the hot end of the superconductor. However, the transverse voltage associated with such motion of a superconducting domain should have the *same sign* as that for regular thermally induced flux flow (from the hot to the cold end of the sample). From these considerations an *enhancement of the Nernst and Seebeck effect* might be expected in a type-I superconductor as one approaches H_c. A shoulder or even a second peak in the Nernst voltage observed in superconducting films of type-I near the critical field H_c rather likely is caused by this effect [9.7].

10. Time-Dependent Theories

10.1 Phenomenological Theories

Following the discovery of flux flow as a dissipative phenomenon in type-II superconductors, a number of phenomenological theories were proposed for treating this subject, namely by BARDEEN and STEPHEN [10.1], VAN VIJFEIJKEN [7.14-16], and NOZIÈRES and VINEN [10.2]. All three models assume a local superconductor and treat the core of a vortex line as being fully normal resulting in dissipation due to quasiparticle scattering by the lattice. The temperature is assumed to be much smaller than T_c such that the normal electrons outside the vortex core can be neglected. The transport current density is assumed to be small but uniform. More specific, the transport current density and the normal current densities generated by the flux motion are taken to be small compared to the supercurrent densities circulating around the vortex cores. This is equivalent to the criterion that

$$v_\varphi \tau \ll \xi_c \, , \tag{10.1}$$

meaning that the distance a vortex line moves in a relaxation time τ is small compared with the core radius ξ_c. Vortices are assumed to move freely without any influence of flux pinning.

In the *Bardeen-Stephen Model* the order parameter goes to zero in the core region and for $r > \xi_c$ gradually increases with the radial distance from the vortex axis to its equilibrium value far away from the core (see Fig.10.1). Outside the core we apply the London equations and inside the core Ohm's law. From the London equation (2.6) and using the relation $\underline{J}_s = n_s e \, \underline{v}_s$ we obtain for the local electric field $\underline{e}$ outside the vortex

$$\underline{e} = \frac{m}{e} \frac{\partial \underline{v}_s}{\partial t} = -\frac{m}{e} \underline{v}_\varphi \cdot \mathrm{grad} \, \underline{v}_s \, , \tag{10.2}$$

where $\underline{v}_s$ is the superfluid velocity. Introducing cylinder coordinates around the vortex axis according to Fig.10.2, we can write

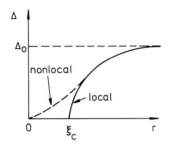

Fig.10.1. Order parameter $\Delta(r)$ as function of the distance from the vortex axis for a local and nonlocal superconductor

Fig.10.2. Coordinates x and θ and vectors $\underline{r}$ and $\underline{\theta}$ to discuss vortex motion

$$\underline{v}_s = \frac{\hbar}{2mr} \underline{\theta} \quad . \tag{10.3}$$

In obtaining relation (10.3) we have used expression (4.8) for the supercurrent density around an isolated vortex line from the London model and the asymptotic approximation for the Hankel function K_1 in the regime $\xi < r < \lambda$. Equations (10.2,3) then yield

$$\underline{e} = -\frac{\varphi_0}{2\pi c} \underline{v}_\varphi \cdot \text{grad}\left(\frac{\theta}{r}\right) \quad . \tag{10.4}$$

Taking the velocity $\underline{v}_\varphi$ along the x-direction we find in the regime $r > \xi_c$

$$\underline{e} = \frac{v_{\varphi x} \varphi_0}{2\pi c r^2} (\underline{\theta} \cos\theta - \underline{r} \sin\theta) \quad . \tag{10.5}$$

The last equation represents the field of a series of electric dipoles as shown in Fig.10.3.

The tangential component of $\underline{e}$ must be continuous at the surface $r = \xi_c$. Hence, from (10.5) we find a uniform field inside the core along the y-direction

$$\underline{e}_{core} = \frac{v_{\varphi x} \varphi_0}{2\pi c \xi_c^2} \hat{y} \quad . \tag{10.6}$$

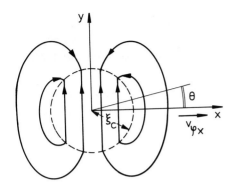

Fig.10.3. Dipole field according to (10.5)

Because of (3.60) we may also write

$$\underline{e}_{core} = \frac{v_{\varphi x} H_{c2}}{c} \hat{y} \quad . \tag{10.7}$$

The radial component of $\underline{e}$ shows a discontinuity at the surface $r = \xi_c$ arising from a surface charge at the core boundary. In a more exact (nonlocal) theory this discontinuity would be smeared out.

From (10.6) we find for the energy dissipation per unit length of core

$$W_{core} = \pi \xi_c^2 \sigma_n e_{core}^2 = \frac{v_\varphi^2 \varphi_0^2}{4\pi c^2 \xi_c^2 \rho_n} \quad . \tag{10.8}$$

According to BARDEEN and STEPHEN, the energy dissipation associated with the normal currents outside the core yields an additional contribution equal to that of (10.8). Hence, the energy dissipation per unit length of vortex is

$$W = \frac{v_\varphi^2 \varphi_0^2}{2\pi c^2 \xi_c^2 \rho_n} = \eta v_\varphi^2 \quad . \tag{10.9}$$

From this we find

$$\eta = \frac{\varphi_0^2}{2\pi c^2 \xi_c^2 \rho_n} = \frac{\varphi_0 H_{c2}}{c^2 \rho_n} \quad . \tag{10.10}$$

By insertion into (7.9) we finally obtain for the flux-flow resistivity ρ_f

$$\rho_f = \rho_n \frac{B}{H_{c2}} \quad . \tag{10.11}$$

In (10.10,11) we have taken the core radius to be approximately equal to the coherence length ξ. In the temperature range $T \ll T_c$ the experimental results [see the empirical relations (7.10,11)] agree reasonably with (10.10, 11).

We emphasize that the above result of the Bardeen-Stephen model is only obtained if free vortex motion is possible. In this case the normal current density in the core is just equal to the applied transport current density. However, if vortex motion is hindered through flux pinning, both current densities are not equal any more. For complete flux pinning and zero flux-flow velocity the current density in the core and the energy dissipation would be zero.

So far in our discussion the vortices have been assumed to move perpendicularly to the transport current, and the Hall effect has been neglected. In the range $\omega_c \tau \ll 1$, where $\omega_c = eH/mc$ is the cyclotron frequency and τ the electron relaxation time, BARDEEN and STEPHEN obtained for the Hall angle θ in the applied field H

$$\tan\theta = \omega_c(H)\tau , \qquad (10.12)$$

i.e., the normal-state value. In this case the vortex moves nearly perpendicular to $\underline{v}_s$, with a velocity $v_\varphi \gg v_s$. In the pure limit, $\omega_c(H_{c2})\tau \gg 1$, BARDEEN and STEPHEN found for the flux-flow velocity

$$v_\varphi = v_s H/H_{c2} , \qquad (10.13)$$

with the vortices moving in the direction of the applied velocity $\underline{v}_s$ (i.e., perpendicular to the Lorentz force).

The (local) Bardeen-Stephen model has been extended to a nonlocal superconductor [10.3-5]. In this case the order parameter gradually approaches zero at the vortex axis (see Fig.10.1), and the effective area of the core must be obtained from microscopic theory. For the viscosity coefficient η the nonlocal model yields an expression not much different from that in (10.10) found for the local case.

The theory by VAN VIJFEIJKEN also assumes the chemical potential to be continuous at the core boundary (resulting in the appearance of a contact potential) similar to the Bardeen-Stephen model. It yields results nearly identical to those contained in (10.11-13).

In the theory by NOZIÈRES and VINEN the continuity of the chemical potential at the core boundary, assumed in the two other models, is abandoned.

It is argued that the concept of a continuous local chemical potential is valid only for thermodynamic equilibrium and becomes questionable in the nonequilibrium situation of the moving vortex lattice. NOZIÈRES and VINEN suggest that the electrostatic potential should be continuous and that there should be no contact potential at the core boundary. In the range $\omega_c \tau \ll 1$ they obtain the same longitudinal resistivity as given in (10.11). However, for the Hall angle they find

$$\tan\theta = \omega_c(H_{c2})\tau \, , \qquad (10.14)$$

i.e., a value independent of H below H_{c2}. For $\omega_c(H_{c2})\tau \gg 1$ they obtain

$$\underline{v}_\varphi = \underline{v}_s \, , \qquad (10.15)$$

indicating that the vortices move with the applied superfluid velocity.

The predictions for the Hall angle of the different theories are summarized in Fig.10.4. Experimentally the Hall effect is clearly more complicated than expected from the three models, as can be seen from Figs.7.7, 8. Furthermore, measurements on dirty type-II alloys yield a Hall angle which increases with decreasing H below H_{c2} and becomes larger than the normal state Hall angle at $H < H_{c2}$ [7.52,60]. Such behavior is difficult to understand from any of the above models. Theories based on the time-dependent Ginzburg-Landau equations are here more promising, as will be discussed in Sect.10.2.

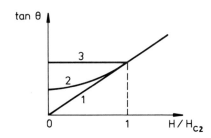

Fig.10.4. Predictions for the Hall angle θ according to the theories of (1) BARDEEN and STEPHEN, (2) VAN VIJFEIJKEN, and (3) NOZIERES and VINEN

10.2 Time-Dependent Ginzburg-Landau Theory

Following the success of the Ginzburg-Landau theory for treating the case of thermodynamic equilibrium, time-dependent extensions of the Ginzburg-Landau theory have been investigated for describing variations of the order parameter with time. Unfortunately, for the time-dependent case no important simplification near the transition temperature is possible as in the static situation. This difficulty is associated with the existence of an energy gap in the energy spectrum and the interconversion between normal excitations and superfluid associated with the time-dependent processes. In the general case an extension beyond the linear term of the Ginzburg-Landau theory is impossible. However, for a gapless superconductor (which may be realized by adding magnetic impurities) these difficulties do not exist, and the time-dependent theory has been developed furthest.

A simple set of time-dependent Ginzburg-Landau equations holds only under highly restricted conditions. It has been shown, that such a set of equations is valid only very close to T_c, where the order parameter is small. Further, as shown by GOR'KOV and ELIASHBERG [10.6], the space and time variation of the order parameter must not be too slow, since the dynamic expansion of those terms proportional to frequency ω proceeds in powers of the order parameter devided by the frequency plus the diffusion constant times the wave vector squared [$\Delta/(\omega + Dk^2)$]. The resulting equation for the order parameter then resembles a diffusion equation, since it contains a first-order time derivative and a second-order space derivative.

Three major lines have been pursued in the derivation and the study of time-dependent Ginzburg-Landau (TDGL) equations. First, one has considered slow variations of small amplitude from the BCS equilibrium state without imposing the requirement that the order parameter be small. For example, ABRAHAMS and TSUNETO [10.7] have shown that TDGL equations exist in two restricted temperature regimes: near absolute zero and near the transition temperature. In the former case the equations are wave-like, and in the latter case diffusion-like. Second, a small order parameter has been assumed by considering the regime just below the second-order phase transition. This work has been done mainly by SCHMID [10.8], CAROLI and MAKI [10.9-11], MAKI [10.12-14], THOMPSON [10.15], TAKAYAMA and EBISAWA [10.16], and EBISAWA [10.17]. Third, one has treated the case of a gapless superconductor which can be achieved, for example, by adding magnetic impurities. As shown by [10.6], a complete set of TDGL equations can be derived for this case. This approach has been utilized to a large extent by THOMPSON and co-workers

[10.18-22]. The main theoretical results to be compared with experiment have been obtained along the second and third line. A more detailed discussion of the problematics associated with the TDGL theory can be found in the review by CYROT [10.23].

For the regime near the transition temperature TDGL equations were first derived by SCHMID [10.8]. Applying the concepts of nonequilibrium thermodynamics to superconductors, one can introduce the functional derivative of the free energy with respect to the order parameter as a generalized force acting on the order parameter. In this way one obtains the transport equation

$$\frac{d\Psi}{dt} + \gamma \frac{df}{d\Psi} = 0 \quad , \tag{10.16}$$

γ being a transport coefficient in the general sense of irreversible thermodynamics. An equation of this type has been used by LANDAU and KHALATNIKOV [10.24] as the starting point for treating superfluid helium. We write the time-dependent order parameter $\Psi(t)$ in the form

$$\Psi(t) = \Psi_0 + \delta\Psi(t) \quad , \tag{10.17}$$

Ψ_0 being its time-independent equilibrium value and $\delta\Psi(t)$ the time-varying departure from equilibrium, with the assumption

$$|\delta\Psi(t)/\Psi_0| \ll 1 \quad . \tag{10.18}$$

Expanding the free-energy density in the form of (3.1) and using (10.16) we obtain

$$\frac{d\delta\Psi(t)}{dt} = -2\gamma\alpha(T)[\Psi_0 + \delta\Psi(t)] - 2\gamma\beta(T)[\Psi_0 + \delta\Psi(t)]^3 \tag{10.19}$$

$$\approx - 2\gamma\alpha(T)[\Psi_0 + \delta\Psi(t)] - 2\gamma\beta(T)\Psi_0^3[1 + 3\frac{\delta\Psi(t)}{\Psi_0}] \quad . \tag{10.20}$$

Inserting relation (3.2) we find

$$\frac{d\delta\Psi(t)}{dt} \approx 4\gamma\alpha(T) \delta\Psi(t) \quad , \tag{10.21}$$

implying that the order parameter relaxes exponentially with a relaxation time

$$\tau = -\frac{1}{4\gamma\alpha(T)} = \frac{1}{4\gamma a(T_c-T)} , \qquad (10.22)$$

where we have used relation (3.3) on the right hand. We note that the temperature dependence of τ is independent of the temperature dependence of Ψ_o.

So far the order parameter has been taken to be spatially constant. Now we allow spatial variations of Ψ and include the presence of magnetic fields. Inserting the expansion of the free-energy density given in (3.8) into (10.16) we have

$$\frac{\partial \Psi}{\partial t} = -2\gamma\left[\alpha\Psi + \beta|\Psi|^2\Psi + \frac{1}{4m}\left(\frac{\hbar}{i}\underline{\nabla} - \frac{2e}{c}\underline{A}\right)^2\Psi\right] . \qquad (10.23)$$

In order to obtain a gauge invariant equation, a term must be added to the left hand yielding the equation

$$\left(\frac{\partial}{\partial t} + \frac{2ie}{\hbar}\tilde{\Phi}\right)\Psi = -2\gamma\left[\alpha\Psi + \beta|\Psi|^2\Psi + \frac{1}{4m}\left(\frac{\hbar}{i}\underline{\nabla} - \frac{2e}{c}\underline{A}\right)^2\Psi\right] . \qquad (10.24)$$

Here the quantity $\tilde{\Phi}$ is the effective potential. A gauge transformation consists of the simultaneous substitutions

$$\Psi \to \Psi \exp\left(\frac{2ie}{\hbar c}\chi\right) ; \quad \underline{A} \to \underline{A} + \nabla\chi ; \quad \tilde{\Phi} \to \tilde{\Phi} - \frac{1}{c}\frac{\partial\chi}{\partial t} , \qquad (10.25)$$

where $\chi(\underline{r},t)$ is an arbitrary function of $\underline{r}$ and t. It has been proposed by SCHMID [10.8] that the effective potential $\tilde{\Phi}$ should be written as

$$\tilde{\Phi} = \Phi - \mu/e , \qquad (10.26)$$

where Φ is the electric potential and μ the chemical potential. In the absence of electro-magnetic fields and for a spatially constant order parameter (10.24) then reduces to

$$\left(\frac{\partial}{\partial t} - \frac{2i\mu}{\hbar}\right)\Psi = -2\gamma[\alpha\Psi + \beta|\Psi|^2\Psi] . \qquad (10.27)$$

The result of (10.24) can be rewritten in the following way. With the quantities $|\Psi_0|^2 = -\alpha/\beta$ from (3.2) and $\xi^2 = -\hbar^2/4m\alpha$ from (3.17), and introducing the normalized order parameter Ψ/Ψ_0 we find from (10.24)

$$\left(\frac{\partial}{\partial t} + \frac{2ie}{\hbar}\tilde{\Phi}\right)\frac{\Psi}{\Psi_0} + \frac{\gamma\hbar^2}{2m\xi^2}\left[\left(\frac{\Psi}{\Psi_0}\right)^2 - 1\right]\frac{\Psi}{\Psi_0} + \frac{\gamma\hbar^2}{2m}\left(\frac{1}{i}\nabla - \frac{2e}{\hbar c}\underline{A}\right)^2\frac{\Psi}{\Psi_0} = 0 \quad . (10.28)$$

The factor $\gamma\hbar^2/2m$ can be identified as the normal-state diffusion constant D

$$\frac{\gamma\hbar^2}{2m} = D = \frac{1}{3}v_F\ell \quad , \tag{10.29}$$

and we obtain finally

$$\left(\frac{\partial}{\partial t} + \frac{2ie}{\hbar}\tilde{\Phi}\right)\frac{\Psi}{\Psi_0} + \frac{D}{\xi^2}\left[\left(\frac{\Psi}{\Psi_0}\right)^2 - 1\right]\frac{\Psi}{\Psi_0} + D\left(\frac{1}{i}\nabla - \frac{2e}{\hbar c}\underline{A}\right)^2\frac{\Psi}{\Psi_0} = 0 \quad . \tag{10.30}$$

Equation (10.30) has been derived by GOR'KOV and ELIASHBERG [10.6] from microscopic theory for a gapless superconductor. The same authors found for the current density,

$$\underline{J}_s = \sigma_n\left(-\nabla\tilde{\Phi} - \frac{1}{c}\frac{\partial\underline{A}}{\partial t}\right) + \text{Re}\left[\frac{\Psi^*}{\Psi_0}\left(\frac{1}{i}\nabla - \frac{2e}{\hbar c}\underline{A}\right)\frac{\Psi}{\Psi_0}\right]\frac{1}{8\pi e\lambda^2} \tag{10.31}$$

and for the charge density

$$\rho = \frac{\tilde{\Phi} - \Phi}{4\pi\lambda_{TF}^2} \quad . \tag{10.32}$$

Here the temperature-dependent coherence length is $\xi = \hbar(6D/\tau_s)^{1/2}/\Psi_0$, where τ_s is the spin-flip scattering time, and the temperature-dependent magnetic penetration depth is $\lambda = \hbar c(8\pi\sigma_n\tau_s)^{-1/2}/\Psi_0$. σ_n is the normal-state conductivity, and λ_{TF} is the Thomas-Fermi static-charge screening length. The set of equations (10.30-32) is completed by the Maxwell equations coupling the scalar and vector potentials Φ and $\underline{A}$ to the charge and current densities ρ and $\underline{J}_s$.

Equations (10.30-32) have been used extensively by THOMPSON and coworkers. A similar set of equations, except for some corrections and additions, was derived earlier by SCHMID [10.8]. As first shown by SCHMID, the general solution of (10.30) for the case of an electric field E in x-direction and a static magnetic field in z-direction has the form

$$\frac{\Psi(\underline{r},t)}{\Psi_0} = \sum_{n=-\infty}^{\infty} C_n \exp[ink(y + ut)]\exp\left[-eH\left(x - \frac{nk}{2eH} - \frac{iu}{4eHD}\right)^2\right]. \quad (10.33)$$

Here n is an integer, u = E/H, and C_n and k are constants chosen to describe the equilibrium solution with the regular vortex lattice. The solution (10.33) satisfies

$$\frac{\partial}{\partial t} \Psi(\underline{r},t) = u \frac{\partial}{\partial y} \Psi(\underline{r},t) , \quad (10.34)$$

indicating that the order parameter moves in y-direction at the uniform velocity -u.

In the following we compare the results on the flux-flow resistivity, the Hall effect, and the transport entropy in type-II superconductors, obtained from the TDGL theory, with the experimental situation.

Flux-Flow Resistivity

A very rough estimate of the flux-flow resistivity in the mixed state can be obtained by considering the fraction of the superconductor which remains superconducting in a perpendicular magnetic field. According to the Abrikosov theory, near H_{c2} the magnetic flux density is given by (3.97). Hence, the fraction of the material remaining superconducting in a magnetic field is $(1-\tilde{h})/(2\kappa^2-1)\beta_A$, where $\tilde{h} = H/H_{c2}$ is the reduced magnetic field. From this we expect for the flux-flow resistivity ρ_f

$$\rho_f/\rho_n \approx 1 - [(1 - \tilde{h})/(2\kappa^2 - 1)\beta_A] . \quad (10.35)$$

We note that for the triangular vortex lattice we have $\beta_A = 1.16$.

Extending SCHMID's calculation [10.8] to all temperatures, CAROLI and MAKI [10.11] and MAKI [10.12] obtained for the flux-flow resistivity in dirty type-II superconductors ($\kappa \gg 1$) near H_{c2}

$$\rho_f/\rho_n = 1 - \{4\kappa_1^2(0)[1 - \tilde{h}]/1.16[2\kappa_2^2(T) - 1] + 1\} . \quad (10.36)$$

Here $\kappa_1(T)$ and $\kappa_2(T)$ are the temperature dependent Ginzburg-Landau parameters introduced by MAKI [10.25], which have the property $\kappa_1(T) \to \kappa$ and

$\kappa_2(T) \to \kappa$ for $T \to T_c$ [1.7]. We see that the result (10.36) is not too different from (10.35) obtained from a rather simple argument. The expressions given here and in the following for the various transport properties refer to a thin superconducting plate placed perpendicularly in a magnetic field, such that the demagnetization coefficient is close to 1.0.

The Caroli-Maki result (10.36) strictly applies only for the case of weak coupling and strong pair breaking. Subsequently the theory has been extended by THOMPSON [10.15], and by TAKAYAMA and EBISAWA [10.16] for the case of weak coupling and weak pair breaking. The theory has further been extended to the case of arbitrary coupling by IMAI [10.26]. For comparison with experiment the dimensionless ratio $\alpha^* = (H_{c2}/\rho_n)(\partial \rho_f/\partial H)_{H_{c2}}$ is often used. According to the various theories this ratio is

$$\alpha^* = \frac{H_{c2}}{\rho_n}\left(\frac{\partial \rho_f}{\partial H}\right)_{H_{c2}} = \frac{4\kappa_1^2(0)}{1.16[2\kappa_2^2(T)-1]+1} \{\text{extensions}\} \quad . \quad (10.37)$$

Here the first term on the right is the Caroli-Maki result, and the factor {extensions} indicates the various temperature-dependent functions arising in the different extensions. The results of the different theories are summarized in Fig.10.5. According to CAROLI and MAKI, $\alpha^*(T)$ increases monotonically with temperature from $\alpha^*(0) = 1.7$ to $\alpha^*(T_c) = 2.5$. The theories by THOMPSON and by TAKAYAMA and EBISAWA yield an increase from $\alpha^*(0) = 1.7$ to $\alpha^*(T_c) = 5$, i.e., to twice the Caroli-Maki value. According to IMAI, the $\alpha^*(T)$ curve for the general case of arbitrary coupling lies between the two other results and can display a pronounced maximum just below T_c.

Experimentally the situation appears somewhat ambiguous [7.9]. Although the data of the Sendai group [7.52,10.28] on Nb-Mo and Nb-Ta alloys agree well with the Caroli-Maki result, there are some discrepancies with the experimental values of other groups [7.63,10.29,30]. Because of considerable roundings of the $\rho_f(H)$ curves near H_{c2}, the slopes α^* cannot always be determined with high accuracy. Recently, HAGMANN et al. [10.27] performed measurements with rather high accuracy through improvements of the experimental technique along various lines. Their results are shown in Fig.10.5 and seem to confirm the generalized theory by IMAI. In comparing their experimental data with theory, these authors used parameter values obtained from tunnel spectra for describing the coupling strength.

For a pure type-II superconductor near H_{c2} MAKI [10.14] obtained the expression

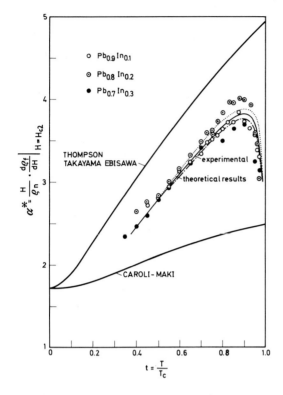

Fig.10.5. Normalized derivative $\alpha^* = (H_{c2}/\rho_n)(\partial\rho_f/\partial H)_{H_{c2}}$ versus reduced temperature according to the theories by CAROLI and MAKI, THOMPSON, and TAKAYAMA and EBISAWA. The data points refer to experimental results obtained for Pb-In alloys as indicated. The theoretical curve drawn through the experimental points is obtained from Imai's theory [10.27]

$$\alpha^* = \frac{H_{c2}}{\rho_n}\left(\frac{\partial\rho_f}{\partial H}\right)_{H_{c2}} = \frac{m}{\pi e n_s}\frac{H_{c2}}{1.16[2\kappa_2^2(T)-1]+1} \tag{10.38}$$

where m and n_s are the mass and the density of the superconducting electrons, respectively. The result (10.38) can be compared with a series of measurements performed on niobium [7.52,19]. The experiments show that the slope α^* for pure Nb is an order of magnitude larger than the theoretical value (10.38) at all temperatures and depends strongly on the electron mean free path or sample purity. It appears that a similar conclusion can be drawn for vanadium [7.55].

Hall Effect

MAKI [10.13,14] calculated the Hall angle for dirty type-II superconductors and obtained the following expression in the vicinity of the upper critical field

$$\tan\theta(H) = \omega_c(H)\tau + \left(\frac{\hbar}{4mD} - \omega_c(H)\tau\right) \frac{4\kappa_1^2(0)}{1.16[2\kappa_2^2(T) - 1] + 1}\left(1 - \frac{H}{H_{c2}}\right) . \quad (10.39)$$

Here D is the diffusion constant of (10.29). Equation (10.39) predicts that just below H_{c2} the Hall angle increases strongly with decreasing magnetic field, in agreement with experimental results on Nb-Ta and Nb-Mo alloys [10.31]. EBISAWA [10.17,32] also calculated the Hall effect in a dirty type-II superconductor and obtained an expression in qualitative agreement with the experiments on type-II alloys.

For the case of a pure type-II superconductor MAKI [10.13,14] found the same result as BARDEEN and STEPHEN, namely (10.12). EBISAWA [10.32] also treated the pure limit of a type-II superconductor. Both microscopic calculations are unable to account for the experimental data (see Fig.7.7,8) on fairly pure vanadium and niobium [7.52,54,55].

Transport Entropy

The transport entropy for dirty type-II superconductors has been calculated by MAKI [10.12-14] yielding the result in the high-field regime

$$S_\varphi(T) = \frac{2(2\pi)^2 \sigma_n}{e^2} \frac{\tilde{\rho}[\kappa_1(0)]^2}{1.16[2\kappa_2^2(T)-1]+1} L_D(T)\left(1 - \frac{B}{H_{c2}(T)}\right) \quad (10.40)$$

with

$$L_D(T) = 1 + \tilde{\rho}\, \psi^{(2)}(\tfrac{1}{2} + \tilde{\rho})/\psi^{(1)}(\tfrac{1}{2} + \tilde{\rho}) \quad (10.41)$$

and

$$\tilde{\rho} = eDH_{c2}/2\pi T \quad . \quad (10.42)$$

Here, $\psi^{(1)}(x)$ and $\psi^{(2)}(x)$ are the higher order derivatives of the di-gamma function. MAKI's expression (10.40) is consistent with the experimental results for alloys [7.62-64,66,10.28]. In particular, the vanishing of the transport entropy as one approaches T_c or H_{c2} is well accounted for by the microscopic theory.

The transport coefficients associated with vortex motion in type-II superconductors have been reexamined by HOUGHTON and MAKI [10.33] from linear response theory. Comparison with the TDGL calculations indicated consistency of the two approaches. The third law of thermodynamics requiring $S_\varphi \to 0$ for $T \to 0$ is satisfied by the result of (10.40-42). However, as pointed out by TAKAYAMA and EBISAWA [10.16] Onsager's reciprocity theorem is violated in Maki's theory. Extensions of the theory to avoid this difficulty have been discussed by HU [10.34,35].

For a pure type-II superconductor MAKI [10.14] obtained an expression similar to (10.40) with $L_D(T)$ replaced by another universal function of order ℓ/ξ_0. Experimental results [9.21] agree qualitatively with Maki's theory. However, a quantitative comparison with experiment suffers from the fact that it is unclear whether the measurements were made for sufficiently large values of the ratio ℓ/ξ_0 for the theory to be rigorously applicable.

11. Flux Pinning

In our discussion of flux motion induced by the Lorentz force or the thermal force we have seen that for flux flow to occur the driving force must exceed the pinning force. The pinning force then determines the critical current and the critical temperature gradient according to (7.8) and (9.3), respectively. Flux pinning can keep the superconductor from reaching thermodynamic equilibrium in its magnetic properties and causes irreversibilities in its magnetic behavior. In the following we shall discuss the critical state of the superconductor and the various mechanisms for flux pinning. The subject of flux pinning in type-II superconductors has been treated in the reviews by CAMPBELL and EVETTS [7.10] and by ULLMAIER [11.1].

11.1 Critical State

In (7.1) and (9.1) for the forces on vortex lines we have introduced the pinning force $\underline{f}_p$ *per unit length* of vortex line or flux tube. It is often convenient to introduce the forces *per unit volume* by multiplying the forces per unit length of vortex line or flux tube with the flux-line density $n = B/\varphi_0$. The pinning force per unit volume (or pinning force density) is then

$$\underline{F}_p = n\underline{f}_p \quad . \tag{11.1}$$

In its critical state the superconductor carries the maximum current density everywhere without the appearance of flux motion because of flux pinning. With the Lorentz force per unit volume $(\underline{J} \times \underline{B})/c$, for each location the critical state of the vortex structure satisfies the equation

$$(\underline{J} \times \underline{B})/c = \underline{F}_p(B) \quad . \tag{11.2}$$

Inserting Maxwell's equation curl $\underline{H}$ = $4\pi \underline{J}/c$ we obtain

$$(\text{curl } \underline{H} \times \underline{B})/4\pi = \underline{F}_p(B) \quad . \tag{11.3}$$

Equation (11.3) is valid for the general case where both an applied transport current and an applied magnetic field can contribute to the quantities $\underline{H}$ and $\underline{B}$. Hence, in general the two quantities on the left hand of (11.3) have components in at least two (if not all three) dimensions. Determining the configuration of the current and of the magnetic field from (11.3) then represents a formidable problem, even if a simple and isotropic behavior of the function $F_p(B)$ is assumed. We note that in (11.2,3) we are referring to the *macroscopic forces* by using the macroscopic quantities $\underline{H}$ and $\underline{B}$ obtained by averaging over the microscopic vortex structure.

We simplify the situation by assuming a one-dimensional geometry where the magnetic field has a component only in one direction and where this component only depends on a single coordinate. This case can be realized by an infinite slab or by a cylinder with circular cross section and infinite length. For a slab of infinite extension along the x- and z-direction and of finite thickness in y-direction placed in an external magnetic field applied in z-direction (11.3) reduces to

$$\frac{1}{4\pi} \frac{\partial H_z}{\partial y} B_z = \frac{1}{4\pi} \frac{\partial H_z}{\partial B_z} \frac{\partial B_z}{\partial y} B_z = F_p(B_z) \quad . \tag{11.4}$$

We see from (11.4) that in the critical state a distinct gradient in the vortex-line density can be sustained within the superconductor. The result of (11.4) has first been obtained by FRIEDEL et al. [11.2] from thermodynamic arguments.

If we know the function H(B), (11.4) can serve for calculating the function $B_z(y)$. In materials where $H_{c1} \ll H_{c2}$ we have $H(B) \approx B$ for $H \gg H_{c1}$ as seen from the magnetization curve in Fig.4.1. In this regime the use of (11.4) is rather straightforward. However, for obtaining the function $B_z(y)$ we need to know the dependence of the pinning-force density on B_z. Here two different models have been proposed. According to the *Bean model* [11.3] $F_p \sim B$, such that $\partial B_z/\partial y$ = const, yielding a *linear variation* of B_z with the coordinate in y-direction. On the other hand in the *Kim model* [11.4, 7.6] F_p has been assumed to be independent of B, such that $B_z \cdot \partial B_z/\partial y$ = const, yielding a *parabolic variation* of the function $B_z(y)$. The profile of magnetic flux density as expected from the Bean model and the Kim model is shown schematically in Fig.11.1. We note that the results obtained for a slab of infinite extension in two directions are identical to those for a cylinder with circular cross section and infinite length. The function $B_z(y)$ then is simply replaced by the identical function $B_z(r)$ of the radial coordinate of the cylinder. Of course, the actual profile of magnetic flux density depends on the previous magnetic history of the superconductor. The curves shown in Fig.11.1 refer to the case where the magnetic field has been raised monotonically from zero to its final value.

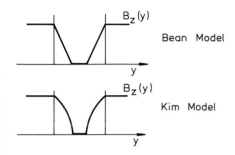

Fig.11.1. Profile of the magnetic flux density $B_z(y)$ for an infinite slab according to the Bean model and the Kim model. (Magnetic field raised monotonically from zero)

In addition to the functions $F_p(B)$ assumed in the Bean model and in the Kim model other functions $F_p(B)$ have been proposed for analyzing experimental results. Experimentally, all methods discussed in Sect.6 can be used in principle for investigating the critical state. Here the primary goal will be the empirical determination of the functional dependence of F_p upon B.

Magnetization measurements with long cylindrical samples in a longitudinal magnetic field have been utilized extensively for studying flux pinning characteristics. Let us consider a long superconducting cylinder of radius R_o and a longitudinal magnetic field H which has been monotonically increased to this value from zero. We assume that magnetic flux has penetrated only a small distance $\Delta r \ll R_o$ into the cylinder. The magnetic flux contained within the cylinder is

$$\varphi = 2\pi R_o \int_{R_o - \Delta r}^{R_o} B(r) dr \quad . \tag{11.5}$$

At the inner edge of the flux ($r = R_o - \Delta r$) we have $B = 0$ and $H = H_{c1}$. At the cylinder surface ($r = R_o$) we have the equilibrium value of $B(H)$ corresponding to the external field H (assuming no surface barrier against flux entry). Using the analog of (11.4) for our case for transforming the coordinate r (11.5) can be written as

$$\varphi = 2\pi R_o \int_{H_{c1}}^{H} [B^2(H)/4\pi F_p(B)] dH \quad , \tag{11.6}$$

yielding the derivative

$$d\varphi/dH = R_o B^2(H)/2F_p(B) \quad . \tag{11.7}$$

We see that from measurements of the derivative $d\varphi/dH$ we can obtain the function $F_p(B)$ if the reversible relation $B(H)$ is known.

In general, magnetization measurements only yield the quantity $F_p(B)$ in the form of an integral. Usually in analyzing the experimental results a reasonable functional form $F_p(B)$ is assumed with some adjustable parameters. The free parameters are then adjusted such that a good fit to the experimental data is obtained.

The very early experiments on flux pinning in type-II superconductors were performed using *hollow cylinders* placed in a longitudinal magnetic field [11.4]. If the external field is gradually increased from zero, the field inside the cylinder first stays strictly equal to zero and starts to rise only when the flux front reaches the inner cylinder surface. If the wall thickness Δr of the tube is sufficiently small the flux-density gradient can be taken as constant within the material. From the analog of (11.4) we obtain

$$F_p = B_m(H - H_i)/4\pi\Delta r \quad . \tag{11.8}$$

Here B_m is the mean flux density and H_i the field inside the hollow tube. Usually, H_i is measured with a small Hall probe.

The advantages of an *ac technique* for investigating the magnetic response of a type-II superconductor were first recognized by BEAN [11.5]. If a small ac component is superimposed on a constant lonitudinal magnetic field, information on the flux-density profile inside the superconducting sample can be derived in various ways from the voltage induced in a pick-up coil wound around the sample [11.6-9].

Most of the experimental methods discussed in Sect.6 have been utilized for studying the flux-pinning characteristics in type-II superconductors. For further details and additional references we refer to the reviews by CAMPBELL and EVETTS [7.10] and ULLMAIER [11.1].

So far we have only dealt with magnetic flux configurations established by an external magnetic field. Next we address ourselves to the situation where the magnetic fields are generated by a *transport current* passing through the superconductor. For simplicity now we assume the external magnetic field to be zero. Again, the problem can be handled analytically only for a one-dimensional geometry. Such a geometry can be realized by a solid or hollow cylinder with circular cross section to which a transport current is applied longitudinally. The magnetic field then only contains an azimuthal component depending only on the radial coordinate. In this case (11.3) yields

$$\frac{1}{r}\frac{\partial}{\partial r}(rH_\varphi) = \frac{\partial H_\varphi}{\partial r} + \frac{H_\varphi}{r} = 4\pi F_p(B)/B \quad , \tag{11.9}$$

where H_φ is the azimuthal magnetic field component.

The left hand of (11.9) consists of two contributions: the term $\partial H_\varphi/\partial r$ arises from the flux-density gradient similar to the case described by (11.4). The second term H_φ/r is associated with the vortex-line energy which tends to shrink a circular vortex line until it reaches zero extension. Because of the r^{-1}-dependence of this force, there exists a critical radius R_c below which the contracting forces are not compensated any more by the pinning force. Hence, in its critical state attained at the maximum transport current I_c a current-carrying solid cylinder will have a flux-free core within the region $r < R_c$. Since $B = 0$ and $H = H_{c1}$ at $r = R_c$, the critical radius is given by

$$R_c = \frac{c}{4\pi} \frac{H_{c1}}{J_c(B=0)} \quad . \tag{11.10}$$

With (11.2) the critical current for a solid cylinder of radius R_0 is

$$I_c = 2\pi c \int_{R_c}^{R_0} [rF_p(B)/B(r)]dr \quad . \tag{11.11}$$

Usually the function $B(r)$ must be obtained by numerical integration of (11.9) using the boundary conditions $H = H_{c1}$ at $r = R_c$ and $H = 2I_c/cR_0$ at $r = R_0$. However, for some special functions $F_p(B)$ analytic expressions for the function $B(r)$ and for the quantity I_c can be derived [11.10]. For example, if F_p is independent of B (Kim model) the critical current is

$$I_c = \frac{2\pi}{\mu_0} R_0^{3/2} \left(\frac{2}{3} \mu_0 F_p\right)^{1/2} \quad . \tag{11.12}$$

From this equation we see that I_c only increases with the square root of the pinning force density and less than proportional to the cross-sectional area of the conductor. Of course, this results from the fact that in the Lorentz force the current enters twice, first by itself and second through the magnetic field associated with the current. (The situation is different for a geometry where a magnetic field, large compared to that generated by the transport current, is applied externally. In this case the critical current increases linearly with the cross-sectional area and the quantity F_p).

The transport-current distribution for zero external magnetic field in type-II superconductors with geometrical shapes different than circular cross section has been analyzed by NORRIS [11.11,12] using conformal mapping methods.

Clearly, flux pinning and the features of the critical state are subjects of great importance for technical applications of superconductors where large currents and magnetic fields are involved (superconducting magnets, transmission lines, etc.). Experiments on the critical current in wires with circular cross section agree well with the critical-state model as expressed in (11.11) [11.13]. Here the relation $F_p(B)$ has been derived from magnetization measurements in longitudinal fields. The distribution of the transport-current density for zero external magnetic field in the critical state of flat tapes carrying large currents has been investigated by MIGLIORI and co-workers [11.14,15]. Their results appear to be inconsistent with a simple critical-state model and can be accounted for by assuming a current-density distribution in form of concentric elliptical shells each with a different current density.

11.2 Summation of Pinning Forces

The pinning forces against the onset of flux motion generally originate from local depressions in the Gibbs free energy of the vortex line. Such depressions can arise through the presence of inhomogeneities in the material such as lattice defects, normal inclusions, etc. The various pinning interactions will be discussed in Sect.11.3. In addition to the presence of localized pinning sites, elastic distortions of the flux-line lattice are necessary for obtaining finite pinning forces, at least for an arbitrary distribution of pinning sites. This is illustrated in Fig.11.2. For a completely rigid flux-line lattice the average force upon the vortex lattice resulting from the various pinning centers is the same for all directions perpendicular to the line orientation and hence results in zero overall flux pinning. On the other hand, in a rather soft flux-line lattice the vortex lines can assume their positions locally, such that the total pinning force in a particular direction is maximized. These aspects for the summation of pinning forces have first been realized by NEMBACH [11.16], ALDEN and LIVINGSTON [11.17], and YAMAFUJI and IRIE [7.32].

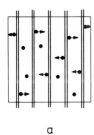
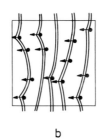

Fig.11.2a and b. Summation of pinning forces for an arbitrary distribution of pinning sites. (a) completely rigid flux-line lattice, (b) soft flux-line lattice. The arrows indicate the forces from the pinning interaction. The driving force is directed to the right

a   b

From Fig.11.2 we see immediately that the elastic properties of the vortex lattice are highly important for the magnitude of the overall pinning force. As pointed out in Sect.7.3 elastic distortions of the vortex lattice during flux flow can also explain the existence of a frictional force $\underline{f}_p$ for a *moving* vortex lattice resulting in the power dissipation $f_p v_\varphi$ per unit length of vortex line. The pinning forces can be obtained from a pinning potential which can be chosen such that it describes pinning by points, lines, or planes. These theoretical developments were carried out by LABUSCH [11.18,19], LOWELL [11.20], GOOD and KRAMER [7.34, 11.21], CAMPBELL and EVETTS [7.10], DEW-HUGHES [11.22] and others. For details we refer to the treatments by CAMPBELL and EVETTS [7.10] and ULLMAIER [11.1].

11.3 Fundamental Pinning Interactions

The energy gain of the vortex lattice achieved from an adjustment of the flux-line configuration to the spatial distribution of the pinning centers can result from various interactions. The pinning mechanism for nearly isolated flux lines is expected to be different from that for a lattice of strongly interacting vortex lines.

11.3.1 Core Interaction

For nearly isolated vortex lines flux pinning can result from the interaction between the normal vortex core and a local inhomogeneity in the material in the following way. As shown in Fig.4.3a, at the center of a vortex line the order parameter drops to zero. This reduction of the order parameter can be described approximately by a normal vortex core of radius ξ. The condensation energy to be supplied for generating the normal core is per unit length of vortex line

$$E_{1c} = \frac{H_c^2}{8\pi} \pi \xi^2 \quad . \tag{11.13}$$

All or part of this energy can be recovered if the core of the vortex line passes through a region in the material where the order parameter is already zero or suppressed below its regular value (through the presence of normal inclusions, voids, etc.). For inclusions larger than the core region the maximum pinning energy per unit length is given by (11.13). The maximum pinning force per unit length is

$$f_{p\,core} = \mathrm{grad}\, E_{1c} \approx E_{1c}/\xi = H_c^2 \xi/8 \quad . \tag{11.14}$$

Inserting (3.59,60) into the last expression we obtain

$$f_{p\,core} \approx \frac{1}{16} \left(\frac{\varphi_0}{2\pi}\right)^{1/2} \frac{H_{c2}^{3/2}}{\kappa^2} \quad . \tag{11.15}$$

Our treatment of the core interaction can easily be generalized by proper geometrical considerations to the case where the spatial extension of the normal inclusions is smaller than the core radius ξ. Further, flux pinning by inclusions, which are not completely normal but where H_c and T_c are only reduced somewhat below the values in the major part of the material, can be treated by straightforward extension of the results given above.

It is interesting to calculate the maximum pinning-force density and the maximum critical current density in a superconductor from the result in (11.14). We assume flux pinning by long normal inclusions extending throughout the whole superconductor in the same direction as the applied magnetic field. For obtaining the maximum pinning-force density we need perfect matching between the flux-line configuration established by the magnetic field and the spatial distribution of the pinning regions. Such a matching arrangement of pinning sites may be achieved by a lattice of holes etched into the superconductor [11.23]. The maximum pinning force density can then be obtained by multiplying $f_{p\,core}$ of (11.14) with the vortex line density $n = B/\varphi_0$. In this way we find

$$F_{p\,core} = nf_{p\,core} = B H_c^2 \xi/8\varphi_0 = J_c B/c \quad . \tag{11.16}$$

With (3.59,60) we obtain for the critical current density from the last equation

$$J_c = cH_c/16\pi \sqrt{2} \lambda \quad . \tag{11.17}$$

This result is rather similar to the expression derived from the Ginzburg-Landau theory and given in (3.43).

11.3.2 Magnetic Interaction

In addition to the energy of the normal vortex core, nearly isolated flux lines contain an energy contribution from the magnetic fields and the circulating supercurrents associated with the lines. This contribution can be calculated from the London model and is given in (4.11,15,16). Sample inhomogeneities will change the distribution of magnetic fields and supercurrents, resulting in spatial variations of the line energy and a magnetic pinning interaction.

A simple example for this interaction is the effect of a large insulating inclusion or void. We assume a planar boundary of the inclusion oriented parallel to the vortex lines and perpendicular to the driving force. As we have seen in Sect.4.6.1, an energy barrier against vortex motion exists near the boundary. According to (4.57) this barrier arises from an attractive and a repulsive interaction near the surface. The attractive force on the vortex can be understood in terms of an image vortex of opposite direction and has the form $C \exp(-2 x_\ell/\lambda)$, where C is a constant and x_ℓ the distance of the flux line from the boundary. The repulsive force results from the interaction between the flux line and the external field H. Assuming the vortex density to be small and $H \approx H_{c1}$, the repulsive force is $\varphi_0 H_{c1} \exp(-x_\ell/\lambda)/4\pi\lambda$. In thermodynamic equilibrium the total work done by moving the vortex from the surface ($x = 0$) to $x = \infty$ must be zero,

$$\int_0^\infty dx \{ [\varphi_0 H_{c1} \exp(-x_\ell/\lambda)/4\pi\lambda] - C \exp(-2 x_\ell/\lambda) \} = 0 , \quad (11.18)$$

yielding

$$C = \varphi_0 H_{c1}/2\pi\lambda . \quad (11.19)$$

The force from the magnetic interaction attains its maximum value at $x = 0$. For the magnetic force per unit length we obtain

$$f_{p\,mag} = \varphi_0 H_{c1}/4\pi\lambda . \quad (11.20)$$

Using (3.59,60), and (4.20) the last result can be cast into the form

$$f_{p\,mag} = \frac{1}{4} \left(\frac{\varphi_0}{2\pi}\right)^{1/2} \frac{H_{c2}^{3/2}}{\kappa^3} \ln \kappa . \quad (11.21)$$

From (11.15,21) we find for the ratio

$$f_{p\,core}/f_{p\,mag} = \kappa/4 \ln \kappa . \quad (11.22)$$

The result of (11.22) indicates that the core interaction dominates for high values of κ. The reason for this can be found in the different range of the two interactions. This range is approximately the coherence length ξ for the core interaction and the penetration depth λ for the magnetic interaction. Since the force is given by the spatial derivative of the interaction energy, $f_{p\,core}$ will be larger than $f_{p\,mag}$ in high-κ materials.

From a comparison of (11.15,21) we see that the field and temperature dependence of the core interaction and the magnetic interaction are rather similar. Hence, the discrimination between the two pinning mechanisms from experimental measurements is quite difficult.

The results on flux pinning through the magnetic interaction can easily be generalized for the case of a boundary between two superconductors with different values of κ and H_{c2}.

11.3.3 Elastic Interaction

In the superconducting state the density and the elastic constants of a material are slightly smaller than in the normal state. For the volume dilatation the fractional changes are typically of the order of 10^{-7} whereas for the elastic constants they are of the order of 10^{-4}. Hence, in the normal core of a vortex line the material is slightly denser and stiffer than in the superconducting region around it. This leads to two pinning interactions in the following way.

First, the contracted normal core of a vortex line is surrounded by a strain field which interacts with the stress field of a defect. The interaction energy is linear in the defect stress. Second, the energy of a defect depends on the elastic constants of the surrounding medium, and will be modified through the presence of the stiff vortex core. The interaction energy is quadratic in the defect stress (for details see [7.10,11.1]).

11.3.4 Ginzburg-Landau Free Energy

So far in our discussion of the various pinning interactions we have restricted ourselves to nearly isolated vortex lines and to materials with a fairly high value of the Ginzburg-Landau parameter. For higher vortex-line densities or for materials with a low κ-value one must proceed differently. In this case the distinction between core interactions and magnetic interactions is not very meaningful. Here the Ginzburg-Landau theory will be more adequate for describing the pinning interactions.

Starting with the expansion of the free-energy density in (3.1) we can express the effect of structural inhomogeneities in the material (acting as pinning centers) in terms of spatial variations of the coefficients α and β. The change in free energy caused by small fluctuations $\delta\alpha$ and $\delta\beta$ is

$$\delta F = \int d\underline{r} \left[\delta\alpha |\Psi|^2 + \frac{\delta\beta}{2} |\Psi|^4 \right] \quad . \tag{11.23}$$

In order to replace the coefficients α and β by measurable quantities we use (3.17,60) and obtain

$$\alpha = -\frac{\hbar e}{2mc} H_{c2} \quad . \tag{11.24}$$

Similarly we find from (3.35,59)

$$\beta = 8\pi\kappa^2 \left(\frac{\hbar e}{2mc}\right)^2 \quad . \tag{11.25}$$

Inserting the last two equations into (11.23) and using the definition (3.2) one finds

$$\delta F = \frac{1}{4\pi} \int d\underline{r} \; H_c^2 \left(-\frac{\delta H_{c2}}{H_{c2}} \left|\frac{\Psi}{\Psi_0}\right|^2 + \frac{1}{2} \frac{\delta\kappa^2}{\kappa^2} \left|\frac{\Psi}{\Psi_0}\right|^4 \right) \quad . \tag{11.26}$$

As an example we calculate the pinning force exerted by a small volume Ω with a different value of H_{c2} [7.10]. The dimensions of this pinning region are assumed smaller than the coherence length. At high magnetic fields $|\Psi|^2$ and h can be represented in good approximation by their first harmonic, and the order parameter varies from $|\Psi/\Psi_0|^2 = 0$ at the vortex cores to its maximum value $|\Psi/\Psi_0|^2 = 3/2 \; \overline{|\Psi/\Psi_0|^2}$ half way between the cores [11.24]. Hence, the maximum pinning energy is gained with the pinning region located halfway between the cores. Noting that for large flux densities we can neglect the second term in (11.26), since $|\Psi|^4 \ll |\Psi|^2$, the maximum pinning energy is given by

$$F_p = \frac{3}{8\pi} \Omega \; H_c^2 \; \frac{\delta H_{c2}}{H_{c2}} \; \overline{\left|\frac{\Psi}{\Psi_0}\right|^2} \quad . \tag{11.27}$$

The spatial derivative of F_p attains its maximum value $\pi F_p/a$ at a distance $x = a/4$ away from the vortex center, where a is the lattice parameter of the vortex lattice. Hence, the maximum pinning force is

$$K_p = \frac{3}{16} \frac{\Omega}{a} \left(\frac{H_{c2}}{\kappa}\right)^2 \left(1 - \frac{B}{B_{c2}}\right) \frac{\delta H_{c2}}{H_{c2}} \quad . \tag{11.28}$$

Here we have used (3.59) and the relation

$$\overline{|\Psi/\Psi_0|^2} \approx 1 - (B/B_{c2}) \quad , \tag{11.29}$$

which follows from (3.96,97) for large κ-values. Inserting expression (4.38) for the vortex-lattice parameter of the triangular lattice into (11.28) we finally obtain

$$K_p = \frac{3}{16}\left(\frac{\sqrt{3}}{2}\right)^{1/2} \frac{\Omega}{\varphi_0^{1/2}} \frac{H_{c2}^{5/2}}{\kappa^2} \frac{\delta H_{c2}}{H_{c2}} \left(\frac{B}{B_{c2}}\right)^{1/2} \left(1 - \frac{B}{B_{c2}}\right) \quad . \tag{11.30}$$

11.4 Some Model Experiments

Because of its importance, in particular with respect to practical applications of superconductors, flux pinning in type-II superconductors and its correlation with the metalurgical microstructure of the material has been the subject of many experiments. Quite often the systems investigated are rather complicated metallurgically, and several pinning interactions may be operating simultaneously. For clarifying the situation, experiments performed with simple model systems and under well controlled variations of the metallurgical microstructure are highly desirable. In the following we describe a few of such experiments.

A regular arrangement of pinning sites can be achieved by a controlled modulation of the surface of the superconductor. The simplest configuration is an array of regular, parallel microgrooves generated in the surface. This idea has been realized by MORRISON and ROSE [11.25] by pressing diffraction gratings into foils of In-Bi alloy. As one would expect, the pinning force exerted by the microgrooves was zero for flux flow parallel to the grooves and reached a maximum value for perpendicular flux flow. Further, due to the asymmetric cross section of the grooves, for perpendicular flux flow the critical current was found to behave asymmetrically with respect to current or magnetic field reversal. Such a rectification effect associated with flux motion in a superconductor is not unusual and always occurs for an anisotropic pinning force [11.26].

Flux pinning by periodic thickness modulation has also been studied in granular aluminum films evaporated in an oxygen atmosphere [11.27]. Here the periodic thickness modulation was achieved by a photolithographic process using the regular interference pattern of two incident He-Cd-laser plane waves at the film surface. The aluminum films were about 5300 Å thick, and the amplitude of the thickness modulation was about 200 Å. The spacing between the grooves at the film surface was 1.9 μm. As a function of the perpendicular magnetic field the critical current density showed characteristic peaks at well-defined field values. These results can be explained by matching of the two-dimensional vortex lattice with the one-dimensional periodic pinning structure. In addition to the static interaction of the vortex lattice with the one-dimensional periodic pinning potential, in the flux-flow regime supercurrent oscillations were observed arising from the periodic thickness modulation. Apparently, under dynamic matching conditions highly coherent Josephson oscillations of the moving vortex lattice take place similar to the behavior of series arrays of resistively shunted Josephson junctions acting in phase and frequency coherence [11.28,29].

Maximum flux pinning will take place if the pinning regions (generated through the absence of superconducting material) show a two-dimensional periodic structure similar to the vortex lattice and extend all the way through the superconductor along the direction of the vortex lines. This situation has been realized by HEBARD and co-workers [11.23,30] who studied the critical currents in granular aluminum films prepared with various lattice arrangements of holes. The oxygen-doped Al films were 0.1 µm thick. The lattice arrays of holes with typically 0.5 µm radius and 3 µm spacing could be generated using an electron beam masking technique in combination with X-ray lithography. In this way hole configurations with dimensions similar to the characteristic dimensions of the vortex lattice have been achieved. For perfect matching between the vortex lattice set up by a perpendicular magnetic field and a lattice of holes with triangular symmetry rather large critical current densities have been observed approaching the maximum theoretical value indicated in (11.17). The strong pinning effect of the hole lattice at the matching perpendicular magnetic field could also be observed for hole lattices deviating considerably from the triangular symmetry, i.e., for a distorted triangular lattice, a square lattice of holes, etc. Apparently, the vortex lattice adjusts itself as much as possible to the configuration of holes, as one would expect from a consideration of the energy gain. This also includes the possibility that each hole may be occupied by more than one flux quantum. It appears that this multiple quantum occupancy of the holes has been observed.

A well-defined pinning structure can also be produced by periodically modulating the composition of the superconductor. This approach has been taken by RAFFY and co-workers [11.31-33] who studied flux flow in Pb-Bi alloy films with a periodic modulation of composition. The modulation was achieved by co-evaporation of the two materials from two sources, keeping the Pb deposition rate constant and varying the Bi deposition rate periodically. For the fresh films, the extremes of the Bi concentration were 2% and 20%. The period of the concentration profile ranged between 700 and 8000 Å. The total film thickness was about 6 µm, corresponding to 60 layers for a period of 1000 Å. The inhomogeneous material could be homogenized by diffusion over a sufficient time period. For a magnetic field parallel to the alloy film and a current direction perpendicular to the field and parallel to the planes of equal composition the critical current density J_c is considerably larger compared to the homogeneous material. Further, J_c shows distinct peaks at certain magnetic field values. The detailed structure of the $J_c(H)$ curves depends on the periodicity length and the amplitude of the modulation of the Bi concentration. For a magnetic field perpendicular to the alloy film the critical current density decreases monotonically with increasing field, and in the inhomogeneous material J_c is very similar to the values in the homogenized film and in a homogeneous film of the same average composition. The peak structure in the $J_c(H)$ curves decreases when the angle θ between the film and the magnetic field is increased from zero and disappears for $θ ≈ 50°$. The results indicate again geometrical matching of the Bi-concentration profile with the vortex lattice.

So far we have discussed model systems for studying flux pinning which have been prepared by periodically modulating the geometry or the composition of the material. Another important way of generating a superconducting material with a reasonably well-defined pinning structure is the controlled precipitation of small clusters of a second phase in a homogeneous matrix. A typical example for this latter technique is a superconducting NbTa alloy containing normal conducting Nb_2N precipitates. This system has been investigated by ANTESBERGER and ULLMAIER [11.33a]. The samples were prepared from a polycrystalline alloy of 75 at.% Nb and 25 at.% Ta. Following an outgassing procedure at high temperatures in ultrahigh vacuum, the samples were loaded with nitrogen in an N_2 atmosphere. Subsequently, they were heat treated in order to form a statistical distribution of Nb_2N precipitates. The concentration of the precipitates could be varied through the nitrogen pressure during the loading procedure. From measurements of the pinning force density the interaction force between a single vortex line and a single defect has been derived. The magnitude and temperature dependence of this force was found to agree well with the theoretical result for the core interaction given in (11.14,15).

In addition to studies of the *macroscopic* pinning behavior through measurements of the critical currents or the critical vortex-density gradients, experiments on the *microscopic* correlations between the arrangement of vortex lines and the spatial distribution of pinning sites are highly interesting. Here the Träuble-Essmann technique, discussed in Sect.6.1, is well suited when combined with an electron-microscopic examination of the same sample area. Such combination experiments have been performed by HERRING [11.34], RODEWALD [11.35], and LISCHKE and RODEWALD [11.36] demonstrating vortex pinning by dislocations and grain boundaries.

12. Flux Creep and Flux Jumps

As we have seen in Sect.11, flux pinning allows the establishment of stationary distributions of magnetic fields and currents which are far away from thermodynamic equilibrium. Relaxation toward equilibrium sets in through flow flow as soon as the vortex-density gradient exceeds its critical value. At finite temperatures an additional relaxation mechanism is possible, namely *thermally activated flux creep*, as first pointed out by ANDERSON [7.4]. Flux creep can reveal itself in two ways: through slow changes of trapped flux and through the appearance of small resistive voltages.

According to Anderson's theory thermally activated flux creep may involve bundles of several flux lines jumping together over the pinning barriers. The jump rate is given by

$$\nu = \nu_0 \exp(-E_{eff}/k_B T) \tag{12.1}$$

where ν_0 is a characteristic frequency and E_{eff} an effective activation energy. This energy E_{eff} is the energy barrier associated with the pinning mechanism reduced by the work contributed from the Lorentz force. In the absence of a vortex-density gradient thermally activated vortex jumps take place in all directions with the same probability, and the net creep velocity vanishes. We note that due to the existence of thermally activated flux creep the critical state only represents a quasi-stationary state in which the creep rate may be extremely small or practically unobservable.

Flux creep can be conveniently investigated by observing the time dependence of trapped flux in a hollow superconducting cylinder. Here the driving force is roughly proportional to the flux density ΔB trapped in the cylinder. Because of the exponential dependence of the creep rate on the driving force, we have

$$\frac{d|\Delta B|}{dt} \approx \mp C_1 \exp(|\Delta B|/B_0) \quad , \tag{12.2}$$

yielding the solution

$$|\Delta B| = C_2 - B_0 \ln t \quad . \tag{12.3}$$

Here C_1 and C_2 are constants. The - sign in (12.2) refers to positive ΔB values and fields decreasing with time, whereas the + sign refers to negative ΔB values and increasing fields. The logarithmic time dependence indicated in (12.3) is in excellent agreement with experiment [7.5,11.4,33,12.1-3]. In most of these experiments the decay of trapped flux in the hollow cylinder was measured using conventional magnetic techniques.

BOERSCH et al. [12.3] detected single flux quanta within a hollow superconducting microcylinder from the phase shift of an electron wave caused by the trapped flux (Bohm-Aharonov effect). Electron waves incident upon the cylinder undergo the analog of Fresnel diffraction and produce interference fringes, which are magnifield electron-optically and observed with the help of an image intensifier. Whether the central fringe is a maximum or a minimum depends upon whether there are an even or an odd number of flux quanta trapped in the cylinder. With this apparatus it is possible to observe the motion, along the cylinder wall, of single flux quanta with velocities down to about 100 Å/s. (For detecting such a velocity of a single vortex by flux-flow voltage measurements, a voltage sensitivity of 10^{-21} V would be required). Figure 12.1 shows the magnetic flux as a function of time for two temperatures in a Pb-In cylinder with only one flux quantum trapped in the bore. The flux change was recorded after switching off the external field. In the upper two diagrams of Fig.12.1 only three curves are shown from a series of many runs. In the lower diagram the mean values of the trapped flux are plotted, demonstrating an enhanced creep rate with increasing temperature.

The logarithmic time dependence shown in (12.3) has been obtained from a rather simple argument. A more careful treatment of flux creep in a cylindrical superconductor placed in a longitudinal magnetic field has been given by BEASLEY et al. [12.2]. They found that the total flux φ in the sample at the time t is

$$\varphi(t) = \varphi(t_0) + k_c \ln(t/t_0) \quad , \tag{12.4}$$

where t_0 is an arbitrary reference time and k_c is the logarithmic creep-rate constant defined by

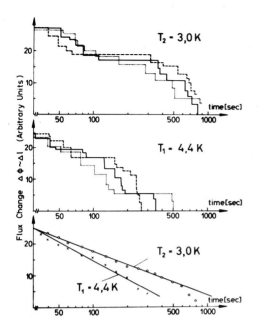

Fig.12.1. Magnetic flux as a function of time at two temperatures in a Pb-In cylinder with only one flux quantum trapped in the bore. The lower diagram contains the mean values of the trapped flux obtained from many runs. For further details see text. A flux change of 10 arbitrary units corresponds to a decrease of one flux quantum in a cylinder of 10 μm length [12.3]

$$k_c \equiv d\varphi(t)/d \ln t \quad . \tag{12.5}$$

For the quantity k_c they obtained

$$k_c = \pm \frac{\pi}{3} k_B T R_0^3 (\partial E_{eff}/\partial |\nabla B|)^{-1} (1 \pm \varepsilon_1) \quad . \tag{12.6}$$

Here R_0 is the cylinder radius and $(\partial E_{eff}/\partial |\nabla B|)$ the change of the activation energy with the flux-density gradient (or the Lorentz force). The positive and negative sign refers to positive and negative values of $\nabla B = \partial B/\partial r$, respectively, corresponding to increasing and decreasing applied fields. The quantity ε_1 represents a correction arising from the fact that in the solution of the flux-creep equation $|\nabla B|$ and $(\partial E_{eff}/\partial |\nabla B|)$ differ as a function of position depending on whether the sample is in the critical state for increasing or decreasing applied fields. Usually ε_1 is small compared to unity.

We see from (12.6) that the logarithmic creep-rate constant is proportional to the absolute temperature. The creep-rate measurements performed by BOERSCH et al. [12.3] tend to confirm this proportionality.

Since the critical state in a superconductor represents a nonequilibrium configuration, its stability against small fluctuations is an open question.

Such a fluctuation may be caused internally (thermally activated flux creep) or externally (change of applied magnetic field or current, mechanical movement). The power dissipation associated with the small perturbation can cause a small temperature rise locally. Since the pinning force density usually decreases with increasing temperature, the perturbation may become larger, resulting in an instability and thermal runaway. This can lead to catastrophic consequences, if the magnetic energy stored in a superconducting magnet is suddenly converted into thermal energy. Hence, the development of stability criteria becomes highly important for applications of superconductors. A properly stabilized material should tolerate individual *flux jumps* without incurring permanent breakdown of superconductivity. We note that in a slowly increasing or decreasing applied magnetic field or transport current the magnetic field within a strong-pinning superconductor may change in form of relatively large flux jumps.

A *stability criterion* with respect to flux jumps can be obtained in the following way. For simplicity we assume a one-dimensional geometry consisting of a superconducting plate of thickness d oriented parallel to the external field H_e. We take the y-coordinate normal to the plate with the origin placed in the center. We assume a critical state within the superconductor which can be described by the Bean model such that B varies linearly in y-direction, see (11.4). The field H_e is taken sufficiently high for complete flux penetration to occur, as shown in Fig.12.2. From Maxwell's equation curl $\underline{H} = 4\pi \underline{J}/c$ we obtain for the magnetic flux density in our geometry

$$B(y) = H_e - \frac{4\pi}{c} J_c \left(\frac{d}{2} - y\right) \quad . \tag{12.7}$$

Here we have restricted ourselves to the half of the plate with $y > 0$. A small local temperature rise $\delta_1 T$ will cause a reduction in the critical current density by the amount

$$\delta J_c = (dJ_c/dT)\delta_1 T \quad , \tag{12.8}$$

leading in turn to the change in magnetic flux density

$$\delta B(x) = -\frac{4\pi}{c} \delta J_c \left(\frac{d}{2} - y\right) \quad . \tag{12.9}$$

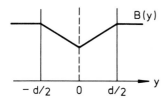

Fig.12.2. Flux profile in a superconducting plate showing complete flux penetration

The change $\delta B(x)$ induces electric fields and causes power dissipation. The average heat per unit volume generated in this way is

$$\delta Q = \frac{J_c}{c} \frac{4\pi}{c} |\delta J_c| \frac{2}{d} \int_0^{d/2} \left(\frac{d}{2} - y\right)^2 dy \qquad (12.10)$$

or

$$\delta Q = \frac{\pi}{3c^2} d^2 J_c |\delta J_c| \quad . \qquad (12.11)$$

This heat δQ causes the temperature rise $\delta_2 T = \delta Q/C$, where C is the specific heat per unit volume. As long as $\delta_2 T < \delta_1 T$ the temperature fluctuation $\delta_1 T$ will not lead to a diverging increase in temperature and the material will return to a stable superconducting state. Hence, we have the stability criterion

$$d^2 < \frac{3c^2 C}{\pi J_c} \left|\frac{dJ_c}{dT}\right|^{-1} \quad . \qquad (12.12)$$

We note that this criterion has been derived for *adiabatic* conditions where no heat energy is removed during the temperature fluctuation. In reality heat conduction will tend to reduce the temperature rise $\delta_2 T$. In order to achieve thermal stability during flux jumps by satisfying condition (12.12), composite conductors have been developed, in which many small superconducting filaments are imbedded in a copper matrix (with high thermal and electrical conductivity). For NbTi filaments according to criterion (12.12) the filament diameter must be smaller than about 10^{-2} cm.

13. Electrical Noise Power

According to the Josephson relation (7.24) the motion of single flux quanta or flux tubes containing many flux quanta generates individual voltage pulses with a characteristic time dependence. Therefore, the time-averaged dc flux-flow voltage contains ac components which are related to the quantized nature of the individual voltage contributions. Microscopic information on the voltage-generating mechanism can be obtained from the electrical noise power spectrum.

13.1 Autocorrelation Function and Power Spectrum

We consider a statistically fluctuating quantity $y(t)$ as function of time t, its fluctuations being statistically independent. Further we assume $y(t)$ to be statistically stationary, i.e., the time average of $y(t)$ is taken as a constant. Useful information on the time behavior of $y(t)$ is contained in the autocorrelation function for correlations in time

$$\chi(\tau) = \overline{y(t) \cdot y(t+\tau)} \quad , \tag{13.1}$$

where $\overline{\ldots}$ denotes time average. We note that $\chi(t)$ has the properties $\chi(0) > 0$ and $\chi(-\tau) = \chi(\tau)$. According to the Wiener-Khintchine theorem the power spectrum $w(\omega)$ and the autocorrelation function are the Fourier transforms of each other and are connected through the relations

$$w(\omega) = 4 \int_0^\infty \chi(\tau) \cos\omega\tau \, d\tau \tag{13.2}$$

and

$$\chi(\tau) = \int_0^\infty w(\omega) \cos\omega\tau \, d\omega \quad , \tag{13.3}$$

where ω is the angular frequency. These equations play an important role in many branches of physics, where the microscopic time behavior of a system is studied with a probe which is periodic in time, such as radio waves, light, X-rays, beams of electrons and neutrons, etc. We note that (13.2,3) represent only the time component of the more general formalism introduced by VAN HOVE [13.1,2] in terms of the dynamical structure factor and the pair correlation function for treating correlations in both space and time.

Some examples for the autocorrelation function $\chi(\tau)$ and the corresponding power spectrum are given in Fig.13.1.

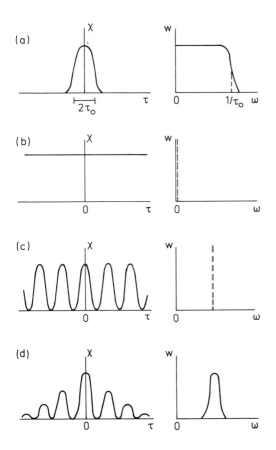

Fig.13.1.a-d. Examples of the autocorrelation function $\chi(\tau)$ and the corresponding power spectrum $w(\omega)$

In utilizing the Wiener-Khintchine theorem one compares electrical noise power data with the results obtained for some model spectrum. In the following we discuss a few important model spectra. In cases a) - c) we assume statistically independent rectangular voltage pulses of height ΔV and average pulse rate ν_p constituting the time-dependent voltage $V(t)$.

a) If the voltage pulses have constant magnitude ΔV and lifetime τ_0 (pure "shot noise") the correlation function is

$$\chi(\tau) = \overline{V(t) \cdot V(t + \tau)} = (\Delta V)^2 \nu_p \tau_0 [1 - (\tau/\tau_0)] \tag{13.4}$$

yielding the power spectrum [13.3]

$$w(\omega) = 2\overline{V} \cdot \Delta V \, \tau_0 \left[\frac{\sin \frac{\omega \tau_0}{2}}{\frac{\omega \tau_0}{2}} \right]^2 . \tag{13.5}$$

b) If the voltage pulses have constant magnitude ΔV and an exponential distribution of their lifetime τ_0 indicated by the autocorrelation function

$$\chi(\tau) = \overline{V(t) \cdot V(t + \tau)} = (\Delta V)^2 \nu_p \tau_0 \exp(-\tau/\tau_0) \tag{13.6}$$

the power spectrum is [13.3]

$$w(\omega) = 4\overline{V} \, \Delta V \, \tau_0 / (1 + \omega^2 \tau_0^2) . \tag{13.7}$$

This case is rather similar to the power spectrum of (13.5), except for the structure in the latter due to the factor $\sin^2(\omega \tau_0/2)$.

c) Next we assume again voltage pulses of magnitude ΔV and an exponential distribution of their lifetime τ_0 in the form $(1/\tau_0) \exp(-\tau/\tau_0)$. However, now the magnitude ΔV and the lifetime τ_0 are coupled through the relation

$$\Delta V \cdot \tau_0 = k_0 , \tag{13.8}$$

where k_0 is a constant. (As we shall see in Sect.13.2 such a coupling between the height and duration of the pulse exists in the case when the voltage pulses are caused by the motion of individual flux quanta. The time integral of each pulse is then equal to the flux quantum φ_0). A consequence of this coupling is an increase of the noise power at high frequencies. In this case the power spectrum is [13.4,5]

$$w(\omega) = 4\nu_p (k_0^2/\omega^2 \tau_0^2) \left[\omega \tau_0 \arctan \omega \tau_0 - \frac{1}{2} \ln (1 + \omega^2 \tau_0^2) \right] . \tag{13.9}$$

We note that the noise power in (13.5,7) is proportional to ω^{-2} at high frequencies, whereas in (13.9) we have ω^{-1} dependence at high frequencies. The three power spectra are shown schematically in Fig.13.2. The noise power is seen to be frequency-independent up to $\omega \approx \tau_0^{-1}$. The low-frequency limit $w(0)$ increases proportional to the lifetime τ_0 of the fluctuation. The behavior of $w(\omega)$ is strongly different if the appearance of a new voltage pulse is delayed for some time after the previous pulse (dead-time effect). The noise power then shows a maximum at a frequency determined by the delay time.

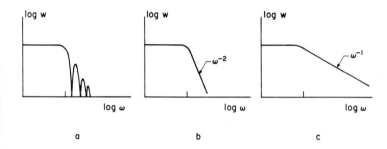

Fig.13.2. (a), (b), and (c) show the noise power spectra described by (13.5, 7,9), respectively. The scale mark on the horizontal axis indicates the frequency $\omega = \tau_0^{-1}$

d) In cases a) - c) we had assumed the individual voltage pulses to be statistically independent. Now we include the possibility that the voltage signal is a mixture of statistically independent pulses and completely correlated pulses. The situation for a sequence of completely correlated voltage pulses is shown in Fig.13.1c. In this case the autocorrelation function may be taken in the form

$$\chi(\tau) = (\Delta V)^2 \nu_p \tau_0 (\cos\omega_0 \tau + 1)/2 \quad , \tag{13.10}$$

and the function $w(\omega)$ consists of a single sharp line at a distinct frequency ω_0. The mixture of the two types of voltage pulses may be described by a new autocorrelation function obtained from the superposition of the correlation functions from (13.6,10) (denoted by $\chi_{(13.6)}(\tau)$ and $\chi_{(13.10)}(\tau)$, respectively) in the form

$$x(\tau) = \alpha x_{(13.6)}(\tau) + (1 - \alpha)x_{(13.10)}(\tau) \quad . \tag{13.11}$$

Here α and $(1 - \alpha)$ are the factors determining the weight of the two correlation functions. From (13.2) we see immediately that the power spectrum $w(\omega)$ is linearly superimposed from the corresponding expressions with the same weighting factors α and $(1 - \alpha)$.

The following are some useful relations applicable to a statistical sequence of identical rectangular voltage pulses of height ΔV, duration τ_0, and average rate ν_p. With $\delta V = V - \overline{V}$ the total mean square noise voltage is

$$\overline{(\delta V)^2} = \int_0^\infty w(\omega)d\omega = \Delta V \cdot \overline{V} \quad . \tag{13.12}$$

For the time-averaged voltage we have

$$\overline{V} = \Delta V \cdot \nu_p \cdot \tau_0 \quad . \tag{13.13}$$

In the case of moving flux quanta the individual voltage pulse is obtained from the Josephson relation (7.24) and a geometrical consideration of the voltage probes and of the measuring circuit (see Sect.13.2). For the moment we consider only relation (7.24). Approximating the time behavior of each pulse by a rectangular pulse of amplitude ΔV and duration τ_0, we can write from (7.24)

$$\frac{\partial \dot{\phi}}{\partial t} = \frac{2\pi}{\tau_0} = \frac{2e}{\hbar} \Delta V \tag{13.14}$$

or

$$\Delta V = \varphi_0/c\tau_0 \tag{13.15}$$

where $\varphi_0 = hc/2e$ is the flux quantum. Inserting (13.15) into (13.5) we have

$$w(0) = 2\overline{V}\varphi_0/c \quad . \tag{13.16}$$

We see that the magnitude of the moving flux unit appears in the quantity $w(0)$. The result (13.16) is analogous to the Schottky relation

$$w(0) = 2\overline{I}e \tag{13.17}$$

for the shot noise associated with the current I in an electron tube. If during flux flow we are dealing with flux tubes of flux $\varphi = N_T \cdot \varphi_0$ (N_T = number of flux quanta) instead of single flux quanta, the quantity φ_0 must be replaced by φ in (13.16). Similarly, for short-range correlation between vortex lines, the total flux contained in the correlated flux lines must appear in (13.16) instead of the flux quantum φ_0 [7.69].

Long-range correlation between flux lines can lead to uniform motion of a monocrystalline vortex lattice through the superconductor. In this case statistical noise in the flux-flow voltage is highly reduced or absent. The autocorrelation function is then periodic in time, and the noise power spectrum consists of a more or less sharp line at some frequency. Such highly correlated vortex motion can lead to various interference phenomena [13.6,7].

13.2 Influence of the Geometry of the Contacts

The voltage pulse associated with the motion of a single vortex line and hence the noise-voltage behavior of a moving assembly of vortex lines depend sensitively on the geometrical arrangement of the voltage probes and of the measuring circuit. A detailed discussion of this point has been given by CLEM [7.69]. We consider the voltage pulse V(t) arising from the current-induced motion of a flux quantum φ_0 through the superconductor between the contact points (1) and (2) (see Fig.13.3). The voltage V(t) measured with an instrument at the ends of the lead wires L is the sum of two contributions

$$V(t) = V_{12}(t) + V_L(t) \quad . \tag{13.18}$$

Here $V_L(t)$ is the voltage induced in the lead wires given by

$$V_L(t) = -\frac{1}{c} \int_{(1)}^{(2)} \frac{\partial A}{\partial t} ds \quad . \tag{13.19}$$

The quantity A(t) is the Coulomb-gauge vector potential associated with the moving vortex line. The integration is performed along the lead wires including the instrument. The voltage $V_{12}(t)$ generated in the superconductor between the points (1) and (2) is obtained from the Josephson relation (7.24). As shown by CLEM [7.69], the contribution $V_L(t)$ can be made arbitrarily small be proper geometrical arrangement of the lead wires above a speciment in the form of an infinite slab. Hence, in the following we only consider the contribution $V_{12}(t)$.

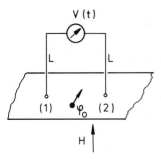

Fig.13.3. Voltage measuring circuit attached to the superconductor at points (1) and (2)

The voltage pulse $V_{12}(t)$ depends only upon the relative positions of the contact points (1) and (2) and the vortex axis if the voltage-lead wires are assumed straight and perpendicular to the sample surface. We identify the surface of the superconductor with the x-y plane and describe the contact points (1) and (2) by the coordinates x_1, y_1 and x_2, y_2, respectively. The superconductor is considered to have infinite extension in x- and y-direction. The vortex is oriented in the z-direction, and the intersection of the vortex axis with the surface is assumed to move along the y-axis. For simplicity we take the vortex velocity v_φ as constant. The voltage $V_{12}(t)$ is then given by

$$V_{12}(t) = \frac{\hbar v_\varphi}{2e}\left[\frac{x_1}{x_1^2+(y_1-v_\varphi t)^2} - \frac{x_2}{x_2^2+(y_2-v_\varphi t)^2}\right] . \qquad (13.20)$$

This function is shown in Fig.13.4. We see that $V_{12}(t)$ is the sum of two Lorentzians. The first and second Lorentzian curves attain their maximum at the time of closest approach of the vortex to point (1) and point (2), respectively. The magnitude of the time integral of each Lorentzian is equal to $\varphi_0/2c = h/4e$. The shape of the Lorentzian becomes narrower and more sharply peaked with decreasing distance of closest approach to the contact point. The magnitude of the time integral of $V_{12}(t)$ is $\varphi_0/c = h/2e$ if the vortex passes between the points (1) and (2). This integral is zero if the vortex passes outside the two points. For the special case where the vortex moves perpendicularly to the straight line passing through the two points (1) and (2) we have $y_1 = y_2$ (by chosing the same coordinate system as before), and the centers of the two Lorentzians in Fig.13.4 coincide. In Fig.13.5 we show a series of pulses $V_{12}(t)$ obtained in this case for different values of the coordinates of the contacts (1) and (2) relative to the path of the vortex. We note that passage of the vortex outside the points

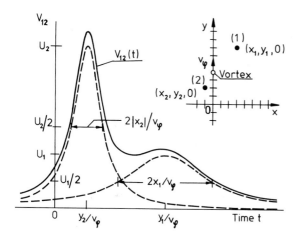

Fig.13.4. Voltage V_{12} between contacts (1) and (2) versus time according to (13.20). The vortex line moves along the y-axis. The relative coordinates of the contacts are shown on the inset. The quantities U_1 and U_2 are defined as $U_1 = \hbar v_\varphi/2ex_1$ and $U_2 = \hbar v_\varphi/2e|x_2|$. The total voltage pulse (solid curve) is the sum of two Lorentzians (dashed curves) as discussed in the text [7.69]

(1) and (2) results in positive and negative contributions to $V_{12}(t)$ such that $\int_{-\infty}^{+\infty} V_{12}(t)dt = 0$.

For the latter case, resulting in voltage pulses such as shown in Figure 13.5, the noise power spectrum is [7.69]

$$w(\omega) = (2\overline{V\varphi}_0/c)\left\{\frac{1}{2} \exp(-\omega x_{12}/v_\varphi) + \frac{v_\varphi}{2\omega x_{12}}[1 - \exp(-\omega x_{12}/v_\varphi)]\right\}, \quad (13.21)$$

where $x_{12} = |x_1 - x_2|$. The low-frequency limit of this expression is $w(0) = 2\overline{V\varphi}_0/c$. At high frequencies, $\omega \gg v_\varphi/x_{12}$, we have

$$w(\omega) \approx \overline{V\varphi}_0 v_\varphi/x_{12} \omega c \ . \tag{13.22}$$

This ω^{-1} dependence at high frequencies arises from the sharply peaked voltage pulses generated by vortices passing very close to the voltage contacts (here assumed infinitesimal in size). We note that this situation is similar to the case of rectangular voltage pulses with a distributed lifetime τ_0 described by (13.8,9), which also leads to the ω^{-1}-dependence of the noise power at high frequencies. Voltage pulses produced by vortex motion and the influence of the arrangement of the voltage leads have also been discussed by PARK [13.8] and JARVIS and PARK [13.9].

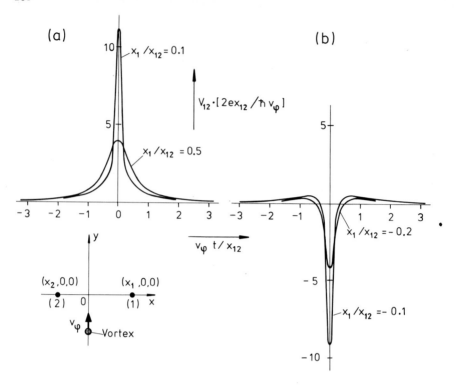

Fig.13.5a and b. Voltage pulses $V_{12}(t)$ between contacts (1) and (2) versus time according to (13.20) for a vortex moving perpendicular to the straight line connecting both contacts. The coordinate system is shown on the inset. The ratio x_1/x_{12}, where x_{12} is the distance between the contacts, indicates the location of the vortex path. For $0 < x_1/x_{12} < 1$ the vortex passes between the two contacts (a); for $x_1/x_{12} < 0$ and $x_1/x_{12} > 1$ the vortex path is located outside both contacts (b)

So far we have assumed well localized voltage probes attached to the superconductor at some points (1) and (2). The extreme opposite case is the geometry of the Corbino disk of Fig.7.5 where the contacts of the voltage leads are spread entirely over the two sides of the specimen. Indeed, VAN GURP [13.10] reported the absence of detectable flux-flow noise for the Corbino-disk geometry. Similarly, the conventional four-probe arrangement of current and voltage leads attached to wide superconducting end sections on both sides of a narrow thin-film specimen is unfavorable for flux-flow noise experiments with high spatial resolution.

13.3 Experiments

Measurements of the electrical noise power arising from vortex motion in a superconductor require the careful elimination of contributions from external sources, such as the operation of the magnet or the supply of the specimen current. Thermal fluctuations due to nucleate boiling of liquid helium can be avoided by operating below the lambda point. By switching the sample current on and off, subtraction of the background noise is possible.

The first systematic experiments on flux-flow noise in type-II and type-I superconductors have been performed by VAN GURP [13.10-12]. The majority of his results on type-II materials (vanadium foils and In-Tl alloy wire) agrees reasonably with the power spectrum indicated in (13.5) or (13.7). The time τ_0 extracted from the experimental $w(\omega)$ curves was found to be close to the transit time τ^* of each vortex obtained from the relation

$$\tau^* = \ell_{12} w_s B / c \overline{V} \quad . \tag{13.23}$$

Here ℓ_{12} is the distance between the voltage probes and w_s the width of the sample. From the low-frequency limit $w(0)$ VAN GURP concluded that short-range correlation can exist for flux bundles containing up to $N_T = 10^5$ flux quanta at low applied magnetic field, low temperature, and low transport-current density. Here the effective bundle size N_T is obtained from extending (13.16) with the definition

$$N_T = w(0)/(2\overline{V}\varphi_0/c) \quad . \tag{13.24}$$

The bundle size N_T was found to decrease toward unity as the field, temperature, and current density increased. Further, the noise power spectra suggested that a fraction p of the vortex lines does not take part in flux motion because of pinning effects, and that the expression (13.23) for the transit time should be multiplied by $(1 - p)$. The fraction p was found to decrease with increasing temperature and current, and to decrease also with increasing magnetic field in the absence of a peak effect in the critical current. Sometimes, at low frequencies, a maximum in the noise power was observed, indicating a delaying mechanism between subsequent voltage pulses.

Extending his flux-flow noise experiments to type-I superconductors (foils of In-2 at.% Pb alloy), at low magnetic fields VAN GURP [13.11] found reasonable agreement with the expressions (13.5) or (13.7), the spectral noise power $w(\omega)$ being independent of ω at low frequencies and proportional to ω^{-2} at higher frequencies. At intermediate fields he found $w \sim \omega^{-\gamma}$ with $0.5 < \gamma < 1$, indicating a distribution of the lifetime τ_0 of the voltage pulses as contained, for example, in (13.8,9). From his results VAN GURP was able to extract the flux-flow component of the resistive voltage. Apparently, the flux-flow component dominates at low magnetic fields, whereas at higher fields ohmic losses become more and more important, in agreement with the behavior expected from other evidence (see Sect.7.3).

VAN GURP analyzed his data in terms of the simple shot-noise model with a single time constant τ_0 and expression (13.5) for the noise power spectrum (showing oscillatory behavior of $w(\omega)$ at high frequencies). This point has been criticized [13.13,14], and the application of expressions (13.6,7) appears to be more realistic. However, the main conclusions by VAN GURP remain unaffected.

Extensive measurements of the flux-flow noise in niobium and vanadium foils have been performed by HEIDEN [13.5,15-17]. In these experiments the location and the geometrical orientation of the pair of voltage contacts could be varied. Using a second pair of voltage contacts, cross correlations could be measured also. HEIDEN observed that the low-frequency limit of the spectral noise power is often much smaller than the value expected from (13.16). Apparently, in the specimens investigated the noise voltages are not generated through the uncorrelated behavior of the individual vortices. Instead, they seem to be caused by local velocity fluctuations in a moving vortex structure with rather strong long-range correlations. Such nonuniform vortex motion can result, of course, from the presence of pinning sites [13.18,19]. Experiments by JARVIS and PARK [13.20] on type-II foils of InPb alloy also indicated strong long-range correlation between the moving vortices.

A model based on the strong influence of flux pinning upon the flux-flow noise characteristics has been proposed also by JOINER and co-workers [13.21-23] for explaining their experimental results on type-II foils of PbIn alloy. They observed the noise power to be independent of ω at low frequencies and to vary proportional to ω^{-1} at higher frequencies. Such behavior can be explained by a distribution of the lifetimes for the individual voltage pulses arising from the fact that during their transit across the sample the vortices are temporarily held up at the pinning centers. In this way vortex motion may take place in some form of a *hopping process*. Direct evidence for such a hopping process has also been obtained from high-resolution magneto-optical studies of current-induced flux-tube motion in type-I superconducting thin films [2.22].

14. Current-Induced Resistive State

In this section we discuss the electric and magnetic phenomena in superconductors caused directly by a transport current and by the magnetic field associated with this current, in the *absence of an external magnetic field*. The dimensions of the superconductor are generally assumed to be large compared to the coherence length and penetration depth.

14.1 Wire Geometry

The destruction of superconductivity by an electric current in a cylindrical type-I superconductor with circular cross section was treated theoretically by LONDON [3.5,14.1] 40 years ago. We consider a wire of radius R_0. Electrical resistance will appear in the wire as soon as the current-generated magnetic field at the circumference of the wire exceeds the critical field, i.e., for currents exceeding the critical value

$$I_c = cR_0 H_c/2 \quad . \tag{14.1}$$

The breakdown of superconductivity in a current-carrying conductor at the current value at which the current-induced magnetic field reaches H_c is often referred to as *Silsbee's rule*. According to London's theory, for currents larger than I_c the superconducting wire splits up into a stationary, cylindrically symmetric mixture of normal and superconducting domains. The configuration proposed by LONDON is shown in Fig.14.1. It consists of a core of alternate layers of normal and superconducting phase arranged in conical shapes along the cylinder axis. This core region of alternate layers extends outward to the radius $R_1 < R_0$. In the normal layers of the core the magnetic field is equal to H_c. The transport-current density along the wire is

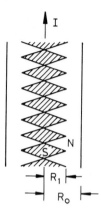

Fig.14.1. Intermediate-state structure of a wire in the current induced resistive state according to LONDON

$$J = \frac{c}{4\pi} \frac{1}{r} \frac{\partial}{\partial r}(rH_\varphi) = \frac{c}{4\pi} \frac{H_c}{r} \quad , \tag{14.2}$$

where r is the radial coordinate and H_φ the azimuthal field component. The spatially averaged magnetic field in the core is

$$H = \frac{\sigma_n E}{J} H_c = \sigma_n E \frac{4\pi r}{c} \quad , \tag{14.3}$$

where σ_n is the normal-state conductivity and E the electric field. The average field increases from $H = 0$ for $r = 0$ to $H = H_c$ for $r = R_1$. From (14.3) we obtain

$$R_1 = cH_c/4\pi\sigma_n E \quad . \tag{14.4}$$

The outer shell between the radii R_1 and R_0 consists of normal phase with $H > H_c$. At the critical current I_c the intermediate state is established such that the core region extends to the surface of the wire ($R_1 = R_0$). As the current is increased above I_c the radius R_1 of the intermediate-state core shrinks, causing a rise in electrical resistance according to

$$R/R_n = \frac{1}{2}\{1 + [1 - (I_c/I)^2]^{\frac{1}{2}}\} \quad . \tag{14.5}$$

Here R_n is the normal-state resistance. From (14.5) at $I = I_c$ we expect the electric resistance to jump discontinuously from zero to 50% of its normal value and to gradually approach R_n as I is increased further. Since the time of London's paper the destruction of superconductivity by a transport current has been studied experimentally from the restoration of electrical resistance in superconducting wires [2.5]. Typical experimental results obtained with In wires are shown in Fig.14.2. We note from (14.1) that the critical current I_c only increases proportional to the square root of the cross-sectional area (πR_0^2) of the wire.

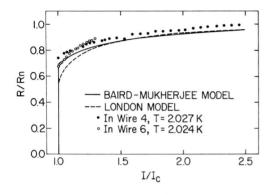

Fig.14.2. Normalized resistance of two indium wires of diameter 0.17 and 0.52 mm, respectively, versus normalized current compared with theoretical curves for the London and the Baird-Mukherjee model [14.2]

In contrast to the static model by LONDON a dynamic model of the current-induced resistive state in a wire of radius R_0 has been proposed by GORTER [7.1,14.3]. In this model electrical resistance is caused by flux flow in the form of continuously collapsing tubes of magnetic flux arranged in cylindrical symmetry along the wire axis. The speed of the flux-tube collapse is regulated by eddy-current damping, and the frequency of this periodic mode is expected to be of the order

$$\omega = c^2 \rho_n / R_0^2 \quad . \tag{14.6}$$

Such periodic flux-tube motion should generate an oscillatory component in the resistive voltage. Assuming that the flux tubes move in phase coherence along the whole length of the wire, GORTER estimated the amplitude of the ac component to be about 2% of the dc voltage. In a search by MEISSNER [14.4] such an oscillatory voltage component could be detected in tin wires. However, the ac component was found to be not larger than 0.015 - 0.018% of the dc voltage, indicating that, if such flux-tube motion exists at all, it would be rather uncorrelated along the wire. Measurements of the electrical noise power in indium wires place an upper limit of only 0.005% on the ratio of the ac and the dc voltage component and raise some doubt about the validity of the Gorter model [14.2].

A whole family of current-induced flux structures in a cylindrical wire with the London and Gorter model as limiting cases has been discussed by ANDREEV and SHARVIN [14.5,6]. A refined version of the static London model has been proposed by BAIRD and MUKHERJEE [14.7,8]. These models account better for the detailed manner in which the resistance appears with increasing current [14.9-11] and they explain the fact that at $I = I_c$ the resistance jumps to about 70% of its normal value, i.e., to a value somewhat higher than predicted by LONDON. In Fig.14.2 the experimental data are compared with the London and Baird-Mukherjee model. The layered configuration of the current-induced intermediate state in a cylinder of type-I superconductor has been confirmed with a Bitter method [14.12]. Detailed measurements on the kinetics of the current-induced transition to the intermediate state have been performed by POSADA and RINDERER [14.13] and by WIEDERICK et al. [14.14] using pulse techniques.

The domain configuration of the current-induced resistive state such as shown in Fig.14.1 for the London model is associated with a rather complicated pattern of the supercurrents and the normal currents flowing in the wire. Since the transport current itself generates the magnetic fields and the normal domains, such a configuration may be subject to various instabilities. As discussed by AZBEL [14.15], magnetic and electric instabilities are a general property of materials with a high value of the magneto-resistance. Here a superconductor represents just the most extreme case. It appears that electric and magnetic instabilities of this kind in the current-induced resistive state of type-I and type-II superconducting wires have been observed by LALEVIC [14.16,17]. He noticed rather abrupt variations of the electric resistance with current and fluctuations of the resistance between metastable resistance levels. Such effects can be understood in terms of abrupt rearrangements of the distribution of currents and magnetic fields within the specimen in the way envisioned by AZBEL for materials with strong magneto-resistive behavior.

In 1937 L.D. LANDAU predicted that the current-induced breakdown of type-I superconductivity in a *hollow cylinder* requires a mixture of superconducting and normal regions in the cylinder wall. This mixture is often referred to as the "two-dimensional mixed state". Some time ago LANDAU and SHARVIN [14.18,19] discovered this state experimentally and measured the V(I) characteristics and the surface impedance of the mixed state. A theoretical study of this state, based on the Ginzburg-Landau theory, has been performed for two limiting cases [14.20]: First, for a very pure type-I superconductor carrying a weak current such that the layer in the mixed

state is nearly completely superconducting [14.21,22], and second, for a
pure and dirty type-I material carrying a large current such that superconductivity is nearly destroyed and exists only in the form of small superconducting fluctuations [14.23,24]. These ideas have been extended to the
"one-dimensional mixed state" in a superconducting cylinder of type-I carrying a large current such that only superconducting fluctuations exist near
the sample axis [14.25].

The stability against current-induced breakdown of superconductivity in
hollow cylinders has been investigated by MEISSNER [14.26,27] using short
current pulses in thin cylindrical films of tin.

14.2 Thin Film Geometry

14.2.1 Dynamic Model: Nucleation of Flux-Tube Trains

Whereas in the wire geometry more than 50% of the normal resistance is abruptly restored when the applied current reaches I_c, in planar thin films
the current-induced transition to the resistive state is more gradual
[14.28]. Even at twice the current value where electrical resistance starts
to appear, the resistance is typically only about 15% of the normal-state
value. Here we have in mind thin-film strips with their width being much
larger than their thickness. Again, all dimensions are assumed larger than
the coherence length and penetration depth. In contrast to the geometry of
a wire or hollow tube with circular cross section, in a planar thin-film
superconductor, *edge effects* become rather important. In a tpye-I superconducting strip carrying a high electric transport current we expect the magnetic behavior shown schematically in Fig.14.3. At small current values the
current flows predominantly along the edges of the strip [14.29,30] such
that the magnetic field is expelled from the specimen except for a surface
layer with the thickness of the penetration depth. The system exhibits the
Meissner effect as indicated in Fig.14.3. When the magnetic field of the
current reaches a value of the order of the critical field at the sample
edge, a normal region will be generated locally at the edge. The transport
current then will try to avoid this region by flowing around it (Fig.14.3b).
Because of the enhancement of the magnetic field at the inside of this current loop, this configuration is magnetically unstable, and the normal
region will grow further until it reaches the center of the strip. This behavior is analogous to the kink instability in magneto-hydrodynamics, as

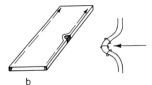

Fig.14.3a-c. Magnetic flux structure generated in a superconducting strip by an electric transport current: (a) Meissner effect; (b) nucleation of a domain at the edge; (c) magnetic flux structure at high currents

indicated in Fig.14.3b. If the current is increased further, more "channels" of normal phase will grow from both sample edges to the center (Fig.14.3c). At the two sides of the strip the magnetic field in these normal channels has opposite sign.

The scheme of the current-induced magnetic structure shown in Fig.14.3 has been verified in high-resolution magneto-optical experiments performed with strips of lead and indium [2.23,14.31]. A typical pattern observed in a Pb strip is presented in Fig.14.4. If the transport current is gradually increased from zero, at a distinct current level the first normal channel appears. Above this critical current value, the number of normal channels in the strip increases about linearly with current, as can be seen from Fig.14.5. This behavior implies a linear increase of the resistance and therefore a quadratic increase of the resistive voltage with increasing current.

In various experiments it could be demonstrated that the channels of normal phase, displayed in Fig.14.4, actually consist of *trains of flux tubes* nucleated at the edge and moving rapidly to the center of the strip. In the center, opposite tubes arriving from the opposite edges annihilate each other. The experimental evidence for this *dynamic model* includes the observation of dynamic magnetic coupling between adjacent films of a sandwich arrangement (see Sect.8.5) and a consideration of the observed resistive

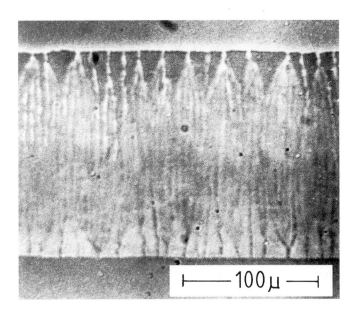

Fig.14.4. Magnetic flux structure in a Pb strip of 4.0 μm thickness and 160 μm width at 4.2 K carrying a transport current of 3.8 A (average current density = 5.9×10^5 A/cm^2). The current flows from left to right. At the top and the bottom the sample edges are clearly visible. The bright and dark "channels" indicate the current-induced flux structure as detected magneto-optically [14.31]

voltage associated with the nucleation of a single "channel" [14.28]. A more direct prove of the dynamic model is provided by photographs of the individual flux tubes in the moving train taken magneto-optically with highly increased time resolution (see Sect.14.2.3).

The abrupt nucleation of the individual flux-tube trains at distinct levels of the transport current leads to structure in the voltage-current characteristic [14.28,32]. Figure 14.6 shows typical voltage steps observed in a Pb strip at the onset of the resistive state. The nucleation of the flux-tube trains and the detail of the voltage-current characteristics are influenced by the metallurgical microstructure of the material, and a complete control of these events in a long strip is rather difficult. However, perfect control of the location where flux-tube nucleation sets in can be achieved in a constricted geometry, as will be discussed in Sect.14.2.3.

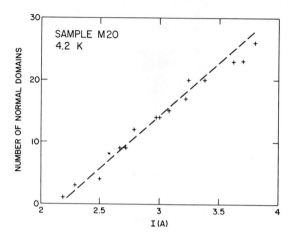

Fig.14.5. Number of current-induced normal domains versus current in a Pb strip of 4.0 μm thickness and 160 μm width at 4.2 K; length of observed section = 320 μm [14.28]

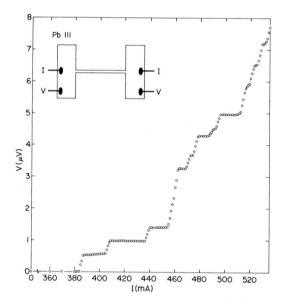

Fig.14.6. Voltage versus current in the current-induced resistive state in a Pb film of 3.2 μm thickness and 100 μm width at 4.2 K [14.33]

The nucleation rate of flux tubes in a single train can be calculated from the resistive voltage steps using the Josephson relation of (7.24) or (13.15), or its equivalent

$$1/\tau_0 V = c/\varphi_0 = 483.6 \text{ MHz}/\mu V \quad . \tag{14.7}$$

Interpreting the voltage steps of about 0.5 µV in Fig.14.6 as being caused by the individual flux-tube trains we see that these steps correspond to a nucleation rate of about 2.4×10^8 flux quanta per second. For the Pb strip yielding the data of Fig.14.6 we expect from magneto-optical observations [2.22] the nucleation of flux tubes containing about 50 flux quanta per tube. [Such a number may also be derived from the flux-spot diameter given in (2.44)]. The voltage steps in Fig.14.6 then correspond to the nucleation rate of about 5×10^6 flux tubes per second.

The periodic nucleation of flux tubes in the current-induced resistive state should lead to a sharp peak in the electrical noise power spectrum at the nucleation frequency. CHURILOV and co-workers [14.34,35] were able to observe directly this high-frequency voltage component in the current-induced resistive state of thin (250 Å) type-II tin films. For detection they applied a strictly passive resonance technique, in which a tunable resonance circuit within the liquid helium bath could be excited by the high-frequency oscillations in the sample. A dynamic model of the current-induced resistive state based on the coherent motion of single-quantum vortex lines perpendicular to the current has been discussed in conjunction with experiments by GUBANKOV et al. [14.36,37] and MUSIRENKO et al. [14.38]. Theoretical calculations yielding a similar model have been carried out by LIKHAREV [14.39] and by ASLAMAZOV and LARKIN [14.40]. Further experiments performed for a constricted geometry and verifying our dynamic model will be discussed in Sect.14.2.3.

In addition to the dynamics associated with the nucleation processes of the individual flux-tube train there exists a second dynamical aspect in the current-induced resistive state of a thin film arising from fluctuations in the *overall magnetic structure* defined by the number and geometrical arrangement of the flux-tube trains. As shown directly by magneto-optical experiments, the configuration of flux-tube trains in a superconducting strip can fluctuate strongly [2.23,7.41]. Apparently, the system is most unstable at a current just above the critical value for the onset of resistance, where only a few resistive "channels" exist. These instabilities can be seen also in the electrical noise power [14.28]. Just above the onset of electric resistance a distinct peak in the low-frequency noise power can usually be observed indicating strong fluctuations in the configuration of currents and magnetic fields. These fluctuations appear to be of similar origin as those observed by LALEVIC [14.16,17] in superconducting wires and seem to be caused by the instabilities arising from a high value of the magneto resistance [14.15]. The rearrangement of currents and magnetic fields in a superconductor in the current-induced resistive state can lead to negative differential resistance [14.41]. The control of such negative differential resistance behavior by proper sample preparation would be an interesting goal for device applications.

14.2.2 Gibbs Free-Energy Barrier

As we have pointed out in Sect.4.6.1, a Gibbs free-energy barrier exists against magnetic flux entry into the superconductor near a planar surface oriented parallel to the magnetic field. This energy barrier arises from the different contributions to the vortex-line energy: 1) the repulsive interaction between the vortex line and the external magnetic field, 2) the attractive interaction between the vortex and its image line, and 3) the energy of the vortex inside the superconductor far away from the surface. (The vortex-line energy as a function of the distance from the surface is shown in Fig.4.12). The same considerations also apply to flux tubes containing the flux $\varphi = n\varphi_0$ in type-I superconductors. For the geometry of a flat superconducting strip with a demagnetization coefficient D close to 1.0 the magnetic self-energy of the vortex or flux tube from the region outside the superconductor and its strong dependence upon the distance of the flux tube from the specimen edge become important. This contribution essentially replaces the second (attractive) term in our calculation of Sect.4.6.1. For the strip geometry this results in a strong Gibbs free-energy barrier against irreversible magnetic flux entry. The critical entry field $H_{en}(edge)$ at the edge of the strip is given by [4.65]

$$H_{en}(edge) = H_c\{1 + [2(a_n w_s)^{1/2}/\pi d]\} \quad . \tag{14.8}$$

Here w_s and d are the width and thickness of the strip, respectively, and a_n is the width of the normal domain entering the strip. In (14.8) the strip and the normal domain placed along the edge are assumed to be infinitely long. We note from (14.8) that for $(a_n w_s)^{1/2} \gg d$ the critical entry field can be much larger than the bulk thermodynamic critical field H_c because of the energy barrier. The magnitude of the energy barrier increases proportional to the flux φ contained in the flux tube entering the superconductor. This is reflected in the fact that the domain width a_n appears in (14.8). Therefore, for type-II superconductors the enhancement from the second term in (14.8) should be less significant.

The result of (14.8) yields the critical current I_c for the onset of irreversible flux-tube motion and electrical resistance [4.65]

$$I_c = (cdH_c/4)\{1 + [2(a_n w_s)^{1/2}/\pi d]\} \quad . \tag{14.9}$$

We note that when $(a_n w_s)^{1/2} \gg d$ the critical current will be much larger than the value expected from Silsbee's rule. In this case I_c is proportional to $w_s^{1/2}$ and independent of d

$$I_c \approx cH_c(a_n w_s)^{1/2}/2\pi \quad ; \quad (a_n w_s)^{1/2} \gg d \quad . \tag{14.10}$$

This value of the critical current is approximately equal to I_c in a wire with the same cross section as the flat strip, as can be seen in the following way. Taking $a_n \approx d$ and setting $d \cdot w_s = \pi R_0^2$, we obtain from (14.10) $I_c \approx cR_0 H_c/2\sqrt{\pi}$, i.e., a value close to the result in (14.1). Apparently, the enhancement of I_c due to the energy barrier is just compensated by the increased current density and current-generated magnetic field at the sample edges due to the nonuniform current-density distribution in the strip geometry with a demagnetization coefficient close to 1.0. The result of (14.9) has been confirmed by experiment [4.65].

It is interesting to compare the critical entry field at the edge of the strip given by (14.8) for the case when $(a_n w_s)^{1/2} \gg d$ with the value $H_e/(1 - D)$ of an external magnetic field H_e enhanced by the factor $(1 - D)^{-1}$. For the strip geometry the demagnetization coefficient is [14.42]

$$D = \left(1 + \frac{d}{w_s}\right)^{-1} \quad . \tag{14.11}$$

For $d/w_s \ll 1$ this yields $H_e/(1 - D) \approx H_e w_s/d$. On the other hand from (14.8) we have $H_{en}(\text{edge}) \approx H_c(2/\pi)(w_s/d)^{1/2}$, taking $a_n \approx d$. We see that the enhancement factor in (14.8) only compensates partly the increase of the magnetic field value due to the demagnetization coefficient.

14.2.3 Constricted Geometry

For investigating the dynamics of a single flux-tube train the localization of the nucleation site and path of the moving flux tubes becomes imperative. Such localization can be achieved in a constricted geometry where a short film strip of only a few μm length is imbedded between end sections where the width and thickness of the film is larger than at the constriction. This geometry can easily be prepared using a multiple-step evaporation procedure [14.43]. The length of the constricted region can be chosen such that here only a single flux-tube train can be nucleated. Nucleation of two, three, or more flux-tube trains in parallel is possible by proper increase of the length of the constriction. Utilizing this geometry in a magneto-optic experiment, the nucleation of a single flux-tube train can be observed directly in conjunction with the structure in the voltage-current characteristic of the sample. With stroboscopic illumination in combination with magneto-optic flux detection, using an arrangement such as shown in Fig.8.1, the individual flux tubes in the current-induced resistive state can be photographed. The result of such an experiment is shown in Fig.14.7. As we have pointed out in Sect.8.1, by varying the phase between the stroboscopic illumination and the periodic sample modulation the complete time behavior during one cycle can be investigated. In Fig.14.8 we show an example obtained by superimposing a small oscillatory current component upon the direct current passing through the sample. Here the time intervals between the successive photographs are 250 ns. Photographs such as shown in Figs.14.7,8 provide perhaps the most convincing evidence for the dynamic nature of the current-induced resistive state of superconducting strips or thin films.

For obtaining a sharp magneto-optical image of the moving flux-tube trains in the current-induced resistive state, synchronization between the flux-tube nucleation and the frequency of the stroboscopic illumination (here a laser beam passing through an acousto-optic shutter) is obviously necessary. This can be achieved by an oscillatory current component of sufficient magnitude. However, synchronization is also possible in the limit of zero alternating current by the modulated illumination alone when the natural frequency of the flux-tube nucleation and the frequency of the light pulses are close to each other. In such a case a sharp magneto-optic image of the flux-tube train at constant frequency of the illumination can be observed over a small interval of direct current. Similarly, at constant direct current a sharp image can be found over a finite bandwidth of the frequency of the illumination. This situation is analogous to the behavior of weak links in the presence of rf or microwave irradiation where the voltage across the superconducting link remains constant for small direct current intervals due to synchronization of the phase-slippage process with the external rf or microwave field.

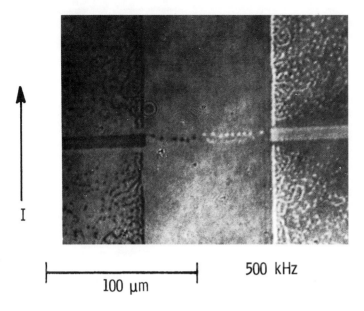

Fig.14.7. Current-induced resistive state of a constricted Pb film of 6.2 μm thickness at 4.2 K. Current direction is vertical. Magneto-optic flux detection with stroboscopic illumination at 500 kHz. The horizontal array of black and bright flux tubes in the center represents the equivalent of a single "channel" such as shown in Fig.14.4 observed under high time resolution (courtesy of D.E. Chimenti)

The constricted geometry has proved to be very useful in combination with the magneto-resistive field probe discussed in Sect.8.6 and shown schematically in Fig.8.6. With such an arrangement the regularity of the flux-tube nucleation process in the current-induced resistive state and its dependence upon the geometrical dimensions and the metallurgical microstructure of the sample can be studied in a *strictly passive* way. Figure 14.9 shows the spectrum of the flux-tube nucleation frequency in an indium constriction at 1.88 K as obtained from the signal amplitude of the field probe. The double peaks in curves A and C represent the sidebands of the mixer output separated by twice the center frequency of the narrowband amplifier (5 - 10 kHz). We see that in the center trace the signal has a bandwidth of only 150 Hz. Here the nucleation frequency is 623 kHz. (For this trace the scale begins at 624 kHz and extends to 634 kHz. Note that in Fig.14.9 we show the sidebands of the mixer output rather than the nucleation fre-

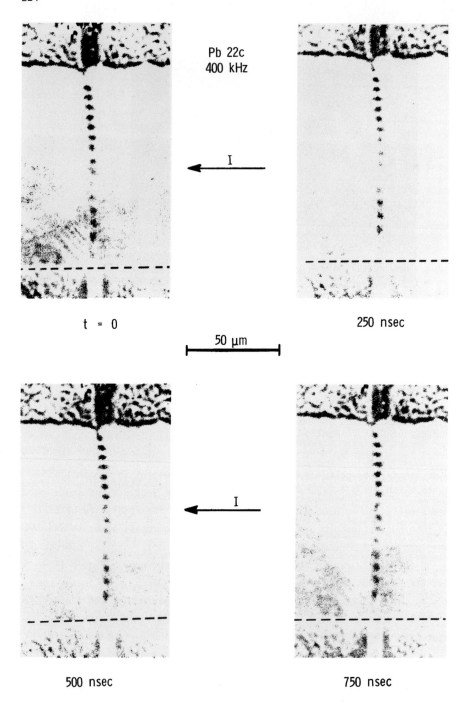

Fig. 14.8 (caption see opposite page)

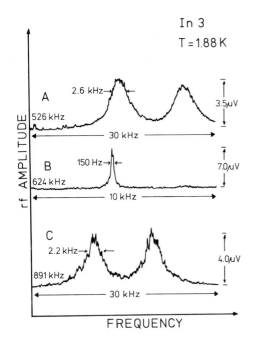

Fig.14.9. RF signal amplitude of the micro-field probe versus frequency for an indium constriction of 2.1 μm thickness, 250 μm width, and 6.0 μm length showing variations in bandwidth. Note scale change in trace B. For further details see text [8.26]

quency directly). In trace A and C the bandwidth of the signal is considerably larger. A combined plot of the nucleation rate and bandwidth vs sample voltage is given in Fig.14.10 for the same specimen. The points labeled A, B, and C refer to the experimental traces in Fig.14.9. We see that the bandwidth of the flux-tube nucleation rate, given on the right-hand side in logarithmic scale, varies by more than an order of magnitude.

These measurements of the flux-tube nucleation rate by counting flux tubes using a micro field probe can be extended further. By frequency demodulation of the probe signal the statistical fluctuations of the nucleation rate and the power spectrum of these fluctuations can be determined. Assuming that the number n of flux quanta per tube remains constant, the power spectrum of the electrical noise can then be obtained directly from the conversion of frequency into voltage using the Josephson relation

---

Fig.14.8. Current-induced resistive state: Stroboscopic sequence observed at 400 kHz for a constricted Pb film of 9 μm thickness and 95 μm width at 4.2 K. The bottom edge of the constriction is indicated by a dashed line. Current direction is horizontal. The dark spots distributed vertically across the constriction represent moving flux tubes containing each about 120 flux quanta. The modulation period is 2.5 μs and the laser pulse width is 250 ns. The time between maximum current and laser pulse is indicated below each frame [8.6]

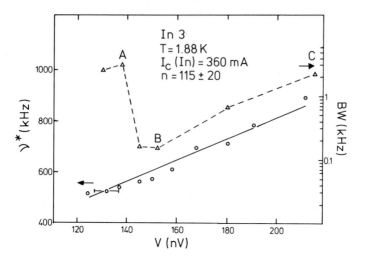

Fig.14.10. Combined plot of the nucleation frequency ν^* (left ordinate) and of the bandwidth BW (right ordinate) versus sample voltage for the same indium constriction as presented in Fig.14.9. The points A, B, and C refer to the traces in Fig.14.9 [8.26]

$$\nu^* = (2e/nh)V \quad . \tag{14.12}$$

Here ν^* is the flux-tube nucleation frequency. We note that the moving flux-tube train in the current-induced resistive state of a superconducting thin film represents a system where the voltage signal can consist of a mixture of well correlated pulses and statistically independent pulses. The autocorrelation function then is of the type given in (13.11), and the power spectrum is composed linearly of the correlated and uncorrelated fraction with the proper weighting factors.

The periodic nucleation of flux tubes in the current-induced resistive state with the frequency given by (14.12) suggests the appearance of structure in the voltage-current characteristic such as constant voltage steps, if the nucleation region is exposed to rf irradiation. Voltage steps can be expected when the radio frequency coincides with the nucleation frequency ν^* of (14.12). Since the expected voltage steps are rather small (typical values are $\nu^* = 100$ kHz; $n = 100$; $V \approx 2 \cdot 10^{-8}$ volt) they are difficult to detect using dc methods. However, the detection becomes much easier if the rf irradiation is frequency modulated and standard ac signal averaging techniques can be employed. An experiment of this kind has been performed by CHIMENTI [14.44] using lead films with a constricted region.

From the signal output arising through synchronization of the flux-tube nucleation rate with the (frequency-modulated) rf irradiation the number n of flux quanta per flux tube could be determined.

So far we have only discussed external sample modulations (irradiation with light pulses or radio frequency) performed close to the nucleation frequency ν^* of the flux tubes. However, in the current-induced resistive state of constricted films of a type-I superconductor steps of constant voltage have been observed under rf irradiation of frequency ν, where the voltage levels are equal to $V = (h/2e)\nu$ and its multiples [14.45]. Examples for such *single-quantum Josephson steps* observed in a constricted In film are shown in Fig.14.11. A model for explaining these results must combine the current-induced nucleation of flux tubes containing many flux quanta with a single-quantum relaxation process. It appears to be the flux-tube nucleation at the sample edge where this single-quantum process takes place.

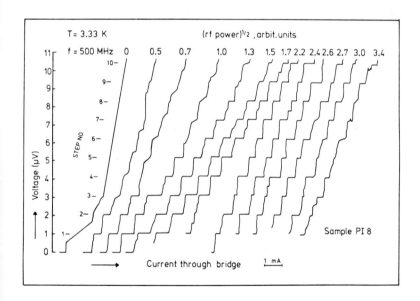

Fig.14.11. Single-quantum Josephson steps in the V-I characteristic of a constricted indium bridge of 2 μm thickness, 30 μm width, and 4.5 μm length for different rf power levels at 500 MHz (T = 3.33 K). The different curves are arbitrarily displaced along the current axis. The positions of the steps are indicated on the left [14.45]

In conclusion we note that the current-induced resistive state in superconducting films with large cross section exhibits interesting features such as the nucleation of moving flux-tube trains proceeding from the film edges via single flux quanta. The dynamic behavior of this state shows up at two levels, first, in the temporal structure of the individual flux-tube train, and second, in the overall arrangement of the total assembly of these trains. Useful insight into these features can be obtained from experiments performed with a constricted geometry.

14.3 Microbridges

If the cross-sectional dimensions of a superconducting wire or film are of the same order or less than the penetration depth or coherence length, no vortex line or flux tube can be accomodated within this configuration. In this case the physical parameters vary only little within the sample cross section and the geometry can be considered as one-dimensional. The Josephson behavior of such one-dimensional structures has been investigated extensively, both experimentally and theoretically. General discussions of weak link Josephson junctions have been given by KULIK and YANSON [14.46], SOLYMAR [14.47], and SILVER and ZIMMERMAN [14.48]. Additional references can be found in a series of recent conference proceedings [14.49-51]. In the following we briefly summarize the important features of nearly one-dimensional microbridges. (Because of the small cross-sectional dimensions, now the geometry deviates from the assumption adopted so far in this section).

For some time voltage instabilities and regular voltage steps in the V-I characteristic have been observed in thin-film microbridges [14.52-58], and whiskers [14.59-61]. The voltage steps found for these "one-dimensional" geometries are analogous to the results discussed in Sect.14.2 (see Figure 14.6) and explained by the nucleation of flux-tube trains.

In Sect.3.3 we have calculated the critical current density in a "one-dimensional" superconductor from the Ginzburg-Landau theory, see (3.41-43). Close to T_c we expect $J_c \sim [1 - (T/T_c)]^{3/2}$. The critical current for the onset of resistance in the microbridges generally shows this $[1 - (T/T_c)]^{3/2}$ behavior. Further experiments using a series of voltage probes along the specimen have shown that in long microbridges the current-induced breakdown of superconductivity is spatially localized and that the voltage steps result from the appearance of isolated and essentially identical phase-slip centers at a current equal to the local critical current [14.62,63]. Apparently because of inhomogeneities in the material and variations in T_c the critical current varies along the length of the microbridges.

The concept of the *phase-slip center*, introduced by SKOCPOL et al., represents a useful model for the understanding of the current-induced resistance in "one-dimensional" microbridges. In particular it explains the two main features of the experimentally observed V-I curves, namely the magnitude of the differential resistance and the zero-voltage intercept of approximately 0.5 - 0.7 times the local critical current. Phase slippage of the wave function $\psi(x)$ occurs at an inhomogeneity (with an infinitesimal reduction of T_c), being localized within the coherence length ξ. In a narrow, one-dimensional bridge the magnetic field can be neglected. Writing $\psi(x)$ in the form of (3.9) the supercurrent density is according to (3.13b)

$$J_s = (2e\hbar/m^*)|\psi|^2 \cdot d\varphi/dx \qquad (14.13)$$

Here the x-coordinate describes the location along the microbridge. As shown in Sect.3.3, at the maximum stable supercurrent density we have $\psi^2 = (2/3)\psi_0^2$, where ψ_0 is the wave function for $J_s = 0$. At higher current densities $|\psi|^2$ collapses to zero. Then also J_s drops to zero and the current is entirely normal, $J = J_n$. At this stage $|\psi|$ can build up again. In this way a relaxation oscillation of the order parameter at the Josephson frequency takes place near the inhomogeneity as soon as the current density exceeds its critical value J_c. The phase difference Φ between both sides of the phase slip center and the supercurrent density exhibit the same oscillation with the relaxation time $\tau = h/2e\bar{V}$. The spatial variation of the wave function ψ and the temporal behavior of the phase difference Φ and of the supercurrent density J_s are shown in Fig.14.12. The time-averaged supercurrent density is approximately

$$\bar{J}_s \approx 0.5 \, J_c \qquad , \qquad (14.14)$$

and the time-averaged normal current density is $\bar{J}_n \approx J - 0.5 \, J_c$. Hence, the resistive voltage is $\bar{V} \approx (L_h \rho_n/S)(I - 0.5 \, I_c)$, where L_h and S are a characteristic healing length and the cross-sectional area of the microbridge, respectively. The healing length L_h can be identified with twice the quasi-particle diffusion length Λ, introduced by PIPPARD et al. [14.64] in connection with the interconversion of quasiparticle currents into supercurrents near a S/N interface,

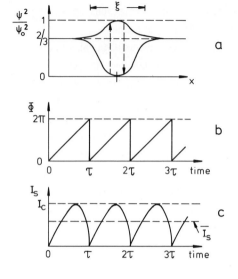

Fig.14.12a-c. Spatial variation of the normalized wave function near the phase-slip center (a), temporal behavior of the phase difference Φ between both sides of the phase-slip center (b), and temporal variation of the supercurrent density J_s at the phase-slip center (c) according to [14.62]

$$L_h = 2\Lambda \quad . \tag{14.15}$$

The length Λ is the diffusion length of quasiparticles between inelastic scattering processes

$$\Lambda \approx (\ell \cdot v_F \cdot \tau_2/3)^{\frac{1}{2}} \quad , \tag{14.16}$$

where ℓ is the ordinary mean free path, v_F the Fermi velocity, and τ_2 an appropriate inelastic scattering time. Finally, we obtain for the differential resistance of a phase-slip center

$$\partial \overline{V}/\partial I = 2\Lambda \rho_n / S \quad . \tag{14.17}$$

The properties of the phase-slip center contained in (14.14-17) are reasonably confirmed by experiment [14.62,65,66]. The healing length L_h has been measured by DOLAN and JACKEL [14.67] using a series of potential probes attached to a microbridge in the nonequilibrium region near a phase-slip center. Their results agree with the model outlined above.

Conceptually, the phase-slip center introduced by TINKHAM and co-workers is associated with an (infinitesimal) perturbation of the sample homogeneity. KRAMER and BARATOFF [14.68] investigated the stability of possible current-carrying solutions of a one-dimensional, perfectly homogeneous, and infinitely long superconductor within the framework of the time-dependent Ginzburg-Landau theory. They found that localized phase slippage occurs *spontaneously* in a narrow range below the critical current density (without the need for an infinitesimal perturbation of sample homogeneity). The dynamical stability of the current in a one-dimensional superconductor and the appearance of soft modes has been analyzed by AMBEGAOKAR [14.69] using time-dependent mean-field theory.

14.4 Thermal Effects

So far in our discussion of the current-induced resistive state we have neglected any thermal effects arising from the power dissipation in the superconductor. However, often heating effects are important and can have catastrophic consequences like thermal runaway and destruction of the superconductor. Hence, the analysis of the thermal behavior represents an important subject, in particular with respect to reliability considerations for practical applications. In the following we briefly discuss self-heating effects in thin-film microbridges.

In a narrow superconducting strip carrying an electric transport current power dissipation can result in the formation of a *localized hotspot*. A detailed theoretical and experimental study of self-heating hot spots in superconducting thin-film microbridges has been performed by SKOCPOL et al. [14.70]. They assumed the microbridge to be sufficiently narrow such that its behavior can be described by a one-dimensional model. Considering a thin-film bridge of length L_s, width w_s, and thickness d, with a normal region (hotspot) of length $2x_0$ generated by Joule heating in the center, SKOCPOL et al. calculated the temperature distribution along the bridge from the heat-flow equations. Heat generated in the hotspot is transferred in two ways: by thermal conduction within the metal film and by surface heat transfer across the boundary with the substrate and with the helium bath, if present. With the boundary condition $T(\pm L_s/2) = T_b$ at the ends of the bridge and the matching condition $T(\pm x_0) = T_c$ at the N/S interface, current and voltage satisfy the following solutions

$$I(x_0) = \left[\frac{\alpha_h w_s^2 d(T_c - T_b)}{\rho_n}\right]^{\frac{1}{2}} \cdot \left[1 + \left(\frac{K_s}{K_n}\right)^{\frac{1}{2}} \coth\left(\frac{x_0}{\eta_n}\right) \coth\left(\frac{L_s}{2\eta_s} - \frac{x_0}{\eta_s}\right)\right]^{\frac{1}{2}} , \quad (14.18)$$

and

$$V(x_0) = I(x_0)\rho_n 2x_0/w_s d \quad . \tag{14.19}$$

Here, T_b and T_c are the bath temperature and the critical temperature of the superconductor, respectively; α_h is the heat-transfer coefficient per unit area to the substrate (and to the helium bath, if present); K_s and K_n is the thermal conductivity of the microbridge in the superconducting and normal state, respectively. The quantities η_s and η_n represent a characteristic length given by

$$\eta = (Kd/\alpha_h)^{\frac{1}{2}} \quad , \tag{14.20}$$

where the subscript refers to the corresponding subscript in K. We see that current and voltage depend on the length x_0 defining the size of the normal region. At low voltages (14.18,19) yield a negative differential resistance. With increasing voltage the current passes through a minimum and at high currents approaches ohmic behavior. Figure 14.13 shows the shape of the V-I characteristic expected from this theory for several values of L_s/η, assuming $\eta_s = \eta_n$. The parameter $R_B = \rho_n L_s/w_s d$ is the resistance of the bridge neglecting end effects. For long microbridges ($L_s \gg \eta$) the V-I curves show a straight vertical portion at the minimum current value

$$I_{min} = [\alpha_h w_s^2 d(T_c - T_b)/\rho_n]^{\frac{1}{2}}[1 + (K_s/K_n)]^{\frac{1}{2}} \quad . \tag{14.21}$$

The negative differential resistance at low voltages can be understood from the fact that here the hotspot remains constant in size and is maintained through constant power dissipation yielding $V \sim I^{-1}$. The vertical branch of the V-I curve is associated with the growth of the hotspot. Finally, at high voltages the behavior becomes ohmic when the size of the hotspot approaches the total length of the bridge and does not grow any further. From (14.21) we note that the value of the minimum current is closely related to the heat-transfer coefficient α_h. The negative differential resistance in combination with an inductive load can lead to relaxation oscillations. The one-dimensional hotspot model by SKOCPOL et al. also accounts reasonably well for the self-heating effects in rather wide (100-200 μm) microbridges where vortex motion perpendicular to the current direction is expected [14.71]. Apparently, in such a two-dimensional geometry the thermal behavior depends only weakly on the coordinate perpendicular to the current direction.

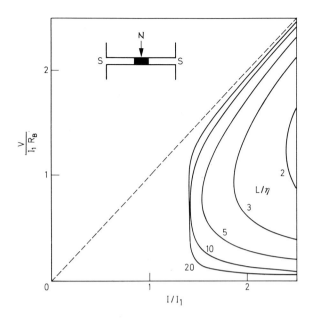

Fig.14.13. Theoretical voltage-current characteristics expected from the hotspot model for different values of the length L_s of the thin-film bridge normalized by the characteristic length η defined in (14.20) and taking $\eta_s = \eta_n$. The scaling current I_1 is $I_1 = [\alpha_h w_s^2 d \cdot (T_c - T_b)/\rho_n]^{1/2}$; R_B is defined in the text. The inset shows the geometry assumed in the model [14.70]

SKOCPOL et al. extended their analysis also to short microbridges where the elevated temperature distribution beyond the actual length of the bridge becomes important. A short microbridge imbedded between wide and thick banks of superconducting material is expected to be rather stable thermally because of the strong heat conduction within the banks.

15. High-Temperature Superconductors: Summary of Recent Developments

15.1 Overview

The discovery of high-temperature superconductors in 1986 by Bednorz and Müller [15.1, 15.2] started an explosive growth worldwide of the research and development activities in the field of superconductivity. In terms of the immediate international scientific response, this event may be compared with the discovery of X-rays in 1895 by W. Röntgen, or with the first observation of nuclear fission in 1938 by O. Hahn and F. Strassmann. The new class of superconductors are cuprate perovskites with electronic properties which depend sensitively on the admixture of specific doping elements. Up to now the highest critical temperature T_c was observed for the cuprate $HgBa_2Ca_2Cu_3O_{8+\delta}$ with $T_c = 133\,K$ at room pressure [15.3]. Many high-T_c superconductors have been found with T_c larger than the boiling point (77,3 K) of liquid nitrogen. Clearly, this fact has important consequences for the technological applications of this class of materials and has already stimulated a host of engineering developments.

In addition to the high values of the critical temperature T_c, the other two unique features of the cuprate superconductors are the small values of the superconducting coherence length ξ and the strong anisotropy. The coherence length ξ reaches values as small as only $1-2\,nm$. With a magnetic penetration depth λ of typically $\lambda = 200-300\,nm$, the Ginzburg–Landau parameter κ attains values as high as $\kappa = 100-200$. The upper critical field B_{c2} reaches $B_{c2} = \frac{\varphi_0}{2\pi\xi^2} = 100-200\,T$. The strong anisotropy results from the layered crystallographic perovskite structure of the cuprates, with the superconductivity residing in the CuO_2 planes.

For the high-temperature superconductors, the subject of magnetic flux structures received a great deal of attention from the very beginning. The experimental verification of the Meissner effect was crucial in the early days for identifying the superconducting phase transition [15.4]. In this chapter we summarize the recent developments in this extremely active and important field. However, to do justice to the many interesting experimental and theoretical investigations would far exceed the intention and scope of this chapter. Instead, we list the important developments and quote key references where further details can be found. A general introduction to the physics of the high-temperature superconductors is given in a number of recent monographs [15.5–15.8].

An important issue in high-temperature superconductivity is the symmetry of the pair wave function. Whereas the classical (low-temperature) superconductors usually display an isotropic energy gap and s-wave pairing symmetry, there is now overwhelming evidence that the hole-doped cuprate superconductors show $d_{x^2-y^2}$ pairing symmetry [15.9] (perhaps with the admixture of another small component). Here, phase sensitive experiments have played a key role [15.10–15.12]. For the electron-doped cuprates the situation remains controversial.

15.2 Static Single Vortex

15.2.1 Pancake Vortices

It is the layered structure of the cuprate superconductors, with the superconductivity arising within the CuO_2 planes, which most strongly affects the properties of a single vortex (or flux line). The orientation of the CuO_2 planes is defined by the crystallographic a- and b-axes, whereas the perpendicular direction is referred to as the c-axis. Between the CuO_2 planes Josephson coupling takes place. A useful phenomenological model for describing such a layered structure with Josephson coupling between the layers was proposed by Lawrence and Doniach (LD) [15.13], see also [15.7, 15.14]. The LD theory contains the anisotropic Ginzburg–Landau and London theories as limiting cases when the coherence length ξ_c in c-direction exceeds the layer spacing s. In this limit the anisotropy is conveniently treated in terms of the reciprocal mass tensor with the principal values $\frac{1}{m_{ab}}$, $\frac{1}{m_{ab}}$, and $\frac{1}{m_c}$. Here m_{ab} and m_c are the effective masses of Cooper pairs moving in the a-b plane and along the c-axis, respectively. If the interlayer coupling is weak, we have $m_{ab} \ll m_c$. Within the anisotropic Ginzburg–Landau limit one finds the following useful relations:

$$\Gamma \equiv \left(\frac{m_c}{m_{ab}}\right)^{1/2} = \frac{\lambda_c}{\lambda_{ab}} = \frac{\xi_{ab}}{\xi_c} = \frac{H_{c2\|ab}}{H_{c2\|c}} = \frac{H_{c1\|c}}{H_{c1\|ab}} \ . \tag{15.1}$$

If the magnetic field is oriented along the c-direction, the flux lines degenerate into stacks of two-dimensional point vortices or *pancake vortices*. A detailed discussion is given by Clem [15.15] and Fischer [15.16] who modelled the layered cuprate superconductor in terms of a stack of thin superconducting films (LD Model). Further references can be found in [15.14]. Energetically, the perfect stacking of the pancake vortices along the c-axis is favorable compared to a more disordered sequence. (We note the similarity to the magnetic coupling effect between two superconducting films discussed in Sect. 8.5). However, compared to a continuous flux line as it appears in the classical

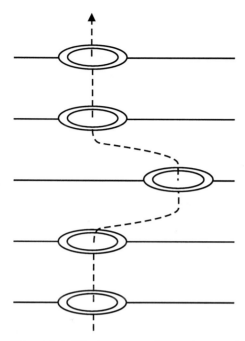

Fig. 15.1. Displacement of a single pancake vortex

superconductors, a stack of pancake vortices contains additional degrees of freedom for thermal excitations. As a result, the vortices in high-temperature superconductors display a rich variety of fluctuating excitations and dynamic properties. This will become clearer in the following sections. As an example we consider the displacement of a single pancake vortex, shown schematically in Fig. 15.1. This displacement is equivalent to the excitation of a vortex–antivortex pair (Kosterlitz–Thouless transition) having the interaction energy $U(r) = \varphi_0^2/\mu_0 \cdot r$, where μ_0 is the vacuum permeability and r the distance between the vortex and the antivortex. With the two-dimensional screening length Λ we obtain the binding energy $U_B \approx \varphi_0^2/\mu_0 \cdot \Lambda$. Interpreting the displacement of a single pancake vortex as an evaporation process, the evaporation temperature T_D is then given by [15.15]

$$T_D \approx \frac{1}{k_B} \frac{\varphi_0^2}{\mu_0 \Lambda} \,. \tag{15.2}$$

For the case of a magnetic field oriented in a direction different from the c-axis, the model discussed above has been extended accordingly. When the applied field is nearly parallel to the a-b plane, the vortex core preferably runs between the CuO$_2$ layers. When the coupling between the layers is weak, the vortex lines along the a-b plane are referred to as *Josephson vortices* or *Josephson strings*. For any magnetic field direction not parallel to the a-b

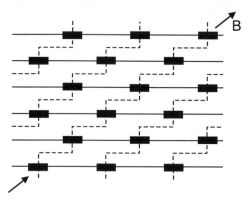

Fig. 15.2. Pancake vortices (*black rectangles*) coupled by Josephson strings (*horizontal dashed lines*)

plane, the pancake vortices existing in the CuO_2 planes are coupled by such Josephson strings, as shown schematically in Fig. 15.2.

15.2.2 Electronic Vortex Structure

In high-temperature superconductors the coherence length ξ is much smaller than in classical superconductors. This has important consequences for the vortex core, since the core radius is given by ξ. As shown by Caroli, De Gennes, and Matricon [15.17] and later by Bardeen et al. [15.18], the electronic structure of the core of an isolated vortex is characterized by discrete energy levels ε_n of the quasiparticles given by

$$\varepsilon_n = \left(n + \frac{1}{2}\right) \frac{\Delta^2}{E_F} = \left(n + \frac{1}{2}\right) \frac{2\hbar^2}{m\xi^2}, \qquad (15.3)$$

where n is an integer and m the quasiparticle mass. The discrete levels ε_n represent quasiparticle bound states originating from Andreev reflection [15.7, 15.19, 15.20] at the core boundary (Andreev bound states). The quantum-mechanical treatment requires the solution of the corresponding Bogoliubov–De Gennes equations. A more detailed discussion can be found elsewhere [15.21, 15.22]. We see from (15.3) that there exists a minigap $\varepsilon_0 = \frac{1}{2}\frac{\Delta^2}{E_F}$ between the Fermi energy and the lowest bound state in the vortex core. Both the minigap and the level separation $\varepsilon_n - \varepsilon_{n-1}$ are proportional to ξ^{-2}. Because of the relatively large coherence length in the classical superconductors (typically $\xi \approx 100$ nm), this electronic quantum structure of the vortex core is negligible, and the core can be treated as an energetic continuum of quasiparticles (cylinder of normal material with radius ξ). However, in the cuprate superconductors the coherence length can be as small as 1–2 nm. Hence, the minigap and the level separation can be up to 10^4 times larger than in the classical superconductors, approaching values of similar order to the energy gap. Clearly, in this case the simple picture of the normal vortex core does not apply any more. A detailed theory of the physics of the quasiparticle excitations in the core of isolated pancake vortices has been presented by Rainer et al. [15.23].

In our discussion of the quasiparticles in the vortex core we must distinguish three important energy scales: the superconducting energy gap Δ, the level spacing Δ^2/E_F of the Andreev bound states, and the energy smearing $\delta\varepsilon = \hbar/\tau$ due to the mean electronic scattering time τ. Correspondingly, we have the following three limits: the "dirty" limit for $\Delta \ll \delta\varepsilon$, the moderately clean limit for $\Delta^2/E_F \ll \delta\varepsilon \ll \Delta$, and the "superclean" limit for $\delta\varepsilon \ll \Delta^2/E_F$. In the classical superconductors we deal almost exclusively with the dirty limit. On the other hand, the cuprate superconductors often approach the superclean limit.

With increasing distance of the quasiparticle energy from the Fermi energy the bound states in the vortex core extend outwards to larger radii, such that the core appears larger for higher quasiparticle energy. Since the states with higher energy become thermally increasingly populated at higher temperature, this leads to an expansion of the core radius with increasing temperature. More than 25 years ago Kramer and Pesch discussed this temperature dependence of the effective core radius in the clean limit for s-wave symmetry of the pair wave function [15.24, 15.25]. For $T \ll T_c$ they showed that the order parameter and supercurrent density increase with the radial distance from the vortex center over a length scale ξ_1 which is much smaller than the BCS coherence length and obeys the relation

$$\xi_1 = \xi_{BCS} \cdot \frac{T}{T_c} \tag{15.4}$$

("Kramer–Pesch effect"). This leads to an increased level spacing of the bound states in the vortex core, and (15.3) must be replaced by

$$\varepsilon_n = \left(n + \frac{1}{2}\right) \cdot \frac{\Delta^2}{E_F} \ln\left(\frac{T_c}{T}\right) . \tag{15.5}$$

Subsequently this behavior was confirmed by Bardeen and Sherman [15.26], and later by Gygi and Schlüter [15.27] and Hayashi et al. [15.28].

The results discussed above and contained in (15.3)–(15.5) are obtained assuming s-wave symmetry of the pair wave function. The strongly increasing evidence obtained for different cuprate superconductors that the pair wave function shows d-wave symmetry [15.9–15.12] has recently stimulated many theoretical studies of the vortex physics for d-wave symmetry of the order parameter. These theoretical discussions were based on the Ginzburg–Landau equations [15.29–15.32] or on the quasiclassical Eilenberger equations [15.33–15.36]. In the latter quasiclassical treatment one assumes $k_F\xi \gg 1$, such that phase coherence can be ignored ("geometric optics approach"). Here, k_F is the Fermi wave vector. In the case $k_F\xi \gg 1$ many bound state energy levels exist between the Fermi energy and the gap energy, and the physics can be adequately described by integration over the quasiparticle energy. In addition to the fourfold symmetry of the pair potential, characteristic of the $d_{x^2-y^2}$ wave function, one obtains a similar symmetry of the local density of states, the supercurrent density, and the magnetic field. The structure of a vortex in a $d_{x^2-y^2}$ superconductor has also been studied for the case where $k_F\xi \approx 1$, in which the quasiclassical approximation is no longer adequate and phase coherence becomes important [15.36–15.40]. Now the self-consistent solution of the Bogoliubov–De Gennes equations is needed. At present, this subject is in a state of rapid development [15.41]. It is hoped that, in combination with local probe experiments (see Sect. 15.7), such theoretical work will contribute significantly to our understanding of the electronic vortex structure. The d-wave symmetry of the pair wave function with the node lines of the order parameter in momentum space also has important consequences for the magnetic penetration depth, resulting in the nonlinear Meissner effect [15.42].

15.2.3 Half-Integer Magnetic Flux Quanta

A spectacular result of the d-wave symmetry of the pair wave function is the spontaneous generation of half-integer magnetic flux quanta observed by Tsuei and coworkers in a frustrated ring geometry. Here frustration arises if the ring contains effectively a Josephson junction with an intrinsic π-phase shift between the two sides of the junction (π-junction) [15.43]. Such a π-phase shift can occur for a Josephson junction between two d-wave superconductors with opposite sign of the pair wave function on the two sides. Up to now the observation of half-integer magnetic flux quanta in a frustrated ring geometry remains the most convincing experimental evidence for d-wave symmetry of the underlying superconducting material. The first such experiments demonstrating d-wave symmetry of the pair wave function were performed with epitaxial c-axis-oriented films of $YBa_2Cu_3O_{7-\delta}$ [15.10, 15.11, 15.44, 15.45]. Recently, half-integer magnetic flux quanta have also been observed in $GdBa_2Cu_3O_{7-\delta}$, $Tl_2Ba_2CuO_{6+\delta}$, and $Bi_2Sr_2CaCu_2O_{8+\delta}$ [15.46]. In all these experiments the half-integer flux quanta were detected using a SQUID microscope. A scanning SQUID microscope image from such an experiment is shown in Fig. 15.3. Four superconducting rings with 48 µm inner diameter and 10 µm width were fabricated in the same epitaxial c-axis-oriented $YBa_2Cu_3O_{7-\delta}$ film. Of the four rings shown, the one in the center contains an effective π-junction, whereas the other three do not. The spontaneously generated half-integer magnetic flux quantum is clearly detected in the center ring, whereas the other rings have zero flux [15.10].

In granular samples of high-temperature superconductors, where the grain boundaries become very prominent, half-integer magnetic flux quanta can also be spontaneously generated, leading to a positive magnetic susceptibility in small applied magnetic fields [15.43]. This is sometimes referred to as the paramagnetic Meissner effect or the Wohlleben effect.

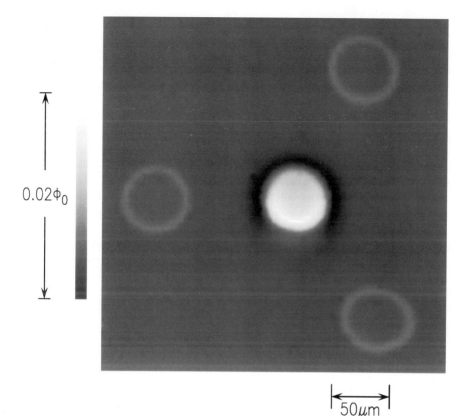

Fig. 15.3. Scanning SQUID microscope image showing four superconducting rings with 48 μm inner diameter and 10 μm width fabricated in the same YBaCuO film. Only the center ring contains an effective π-junction and spontaneously generates a half-integer magnetic flux quantum [15.10]

15.3 Static Vortex Lattice

15.3.1 Vortex Matter

The layered crystallographic structure and the resulting strong anisotropy of the cuprate superconductors in an applied magnetic field give rise to the pancake vortices attached to the individual CuO_2 planes, as we have discussed in the last section. This has important consequences for the static and dynamic properties of the vortex lattice, yielding new physics not encountered in the classical superconductors. This new physics was already noticed immediately following the discovery of high-temperature superconductivity, as described in another famous paper by Müller, Takashige, and Bednorz [15.47]. Studying the temperature and magnetic field dependence of the magnetization in powder samples of BaLaCuO, these authors discovered an irreversibility line in the H–T phase space (H = magnetic field; T = temperature). Above this line the magnetization is perfectly reversible with no detectable flux pinning. On the other hand, below this line the magnetization becomes hysteretic, and the equilibrium vortex distribution can no longer be established because of flux pinning. Soon after this discovery of the irreversibility line, a similar line was observed by Yeshurun and Malozemoff in a $YBa_2Cu_3O_{7-\delta}$ single crystal [15.48]. From these and similar observations within the last decade, the new concept of "vortex matter" evolved, with a liquid, glassy, and crystalline state, displaying a complex phase diagram. These new features have important consequences for the transport processes associated with vortex motion, as will be discussed in Sects. 15.4 and 15.5.

The vortex lattice originally proposed by Abrikosov in 1957 [1.3] consists of a configuration where the vortices (magnetic flux lines) occupy regular positions on a hexagonal or quadratic lattice, in this way minimizing their mutual interaction energy (see Sect. 3.7). In high-temperature superconductors thermal energies are large enough to melt the Abrikosov vortex lattice, forming a vortex liquid over a large portion of the phase diagram. In addition to the high temperatures, it is the structure of a flux line consisting of individual, more or less strongly coupled pancake vortices which promotes this melting transition.

The simplest theoretical description of a melting transition is based on the *Lindemann criterion*, according to which a crystal melts if the thermal fluctuations $\langle u^2 \rangle^{1/2} = c_L \cdot a$ of the atomic positions are of the order of the lattice constant a. The Lindemann number $c_L \approx 0{,}1 - 0{,}2$ depends only slightly on the specific material. This concept has been used to determine the melting transition of vortex lattices. In particular, expressions for the three-dimensional and two-dimensional melting temperature have been derived [15.7, 15.14, 15.49–15.51]. A schematic phase diagram for melting of the vortex solid for a three-dimensional material such as $YBa_2Cu_3O_{7-\delta}$ with H applied parallel to the c-axis is shown in Fig. 15.4.

The remarkable variety in the vortex phase diagram results from the competition of four energies: thermal, vortex interaction, vortex coupling between

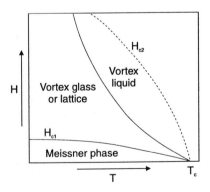

Fig. 15.4. Schematic phase diagram for a three-dimensional material such as $YBa_2Cu_3O_{7-\delta}$

layers, and pinning. The thermal energy pushes the vortex system towards the liquid state, the interaction energy favors the lattice state, the coupling energy tends to align the pancake vortices in form of linear stacks, and the pinning energy generates disorder. The interplay of these energies, whose relative magnitudes vary strongly with magnetic field and temperature, results in the complex phase behavior displayed by the vortex matter. The irreversibility line mentioned above is then interpreted as the melting line, above which the state of the vortex liquid and below which the state of the vortex glass or the vortex lattice is attained. The vortex glass state is intimately connected with flux pinning in the sample disrupting any vortex motion. We will return to this subject in Sects. 15.4 and 15.6. The first clear evidence for vortex lattice melting came from transport measurements (electrical resistance) for detwinned single crystals of $YBa_2Cu_3O_{7-\delta}$ with H parallel to the c-axis. At a well-defined freezing temperature T_m which depends on the magnetic field a sudden drop to zero of the resistivity was observed, signaling the onset of strong pinning in the vortex solid. The sharp drop of the resistivity at T_m indicates a first-order freezing transition [15.52]. Subsequently, the first-order vortex-lattice melting transition was also observed in a thermodynamic measurement using a high-quality single crystal of $Bi_2Sr_2CaCu_2O_8$ with H again parallel to the c-axis [15.53]. This thermodynamic observation was achieved by detecting the local magnetic flux density at the sample surface using an array of microscopic Hall sensors. Excellent reviews of the vortex matter and its melting transition have been presented recently by Crabtree and coworkers [15.52, 15.54].

The high values of the upper critical field B_{c2} in many of the cuprate superconductors often exclude the use of theoretical formulas obtained in the high-field limit for analyzing experimental results such as magnetization data. In order to overcome this difficulty, Hao and Clem developed an extended London model for the reversible magnetization in the whole field regime between B_{c1} and B_{c2}, which provides a useful theoretical basis [15.55, 15.56]. By taking into account the kinetic energy and the condensation energy terms arising from the suppression of the order parameter in the vortex core, in addition to the supercurrent kinetic energy and the magnetic field energy, Hao and Clem derived expressions yielding the implicit magnetic field dependence of the magnetization. They used a specific fitting procedure to obtain the Ginzburg–Landau parameter κ and the function $B_c(T)$ or equivalently $B_{c2}(T) = \kappa \cdot \sqrt{2} \cdot B_c(T)$.

15.3.2 Electronic Structure

Having discussed the geometric aspects of the different forms of vortex matter, next we turn to the quasiparticle electronic structure within the vortex system. Motivated by the physics valid for the classical superconductors, the theoretical treatments of the electronic vortex structure have, to a large part, been based on the quasiclassical approximation assuming a sufficiently small distance between the bound-state energy levels such that the energy dependence of the density of states (DOS) can be obtained by integration [15.57–15.60]. Going beyond the quasiclassical approximation, the discrete energy levels ε_n of the quasiparticle Andreev bound states in the core of an isolated vortex [discussed in the last Section (see (15.3) and (15.5))] are expected to develop into subbands in a perfect vortex lattice. Here the finite overlap between the bound state wave functions of two neighboring vortices can be treated in the tight-binding approximation (Bloch electrons). Theoretical discussions along these lines have been given by several authors [15.61–15.64]. As shown by Pöttinger and Klein [15.64], the development of these subbands must be taken into account if the intervortex distance a becomes smaller than about 7ξ. Clearly, for these concepts to apply, the quasiparticle excitations must propagate coherently through many unit cells of the vortex lattice, approaching the superclean limit. As we have discussed in Sect. 15.2, in classical superconductors this electronic quantum structure remains unobservable, in contrast to the expectation for high-temperature superconductors. The arguments for the development of these subbands between the Fermi energy and the gap energy appear straightforward and convincing in the case of s-wave symmetry of the pair wave function. Experimental spectroscopic evidence of a subband structure in the mixed state of high-temperature superconductors has not yet been reported.

Experimentally, the establishment of a perfect vortex lattice in high-temperature superconductors is extremely difficult because of flux pinning. However, in the flux-flow regime under the influence of an electric transport current of sufficient magnitude, the quality of the vortex crystal can improve considerably (dynamic correlation, see Sect. 6.4), such that the ansatz with Bloch wave functions may be justified. We will return to this subject in Sect. 15.4.5.

For s-wave symmetry of the pair wave function, the quasiparticle DOS in the mixed state scales proportional to B, since the contribution from each vortex must be counted independent of B. However, in the case of d-wave symmetry the situation is completely different. Now in the direction of the node lines the energy gap vanishes, and quasiparticles with wave vector along these directions and zero excitation energy exist even in the absence of a magnetic field. As first discussed by Volovik [15.65] and Simon and Lee [15.66], because of the node lines the quasiparticle DOS in the mixed state scales proportional to $T \cdot B^{1/2}$. This $B^{1/2}$-behavior results from the delocalized quasiparticles with wave vector parallel to the node lines outside the vortex cores.

Experimental evidence for a $T \cdot B^{1/2}$ term in the specific heat has been obtained by different groups for single crystals of $YBa_2Cu_3O_{7-\delta}$ and some other cuprate superconductors in the mixed state [15.67–15.71]. Since this term is usually only a small fraction of the total specific heat, the curve fitting procedure used for the data analysis represents an important issue.

Motivated by the impressive recent advances in vortex imaging experiments (see Sect. 15.7), theoretical investigations of the vortex lattice structure have been carried out by many groups. The calculations include s-wave and d-wave symmetry of the pair wave function. They are based on the Ginzburg–Landau equations [15.30, 15.31, 15.72] or on the quasiclassical Eilenberger equations [15.73–15.75]. The low energy quasiparticle excitations in the mixed state of d-wave superconductors have also been calculated using the Bogoliubov–De Gennes equations [15.76–15.78]. Often these papers present useful contour plots of the local order parameter, supercurrent density, magnetic field, and quasiparticle DOS, with the magnetic field oriented along the c-axis.

15.4 Lorentz Force and Vortex Motion

As one would expect, vortex motion due to the Lorentz force and the resulting flux-flow resistivity are strongly influenced by the three unique properties of the cuprate superconductors: high critical temperature, small coherence length, and layered structure. All three features combined highly facilitate vortex motion, leading to electric resistive losses and breakdown of superconductivity. As a consequence the resistive transition is broadened significantly in a magnetic field. It is also this aspect which has stimulated a great technological and materials-science effort in order to reduce vortex motion by means of flux pinning. The issue of vortex motion may become even more crucial if new superconducting materials with critical temperatures higher than those presently known (room temperature superconductivity) should become available. We note that the Lorentz force can be generated both by an electric transport current and by a shielding current associated with the magnetization of the sample.

15.4.1 Thermally Assisted Flux Flow

In the cuprate superconductors, because of their high T_c values, thermally assisted flux flow (TAFF) represents an important process. The underlying concepts go back to ideas of Anderson and Kim [7.4, 15.79]. We consider a flux line or flux-line bundle attached to a pinning potential well (which represents a local minimum in the Gibbs free energy density landscape). We denote the depth of the potential well by U_0. For simplicity, in our discussion we refer to single flux lines carrying the flux quantum φ_0, but it can easily be extended to flux-line bundles. By thermal activation the flux line can jump out of the potential well with the jumping rate R_j given by

$$R_j = \nu_0 \exp\left(-\frac{U_0}{k_B T}\right). \tag{15.6}$$

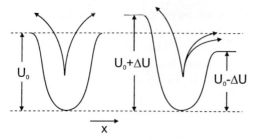

Fig. 15.5. Thermally assisted flux flow. *Left*: Without an external force the flux jumps have no preferred direction. *Right*: In the presence of an external force the flux jumps show a preferred direction

Here ν_0 is a characteristic attempt frequency, and $U_0 \gg k_B T$. In the absence of an external force acting on the flux line, the jumps have no directional preference, and the resulting overall vortex motion remains zero. However, if an external force acts on the flux line breaking the symmetry, the wall height of the potential well is reduced by ΔU along the direction of this force, whereas in the opposite direction it is increased by ΔU. A schematic representation is shown in Fig. 15.5. The jumps now possess a preferred direction, and by subtraction of the jumping rates in forward and backward direction we obtain

$$R_j = \nu_0 \exp\left(-\frac{U_0}{k_B T}\right)\left(e^{\Delta U/k_B T} - e^{-\Delta U/k_B T}\right) \tag{15.7}$$

$$= 2\nu_0 \exp\left(-\frac{U_0}{k_B T}\right) \sinh\frac{\Delta U}{k_B T} . \tag{15.8}$$

In the limit $\Delta U/k_B T \ll 1$, (15.8) yields

$$R_j = 2\nu_0 \exp\left(-\frac{U_0}{k_B T}\right) \frac{\Delta U}{k_B T} . \tag{15.9}$$

In the case of the Lorentz force $(\boldsymbol{J} \times \boldsymbol{\varphi}_0)/c$ we have

$$\Delta U = \frac{1}{c}(\boldsymbol{J} \times \boldsymbol{\varphi}_0) \cdot \boldsymbol{s} \cdot l_c , \tag{15.10}$$

where s is the jumping distance and l_c a characteristic correlation length along the magnetic field direction. As a limiting case, l_c refers to an individual pancake vortex. Noting that the directionally oriented flux-line jumping rate of (15.8) and (15.9) yields an overall electric field $\boldsymbol{E}$ according to (7.5) (see also Sect. 7.7), we find from (15.9) and (15.10) the proportionality $\boldsymbol{E} \propto \boldsymbol{J}$. Hence, as an important result in the limit of small electric current densities, TAFF yields Ohm's law.

At this stage it is convenient to introduce some parameters which are experimentally directly accessible. We define the critical current density

$$J_{\rm c} \equiv \frac{U_0}{\Delta U} J \tag{15.11}$$

as the current density at which the pinning potential well is completely compensated from the energy gain due to the Lorentz force ($\Delta U = U_0$). Assuming $\boldsymbol{J}$ and $\boldsymbol{\varphi}_0$ are oriented perpendicular to each other and, for simplicity, ignoring vector notation, with $B = n\varphi_0$ we obtain from (15.10) and (15.11)

$$J_{\rm c} = \frac{cU_0}{Bs(l_{\rm c}/n)}, \tag{15.12}$$

where $(l_{\rm c}/n)$ represents some jumping volume. Using (7.5) and with the vortex velocity $v_\varphi = R_{\rm j} s$ we have

$$E = 2\nu_0 \exp\left(-\frac{U_0}{k_{\rm B}T}\right) \sinh\left(\frac{JU_0}{J_{\rm c} k_{\rm B}T}\right) sB/c. \tag{15.13}$$

Introducing the quantities $E_{\rm c}$ and $\varrho_{\rm c}$ for $J = J_{\rm c}$

$$E_{\rm c} \equiv \varrho_{\rm c} J_{\rm c} = \nu_0 sB/c, \tag{15.14}$$

finally we obtain

$$E = 2\varrho_{\rm c} J_{\rm c} \exp\left(-\frac{U_0}{k_{\rm B}T}\right) \sinh\left(\frac{J}{J_{\rm c}}\frac{U_0}{k_{\rm B}T}\right). \tag{15.15}$$

We distinguish two important limits. In the TAFF limit, $J \ll J_{\rm c}$,

$$E = 2\varrho_{\rm c} \exp\left(-\frac{U_0}{k_{\rm B}T}\right) \frac{U_0}{k_{\rm B}T} J \tag{15.16}$$

and in the limit $J \approx J_{\rm c}$,

$$E = \varrho_{\rm c} \exp\left[-\frac{U_0}{k_{\rm B}T}\left(1 - \frac{J}{J_{\rm c}}\right)\right] J_{\rm c}. \tag{15.17}$$

The TAFF limit is seen to produce Ohm's law. However, the resistivity E/J is strongly reduced because of the factor $\exp(-U_0/k_{\rm B}T)$. The limit $J \approx J_{\rm c}$ of (15.17) is referred to as *flux creep* with the current dependent effective barrier energy $U_{\rm eff} = U_0(1 - J/J_{\rm c})$. This current dependent exponent in (15.17) often results in a strong increase of E with J over many orders of magnitude.

The case $J \gg J_{\rm c}$ is referred to as *flux flow* where flux-pinning effects become negligible (and where the phenomenological theories discussed in Sect. 10 may perhaps be applicable). The different regimes of vortex motion we have discussed are summarized in Fig. 15.6.

A detailed discussion of the importance of TAFF has been presented by Kes et al. [15.80] and further references can be found elsewhere [15.14, 15.50, 15.81, 15.82].

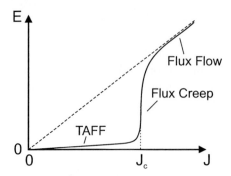

Fig. 15.6. Electric field E versus the current density J for the different regimes of vortex motion

The linear TAFF regime has been observed for sufficiently large values of B and T. It appears to be established above the irreversibility line of the magnetization [15.47, 15.48]. Early observations of the TAFF ohmic behavior were reported by Palstra et al. for single-crystals of $Bi_{2.2}Sr_2Ca_{0.8}Cu_2O_{8+\delta}$ [15.83] and $YBa_2Cu_3O_7$ [15.84, 15.85], and the magnetic field and orientation dependence of the activation energy U_0 were determined. These experiments clearly indicated that flux pinning is stronger in YBCO than in the Bi-system, a fact that is now understood in terms of the stronger coupling between the CuO_2 layers in the former compared to the latter material.

The early transport experiments on high-temperature superconductors already indicated a novel power-law behavior of the I-V curves [15.86, 15.87] which was then taken as evidence for a freezing transition into a superconducting vortex-glass phase (see Sect. 15.6.1). In a different interpretation, a distribution of activation energies was suggested [15.88].

Melting of the vortex lattice and its effect on flux pinning represents an interesting question. Weakening of flux pinning by melting of the vortex lattice is expected only if there are many more flux lines or pancakes than pinning centers. However, in the opposite case we must note that softening of the vortex lattice usually leads to stronger pinning compared to a rigid vortex lattice (see Sect. 11.2). Since atomic-scale defects (such as oxygen vacancies) can act as pinning centers for high-temperature superconductors, this second case is often realized, and flux-lattice melting does not necessarily result in a reduction of pinning effects. However, because of the complexity of this issue, a simple answer cannot be given, as has been discussed in the reviews by Brandt [15.14, 15.81]. We will return to the subject of flux pinning in Sect. 15.6.

Because of the extremely small values of the electric resistivity in the TAFF regime, ac susceptibility measurements as a function of frequency turned out to be convenient due to their high sensitivity [15.89, 15.90]. As an additional advantage, the attachment of electrical wiring to the sample can be avoided. This technique appears useful particularly for exploring the limit of small electric current densities.

15.4.2 Broadening of the Resistive Transition

An interesting new feature of many high-temperature superconductors, which quickly became well established, is the broadening of the resistive transition in a magnetic field. Whereas in the classical superconductors the drop in the electric resistivity due to the onset of superconductivity is usually shifted to lower temperatures with increasing magnetic field, in high-temperature superconductors the resistively measured onset temperature of superconductivity is only weakly magnetic field dependent, but the transition curve is broadened, and the broadening increases strongly with increasing magnetic field. A typical case is shown in Fig. 15.7a for a $YBa_2Cu_3O_7$ single crystal with the magnetic field applied along the c-axis [15.91]. Additional references have been collected by Freimuth [15.82]. An exception to this broadening of the resistive transition in a magnetic field has been observed for the electron doped cuprate superconductor $Nd_{2-x}Ce_xCuO_{4\pm y}$ in the whole doping regime from slightly underdoped to slightly overdoped [15.92]. Here the whole resistive transition curve is shifted to lower temperatures with increasing magnetic field, more like in classical superconductors.

Although no simple general theory exists for explaining all experimental observations, the three unique features of the high-temperature superconductors listed above facilitating vortex motion clearly play a role. The reversible magnetization behavior above the irreversibility line in the H–T plane, we have discussed in Sect. 15.3.1, is consistent with an intermediate regime in the transition to zero resistance. In this context TAFF must be taken into account. Furthermore, the three unique features strongly promote *fluctuations* contributing significantly to the broadening of the transition. The study of fluctuation effects in superconductors has a long history with much of the early work focusing on fluctuation effects in zero magnetic field (see, e.g., [15.93]). Ullah and Dorsey pointed out that for treating the fluctuations in high-temperature superconductors one must go beyond the Gaussian approximation [15.94, 15.95]. They included the interaction between the fluctuations using the Hartree–Fock approximation. In the fluctuation regime they derived universal scaling functions of various thermodynamic and transport properties allowing discrimination between two-dimensional and three-dimensional scaling. An analysis of resistivity and magnetization measurements performed with twinned and untwinned crystals of $YBa_2Cu_3O_{7-\delta}$ in terms of this universal scaling behavior was reported by Welp et al. [15.96]. In this analysis the theoretical procedure introduced by Hao and Clem [15.55, 15.56] also turned out to be crucial for obtaining correct values of the derivative $\partial H_{c2}/\partial T$ near T_c.

The vortex motion induced by the Lorentz force always leads to the *Ettinghausen effect*: the appearance of a temperature gradient perpendicu-

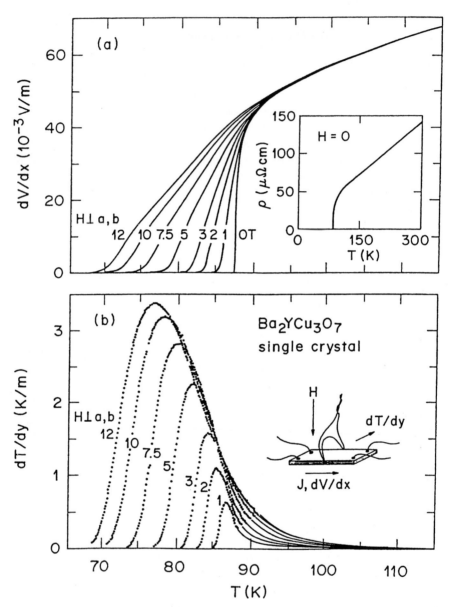

Fig. 15.7. (a) Longitudinal electric field versus temperature for the current density $J = 14\,\text{A/cm}^2$ in a $YBa_2Cu_3O_7$ single crystal for magnetic fields parallel to the c-axis from 0 to 12 T. The *inset* shows the resistivity in zero magnetic field up to 300 K. (b) Transverse temperature gradient versus temperature for the same crystal and the same current density and magnetic fields. The *inset* shows the lead configuration [15.91]

lar to the magnetic field and the electric transport current. As one expects, a broadening effect similar to that observed in the resistive transition also appears in the Ettinghausen effect associated with vortex motion. In Fig. 15.7b we show the transverse temperature gradient dT/dy as a function of temperature for the same crystal, electric current, and magnetic fields as in Fig. 15.7a. The experimental demonstration of the Ettinghausen effect in the superconducting mixed state represents an important confirmation that vortex motion is the dissipative mechanism in this regime of temperature and magnetic field. Similar to the Ettinghausen effect, the mixed state regime of the other thermomagnetic and thermoelectric effects in the high-temperature superconductors is strongly broadened, all effects being proportional to the resistivity due to flux motion. We will return to this point in the subsequent sections.

15.4.3 Damping of the Vortex Motion and Hall Effect

The phenomenological force equation (7.1) is now often written in the form

$$(\boldsymbol{J} \times \boldsymbol{\varphi}_0)/c - \eta \boldsymbol{v}_\varphi - \alpha(\boldsymbol{v}_\varphi \times \boldsymbol{n}) = 0 \ . \tag{15.18}$$

Here, for simplicity, we have ignored the pinning force. $\boldsymbol{n}$ is a unit vector in magnetic field direction. The current density $\boldsymbol{J}$ and the electric field $\boldsymbol{E}$ are perpendicular to the vector $\boldsymbol{n}$. Equation (15.18) represents a convenient form for discussing the role of scattering. It was proposed many years ago by Vinen and Warren [15.97]. A microscopic derivation was presented by Kopnin and coworkers [15.98–15.100]. A simple heuristic argument for (15.18) is given by Blatter and Ivlev along the following lines, based on the requirement that the resulting vortex velocity is consistent with the charge carrier motion inside the vortex core [15.101]. We start with the generalized conductivity law in the presence of both an electric and magnetic field:

$$\boldsymbol{J} = \sigma_\| \boldsymbol{E} + \sigma_\perp \boldsymbol{n} \times \boldsymbol{E} \ , \tag{15.19}$$

where $\sigma_\| = \sigma_n/\left(1+\omega_c^2\tau^2\right), \sigma_\perp = \sigma_n \omega_c \tau/\left(1+\omega_c^2\tau^2\right), \sigma_n = e^2 n^* \tau/m, \omega_c = eB/mc =$ cyclotron frequency, $n^* =$ charge carrier density and $\tau =$ mean electronic scattering time. $\sigma_\|$ and $\sigma_\perp$ are referred to as ohmic and Hall conductivity, respectively. Using (7.5) and taking the cross product of (15.19) with $\boldsymbol{\varphi}_0 \cdot \boldsymbol{n}/c$ we obtain (15.18), with

$$\eta = \frac{\pi \hbar n^* \omega_c \tau}{1+\omega_c^2\tau^2} \tag{15.20}$$

and

$$\alpha = -\frac{\pi \hbar n^* \omega_c^2 \tau^2}{1+\omega_c^2\tau^2} \ . \tag{15.21}$$

Here we have used $\sigma_n \varphi_0 B/c^2 = \pi \hbar n^* \omega_c \tau$. $\hbar \omega_c$ is equal to the level spacing in the core given by (15.3): $\hbar \omega_c = \Delta^2/E_F$. Expressions (15.20) and (15.21) are also obtained from microscopic theory [15.98–15.102].

Assuming the current density $\boldsymbol{J}$ is applied in the x direction and the magnetic field $\boldsymbol{B}$ in z direction, the x component of the force equation (15.18) yields the Hall angle θ

$$\tan \theta = \frac{v_{\varphi x}}{v_{\varphi y}} = -\frac{\alpha}{\eta} = \omega_c \tau . \tag{15.22}$$

Then from the y component we obtain the flux-flow resistivity

$$\varrho_f = \frac{E_x}{J_x} = \frac{\varphi_0 B}{\eta(1 + \alpha^2/\eta^2)c^2} . \tag{15.23}$$

For strong electron scattering ($\omega_c \tau \ll 1$) the Hall angle is small, and the vortices move nearly perpendicular to the current. In the opposite limit ($\omega_c \tau \gg 1$) the Hall angle approaches $\pi/2$, the vortices move nearly parallel to the current, and the dissipation is strongly reduced. In this case an effective damping coefficient $\eta_{\text{eff}} = \eta(1 + \alpha^2/\eta^2)$ appears in (15.23), which is strongly enhanced above the value of η valid in the limit $\omega_c \tau \ll 1$ [15.103]. It appears that this enhanced value η_{eff} has been observed recently combined with a large Hall angle in $YBa_2Cu_3O_{7-\delta}$ at low temperatures [15.104, 15.105]. We emphasize that due to the force equation (15.18) large values of the damping coefficients η and/or α always lead to small values of the corresponding electric field according to (7.5).

So far in our discussion we have neglected flux pinning. In the presence of pinning our conclusions about the dependence of the flux-flow Hall angle on the quasiparticle scattering rate may have to be modified. According to recent calculations by Kopnin and Vinokur, flux pinning is expected to reduce the Hall angle compared to the case where pinning effects are absent [15.106].

The expressions for the damping coefficients η and α in (15.20) and (15.21), respectively, show the prominent role of the quasiparticle scattering time τ. Since in the cuprate superconductors this scattering time is dominated by the electron–electron interaction [15.107–15.110], the quasiparticle electronic structure providing the phase space for scattering will be crucial. In the mixed state of a superconductor, where the quasiparticle spectrum consists of subbands between the Fermi energy and the gap energy, perhaps with the appearance of a minigap (see Sect. 15.3.2), the phase space available for scattering can be strongly restricted, particularly in the low-temperature limit. This leads to interesting consequences for the flux-flow resistivity [15.111, 15.112]. In this case the Bardeen–Stephen model [10.1], representing a reasonable description for the classical superconductors (see Sect. 10.1), is not valid any more, and the flux-flow resistivity may be expressed in the form

$$\varrho_f = \frac{\tau_s^{-1}}{\tau_n^{-1}} \varrho_n \tag{15.24}$$

instead of (10.11). Here τ_{s}^{-1} and τ_{n}^{-1} are the quasiparticle scattering rates in the superconducting mixed state and in the normal state, respectively.

Recently, vortex dynamics and the underlying force equation have been the subject of many theoretical discussions and also some controversy. As first pointed out by Volovik [15.113] and subsequently discussed by others [15.114–15.118], vortex motion induces a flow of the fermionic energy levels through the gap nodes in the center of the vortices. This spectral flow carrying particles from negative energy states of the vacuum to positive energy states transfers linear momentum from the coherent superfluid motion into the normal component of the electron system. Hence, it results in a mutual friction force between the superfluid and normal components. The theoretical discussions assumed the validity of the quasiclassical limit, where the discrete energy spectrum can be treated as a quasicontinuum. For details we refer to the publications cited above. An excellent discussion of the Magnus force in superfluids and superconductors has recently been presented by Sonin [15.41, 15.119].

The early experiments on the *flux-flow Hall effect* in the cuprate superconductors already revealed an unexpected reversal of the sign of the Hall voltage at temperatures below the superconducting transition [15.120, 15.121]. An early summary of the literature was given by Hagen et al. emphasizing the intrinsic nature of the anomalous Hall effect [15.122]. Recently, the flux-flow Hall effect has been studied by many groups, and further references can be found in [15.82]. Experimentally the free flux-flow limit has been approached using high pulsed current densities [15.123] or measuring the complex Hall conductivity at radio frequencies [15.124]. From an investigation of the flux-flow Hall effect in various high-temperature superconductors Matsuda and coworkers concluded that the sign of the Hall voltage is universal and is determined by the doping level: the sign is electron-like in the underdoped regime and hole-like in the overdoped regime [15.125].

From the conventional models of vortex dynamics no sign change of the flux-flow Hall voltage is expected (see Sect. 10.1). The theoretical models which have been proposed for explaining the sign reversal include the following ideas: flux pinning, motion of vacancies in the pinned vortex lattice, melting of the vortex lattice, thermal fluctuations, unbinding of vortex–antivortex pairs, moving segments of Josephson vortices, upstream vortex motion, spectral flow, breaking of the particle–hole symmetry, backflow of quasiparticle excitations, thermoelectric effects, or charging of the vortex cores. An early theoretical discussion was given by Dorsey [15.126]. It is interesting that more than 20 years ago the sign reversal of the flux-flow Hall voltage had already been observed for the classical superconductors (see Sect. 7.5 and Fig. 7.7), but had received only little attention.

An interesting scaling behavior between the Hall resistivity ϱ_{xy} and the longitudinal resistivity ϱ_{xx} in the mixed state has been observed as a function of temperature: $\varrho_{xy} \sim \varrho_{xx}^{\beta}$ with $\beta \approx 2$. The first results were reported by Luo et al. [15.127], and subsequently a similar scaling relation was found in most high-temperature superconductors [15.128]. A theoretical explanation based on the vortex glass transition was proposed by Dorsey and Fisher [15.129]. Another theory emphasized the role of flux pinning in the TAFF regime [15.130].

The results of (15.22) and (15.23) need no further discussion if the electronic structure of the superconductor consists of a single band with only one type of charge carrier. However, an interesting situation arises if two types of charge carriers of opposite sign are present. As an extreme example, we take the case where electrons and holes exist with equal concentration. The coefficient α then has two contributions of opposite sign, α_1 and α_2, cancelling each other: $\alpha = \alpha_1 + \alpha_2 \approx 0$. On the other hand, this compensation does not occur in the effective damping coefficient η_{eff}, since α^2 appears in this quantity and we have $\eta_{\text{eff}} = \eta \left[1 + \left(\alpha_1^2 + \alpha_2^2 \right) / \eta^2 \right]$. We conclude that in this case the damping coefficient can be strongly enhanced because of the large values of $|\alpha_1|$ and $|\alpha_2|$, whereas the observed Hall angle can remain very small. It appears that this is approximately the situation in the cuprate $Nd_{1.85}Ce_{0.15}CuO_x$ where a two-band model for describing the electronic structure has been in discussion for some time [15.92, 15.131].

In the resistive transition regime, flux motion leads one to expect the *Peltier effect* which is proportional to the resistivity. If the Hall angle remains negligible, the Peltier effect results from the quasiparticles in the mixed state [15.132]. A detailed discussion has confirmed the validity of the Thomson relation between the Peltier and Seebeck coefficient [15.133]. For finite flux-flow Hall angle a contribution to the Peltier effect also arises from the component of the vortex motion parallel to the applied transport current. The experimental observation of the mixed state Peltier effect has been reported for $Tl_2Ba_2CaCu_2O_x$ [15.134].

15.4.4 Flux-Flow Instabilities at High Velocities

In the phenomenological force equation (15.18) the damping coefficients η and α are usually taken as constants, being independent of the vortex velocity v_φ and the electric field E. If we continue to ignore flux pinning, (15.18) yields the proportionality $E \propto J$, i.e., we recover Ohm's law. However, the constancy of the coefficients η and α requires that the vortex structure remains unchanged during its motion, which appears reasonable only in the limit of small values of v_φ and E. On the other hand, at high values of v_φ and E electronic nonequilibrium effects occur, leading to strong deviations from Ohm's law including electronic instabilities. We emphasize that here and in the following we are only concerned with the electronic nonequilibrium and not with the effects due to Joule heating (as they are discussed in Sect. 14.4). In a discussion of the electronic nonequilibrium effects two limits must be distinguished: the high-temperature limit ($T \approx T_c$) and the low-temperature limit ($T \ll T_c$).

In the *high-temperature limit* the vortex core is filled with quasiparticles, and its description in terms of a cylinder of normal phase with radius ξ represents a valid approximation. Because of the electric field generated by vortex motion, the quasiparticle distribution is shifted upwards in energy. During this process the quasiparticles traversing the vortex core experience Andreev reflection at the core boundaries, alternating between electrons and holes and always picking up energy from the electric field. Figure 15.8 presents a schematic visualization of this process. At the end a fraction of the quasiparticles leaves the vortex core, the core shrinks, and the damping coefficient η of the friction force $\eta \cdot v_\varphi$ decreases with increasing vortex velocity. A theoretical treatment of this effect has been given by Larkin and Ovchinnikov [15.135], who derived the following relation for the damping coefficient

$$\eta(v_\varphi) = \frac{\eta(0)}{1 + \left(v_\varphi/v_\varphi^*\right)^2} = \frac{\eta(0)}{1 + (E/E^*)^2} \,. \tag{15.25}$$

Here $\eta(0)$ is the damping coefficient in the limit $v_\varphi \to 0$ and v_φ^* a critical vortex velocity at which the damping force $\eta \cdot v_\varphi$ passes through a maximum and negative differential resistivity sets in. Because of (7.5) we have the equality $v_\varphi/v_\varphi^* = E/E^*$, where E^* is the electric field corresponding to v_φ^*. From (15.25) we find for the flux-flow resistivity

$$\varrho_f = \varrho_f(0)\left[1 + \left(\frac{E}{E^*}\right)^2\right] \,, \tag{15.26}$$

where $\varrho_f(0)$ refers to the limit $E \to 0$. The form of the electric field dependence of ϱ_f shown in (15.26) is also obtained from general symmetry arguments. Furthermore, we note that an electric-field-dependent resistivity produces a positive feedback effect leading to a contribution to ϱ_f proportional to E^2, since the field acts on the resistivity and vice versa. The critical velocity v_φ^* is given by [15.135]

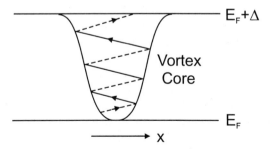

Fig. 15.8. Energy increase of the quasiparticles in the vortex core in the electric field generated by vortex motion. The quasiparticles experience Andreev reflection at the core boundaries and alternate between electrons and holes

$$v_\varphi^{*2} = \frac{D[14\zeta(3)]^{1/2}(1-T/T_c)^{1/2}}{\pi\tau_\varepsilon}.\qquad(15.27)$$

Here D is the quasiparticle diffusion coefficient ($D = v_F\ell/3$ with v_F the Fermi velocity and ℓ the electron mean free path); $\zeta(x)$ is the Riemann zeta function, and τ_ε is the quasiparticle energy relaxation time. The rate τ_ε^{-1} tends to move the quasiparticle system back to its equilibrium distribution.

Because of the effect expressed in (15.25)–(15.27) the IV characteristic is curved upwards, and at the velocity v_φ^* (corresponding to the field E^*) under current control the sample switches to another state with higher resistivity. In a voltage controlled measurement the IV branch with negative differential resistance can also be explored explicitly. As we see from (15.27), from an experimental study of the instability at v_φ^* or E^* we can obtain information on the quasiparticle energy relaxation rate τ_ε^{-1} and its magnetic-field and temperature dependence. Experiments of this kind have already been performed for classical superconductors, and reasonable agreement with the Larkin–Ovchinnikov theory has been found [15.136, 15.137].

Recently, similar experiments have been carried out for epitaxial c-axis-oriented films of $YBa_2Cu_3O_{7-\delta}$ and $La_{1.85}Sr_{0.15}CuO_{4-x}$ with the magnetic field oriented along the c-axis [15.138, 15.139]. In order to minimize effects from Joule heating the films were deposited on substrates with high thermal conductivity, and the film thickness was kept small (60–100 nm). In addition, the IV characteristic was measured using a rapid single-current pulse method. The quasiparticle energy relaxation rate τ_ε^{-1} was determined as a function of temperature in the range above $T/T_c \geq 0{,}4-0{,}5$ using (15.27). Noting that the energy relaxation rate τ_ε^{-1} refers to processes removing excess energy from the quasiparticle system, this rate is expected to be considerably smaller than the elastic or inelastic scattering rate obtained from other experiments such as microwave absorption. In the cuprates $YBa_2Cu_3O_{7-\delta}$ and $La_{1.85}Sr_{0.15}CuO_{4-x}$ the relaxation rate τ_ε^{-1} was well fitted by the proportionality $\tau_\varepsilon^{-1} \sim \exp[-2\Delta(T)/k_BT]$, suggesting that three-quasiparticle interaction, with two quasiparticles recombining to a Cooper pair and one quasiparticle generated at a higher energy, is the underlying dominant process. Here $\Delta(T)$ is the temperature-dependent energy gap. A detailed discussion of the quasiparticle energy relaxation is given in [15.139]. Experiments on the flux-flow instability in the high-temperature regime have also been reported by Xiao et al. for thin films of $YBa_2Cu_3O_{7-\delta}$ and $Bi_2Sr_2CaCu_2O_{8+\delta}$ [15.140–15.142].

From (15.27) the critical velocity v_φ^* is expected to be independent of the magnetic flux density B. This independence has, indeed, been observed, except at low magnetic fields, where a crossover to the behavior $v_\varphi^* \sim B^{-1/2}$ has been found [15.140, 15.143, 15.144]. For explaining this crossover phenomenon we note that in the Larkin–Ovchinnikov theory spatial homogeneity of the nonequilibrium quasiparticle distribution is assumed. However, this assumption is only satisfied if the quasiparticle energy relaxation length

$\ell_\varepsilon = (D\tau_\varepsilon)^{1/2}$ is larger than about the intervortex distance a. Hence, the crossover at low magnetic fields is expected for

$$\ell_\varepsilon = (D\tau_\varepsilon)^{1/2} = a \ . \tag{15.28}$$

Using (15.27) one can show that the length ℓ_ε is approximately equal to the distance $v_\varphi^* \tau_\varepsilon$ and that (15.28) can be replaced by

$$v_\varphi^* \tau_\varepsilon = a \propto B^{-1/2} \ . \tag{15.29}$$

From this we expect that beyond the crossover point at low magnetic fields v_φ^* is proportional to $B^{-1/2}$. These ideas are well supported by experiments [15.143, 15.144].

In the *low-temperature limit* ($T \ll T_c$), in contrast to the case discussed above, the equilibrium Fermi distribution function for the quasiparticle energy drops abruptly from one to zero even on the scale of the energy gap. Now, at equilibrium the states in the vortex core above (below) the Fermi energy are hardly occupied (unoccupied). It is only in the presence of an electric field, generated by vortex motion and resulting in a strong deviation from equilibrium, that the states far above (below) the Fermi energy become occupied (unoccupied). In the quasiclassical limit and for $B \ll B_{c2}$ according to Larkin and Ovchinnikov the resistivity due to vortex motion is

$$\varrho_f = \tilde{\varrho} \cdot 4.35 \cdot \frac{B}{B_{c2}} \frac{1}{\ln(\Delta/k_B T)} \ , \tag{15.30}$$

where $\tilde{\varrho}$ is the core resistivity [15.145, 15.146]. An expression similar to (15.30) was derived by Bardeen and Sherman [15.26]. The logarithmic singularity on the r.h.s. of (15.30) is a result of the shrinkage of the vortex core diameter in the limit $T \to 0$ that we discussed in Sect. 15.2.2 (Kramer–Pesch effect). The quasiparticle nonequilibrium due to the electric field can be expressed as an effective quasiparticle temperature T^* given by

$$k_B T^* = \Delta \left[k_B T_c \frac{\tau_\varepsilon}{\hbar} \left(\frac{J}{J_0}\right)^2 \right]^{1/5} \ . \tag{15.31}$$

Here J_0 is the critical pair-breaking current density. The quasiparticle heating is now taken into account by replacing T in (15.30) by the effective temperature T^* from (15.31). In a sufficiently strong electric field we can have $T^* \gg T$, and the diameter of the vortex core increases in proportion to T^*. Inserting T^* into (15.30), we see that the resistivity diverges for $k_B T^* = \Delta$. The quasiparticles within the vortex core are energetically excited to the value of Δ and leave the core region with subsequent recombination and emission of phonons. Denoting the current density $J = J^*$ at which $k_B T^* = \Delta$, we obtain from (15.31)

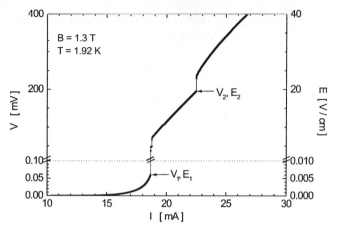

Fig. 15.9. Two voltage steps in the IV curve at the onset voltages V_1 and V_2, respectively, in an epitaxial c-axis-oriented $Nd_{2-x}Ce_xCuO_y$ film. $x \approx 0.15, T = 1{,}92\,K, B = 1300\,mT$. Current bias for increasing current. The corresponding electric field is indicated on the right vertical axis. Note the change of scale on the voltage axis at $V = 0{,}10\,mV$ [15.147]

$$\frac{J^*}{J_0} = \left[\frac{\hbar}{k_B T_c \tau_\varepsilon}\right]^{1/2}, \qquad (15.32)$$

and for the electric field

$$E = 10.9 \tilde{\varrho} \frac{B}{B_{c2}} \frac{J}{\ln(J^*/J)} . \qquad (15.33)$$

We see that the electric field diverges at $J = J^*$ and that $J^* \propto (\tau_\varepsilon)^{-1/2}$. The factor $\ln(J^*/J)$ in the denominator of (15.33) originates from the Kramer–Pesch effect in close analogy to the influence of the temperature in the logarithmic factor in the denominator of (15.30). If this vortex contraction with decreasing temperature were absent, $\ln(J^*/J)$ would be replaced by a number of order unity.

A clear experimental observation of the electric field divergence predicted in (15.33) is difficult due to the influence of flux pinning in the low-temperature limit. On the other hand, in the cuprate superconductors the quasiclassical limit $k_F \xi \gg 1$ is not established generally because of the small values of ξ. Instead, we deal with the quantum limit and the possible appearance of subbands between the Fermi energy and the gap energy (see Sects. 15.2.2 and 15.3.2).

Interesting flux-flow instabilities at $T \ll T_c$ have been observed recently in epitaxial c-axis-oriented films of $Nd_{2-x}Ce_xCuO_y$ with the magnetic field oriented along the c-axis [15.147]. The main features of this novel phenomenon are two hysteretic flux-flow resistance steps under current control and negative differential resistance and spontaneous oscillations under quasi-voltage bias [15.41, 15.111]. During the measurements the samples were embedded in superfluid helium, in this way reaching temperatures in the range $T/T_c \approx 0.08$ and minimizing effects from Joule heating. A typical IV curve for current bias (current increasing) is shown in Fig. 15.9. We note two voltage steps at the threshold voltages V_1 and V_2, respectively. V_2 is about thousand times larger than V_1. For explaining the appearance of two intrinsic flux-flow resistance steps and the associated negative differential resistance, an electronic structure in the mixed state, consisting of two subbands between the Fermi energy and the gap energy, has been proposed [15.111, 15.112, 15.148]. The subbands originate from the Andreev bound states in the core of an isolated vortex because of the interaction between vortices, as discussed in Sect. 15.3.2. The quasiparticles then perform Bloch oscillations in the narrow subbands, leading to localization and the onset of negative differential resistance or the corresponding voltage steps. Such a model can explain qualitatively nearly all important features of the observed IV curves and in particular their magnetic field and temperature dependence.

The Bloch oscillation of the quasiparticles resulting from the Bragg reflection at the Brillouin zone boundary $\pm \pi/a$ of the vortex lattice is governed by the equation

$$e^* E = \hbar \dot{k} = \hbar \frac{2\pi}{a \tau_B} . \tag{15.34}$$

Here e^* is the effective quasiparticle charge, and τ_B is the cycle time of the oscillation. Defining the Bloch frequency $\omega_B = 2\pi/\tau_B$, we obtain from (15.34)

$$\omega_B = \frac{e^* E a}{\hbar} . \tag{15.35}$$

The resistivity is then given by

$$\varrho = \varrho_0 \left(1 + \omega_B^2 \tau^2\right) = \varrho_0 \left(1 + \frac{E^2}{E^{*2}}\right) , \tag{15.36}$$

where τ is the quasiparticle scattering time for the particular subband, and

$$E^* = \frac{\hbar}{e^* a \tau} . \tag{15.37}$$

It is interesting to compare the dynamics of the quasiparticles with the dynamics of the pair wave function during the flux-flow process. The oscillation of the pair wave function is given by the Josephson relation applied to a single unit cell of the moving vortex lattice and yielding the Josephson frequency

$$\omega_J = \frac{2eEa}{\hbar} . \tag{15.38}$$

We see that $\omega_B = \omega_J$ if we set $e^* = 2e$. In this context we note that the quasiparticle dynamics in a narrow subband between the Fermi energy and the gap energy and the validity of (15.34)–(15.37) are still speculative, since a rigorous theoretical treatment of the underlying physics is lacking. Here the correct handling of the symmetry of the pair wave function represents an additional challenge. However, we know that electrons and holes always appear together as quasiparticles and that they are coupled by means of the Bogoliubov–De Gennes equations. Therefore, setting $e^* = 2e$ appears reasonable, and we obtain, indeed, the same value for ω_B and ω_J.

The concepts discussed above can be justified only for a reasonably perfect vortex lattice and provided phase coherence of the wave functions of the quasiparticles and of the pairs is established over many unit cells of the vortex lattice. This phase coherence is strongly improved at high flux-flow velocities, as we will discuss in Sect. 15.4.5. Excitations of quasiparticles from a lower to a higher subband due to an electric field are expected to take place by Zener breakdown. A discussion relating this to the flux-flow instabilities in $Nd_{2-x}Ce_xCuO_y$ is given in [15.112].

This model based on the existence of two subbands can be generalized by assuming instead a strongly energy dependent quasiparticle DOS in the mixed state, with two pronounced steps appearing between the Fermi energy and the gap energy. The electric-field-induced shift of the quasiparticle distribution to higher energies then results in a corresponding electric-field dependence of the flux-flow resistance. This, in turn, can lead to voltage steps or negative differential resistance and spontaneous resistance oscillations, depending on the bias conditions [15.41, 15.149].

15.4.5 Dynamic Correlation

The development of a single-crystalline vortex lattice is usually impossible because of flux pinning. However, the quality of the vortex lattice is considerably improved in the flux-flow state, an effect which is referred to as dynamic correlation. As discussed in Sect. 6.4, this effect was already observed nearly 30 years ago by neutron diffraction. Recent neutron diffraction experiments in single crystals of $NbSe_2$ confirmed these early results showing that the correlation length of the vortex lattice strongly increases if the vortices are depinned during flux flow [15.150]. A similar effect was also found in magnetic decoration studies of the pinned and the moving vortex structure in $NbSe_2$ single crystals, when the decoration image was taken after the current had been turned off abruptly (current annealing): for high vortex velocities a reduced defect concentration was observed [15.151, 15.152]. The issue of a dynamic phase transition to a more ordered vortex state has also been investigated in a combination of IV measurements on $YBa_2Cu_3O_{7-\delta}$ single crystals and numerical simulations [15.153].

The motion of a coherent vortex crystal in $YBa_2Cu_3O_{7-\delta}$ single crystals has been demonstrated for the Corbino disk geometry with a spatially varying Lorentz force density. In this experiment the in-plane vortex correlation was found to extend up to about 10^4 vortex lattice constants [15.154, 15.155].

The dynamic correlation of the vortex lattice discussed in this short section represents an important input for the physics dealing with the spatial phase coherence of the quasiparticle and condensate wave functions.

15.4.6 Quantum Tunneling

When the early magnetic relaxation experiments in the TAFF regime of high temperature superconductors were extended to lower temperatures, below about 1 K, an anomalous behavior was discovered, in which the magnetic relaxation rate becomes temperature independent [15.156–15.159]. The growing amount of data suggested that this anomaly can be interpreted in terms of dissipative macroscopic quantum tunneling of vortices. A summary of the early experiments is given in [15.159, 15.160].

The effect of dissipation on quantum tunneling in macroscopic systems has been studied theoretically by Caldeira and Leggett [15.161] and by Larkin and Ovchinnikov [15.162]. They showed that dissipation reduces the probability of the quantum tunneling process. Quantum tunneling of vortices in bulk superconductors was treated by Blatter et al. [15.163], and a detailed discussion including further references is given in [15.50] (see also [15.14]).

In quantum tunneling the thermal depinning rate $\sim \exp(-U_0/k_\mathrm{B}T)$ is replaced by the tunneling rate $\sim \exp\left(-S_\mathrm{E}^\mathrm{eff}/\hbar\right)$ where $S_\mathrm{E}^\mathrm{eff}$ is the effective Euclidean action of the tunneling process. For single vortex tunneling in the limit of strong dissipation, one obtains for the normalized relaxation rate of the magnetization M

$$\left|\frac{\partial \ln M}{\partial \ln t}\right| = \frac{\hbar}{S_\mathrm{E}^\mathrm{eff}} = \frac{e^2}{\hbar}\frac{\varrho_\mathrm{n}(0)\Gamma}{\xi}\left(\frac{J_\mathrm{c}}{J_0}\right)^{1/2}. \tag{15.39}$$

Here Γ is the anisotropy parameter of (15.1). The effective action $S_\mathrm{E}^\mathrm{eff}$ depends neither on the angle between the magnetic field and the ab planes nor on the direction of the flux motion. In (15.39) the resistivity $\varrho_\mathrm{n}(0)$, the coherence length ξ, the critical current density J_c and the depairing current density J_0 refer to the in-plane values. From (15.39) we see that in the high-temperature superconductors, because of their small values of ξ and large values of $\varrho_\mathrm{n}(0)$, a large quantum tunneling rate is expected.

Highly sensitive measurements of quantum creep in a series of high-temperature superconductors, using capacitance torque magnetometry down to the mK temperature regime, were performed recently by Hoekstra et al. [15.164, 15.165].

15.5 Thermal Force and Vortex Motion

As discussed in Chap. 9, the thermal force $-S_\varphi \,\mathrm{grad}\, T$ acting on the magnetic flux quanta results in the Nernst and Seebeck effect in the superconducting mixed state. In order to deal with these effects, in the phenomenological force equation (15.18) the Lorentz force on the left must be replaced by the thermal force. It is now clear that the dominant contribution to the Seebeck effect in the mixed state has a different origin, namely the thermal force acting on the quasiparticles. The broadening effect and the irreversibility line, observed in the resistive transition of the high-temperature superconductors in the mixed state, also appear in the vortex dynamics responding to a temperature gradient. Because of the high sensitivity of the properties of a superconductor to the temperature, the application of a large temperature gradient to the sample often causes complications. The only exception is a geometry where the sample dimension in the direction of the temperature gradient is kept small, say, about 100 μm. In this case a large gradient can be achieved even for a small temperature difference between the hot and cold side of the sample [15.166, 15.167]. It is important to note that in the superconducting mixed state thermal diffusion due to a temperature gradient is experienced by the two species present: the quasiparticles and the magnetic flux quanta. Therefore, we discuss each contribution separately. This subject has been treated in previous reviews [15.168, 15.169].

15.5.1 Thermal Diffusion of the Quasiparticles

In the *normal state* the thermal diffusion of charge carriers down the applied temperature gradient $\mathrm{grad}_x T$ (assumed in the x direction) results in the accumulation of space charges and a longitudinal electric field E_x (Seebeck effect). From the balance between the thermal force $f_{\mathrm{th}} = -S_q \, \mathrm{grad}_x T$ and the electrostatic force $f_{\mathrm{el}} = q \cdot E_x$ acting on the electric charges q one obtains the normal state Seebeck coefficient $s_\mathrm{n} = E_x / \mathrm{grad}_x T = S_q/q$. Here S_q is the transport entropy of the charge carriers, which is related to the Seebeck coefficient.

In the *superconducting state* the thermal diffusion of the quasiparticles generates the current density

$$J_\mathrm{n} = -\frac{E_x}{\varrho_\mathrm{n}} = -\frac{s_\mathrm{n}}{\varrho_\mathrm{n}} \mathrm{grad}_x T \ . \tag{15.40}$$

Now, however, electric fields (such as generated by the space charges in the normal state) can no longer exist. As discussed by Ginzburg [15.170, 15.171], in *zero magnetic field* the thermal diffusion current density J_n of (15.40) is locally compensated by a counterflow of supercurrent density J_S ($J_\mathrm{n} + J_\mathrm{S} = 0$). This phenomenon is the exact analog of the counterflow in superfluid helium, leading to the fountain effect. Below T_c the quasiparticle concentration decreases rapidly with decreasing temperature, and the counterflow eventually vanishes. From a simple two-fluid model we expect $J_\mathrm{n} = -J_\mathrm{S} \sim (T/T_\mathrm{c})^4$. For the classical superconductors the counterflow effect has been observed in a series of experiments [15.172]. An experimental demonstration has also been reported for a polycrystalline YBaCuO film [15.173].

In the *superconducting mixed state* we have an inhomogeneous situation, and the counterflow model of Ginzburg must be extended [15.174, 15.175]. For temperatures not much below T_c, where the quasiparticle concentration is still high everywhere, far from the vortex cores the two-fluid counterflow picture of Ginzburg is valid. However, near the vortex cores the supercurrent density redistributes itself, and exact local compensation of the two counterflow components of the current no longer exists. The supercurrent flows around the vortex cores, whereas the normal current does not. Together with the circulating supercurrent generating the vortex, a driving force acting on the vortex is developed similar to the hydrodynamic lift force (see Fig. 15.10a). At temperatures $T \ll T_c$, where the quasiparticle concentration away from the vortex cores vanishes, the quasiparticle thermal diffusion current is restricted to the core region and is compensated again by superfluid backflow (see Fig. 15.10b). In both cases shown in Fig. 15.10 the supercurrent backflow generates a driving force acting on the vortices and causing vortex motion in the y direction assuming that the magnetic field is oriented in z direction. As a result a longitudinal electric field E_x develops. It is this mechanism that generates the dominant component of the Seebeck effect in the superconducting mixed state.

The concept of the driving force acting on the vortices perpendicular to the applied temperature gradient and caused by the supercurrent backflow, even in the presence of a compensating quasiparticle current in the opposite direction, is strongly supported by experiments studying the effect of a temperature gradient on the dynamics of a Josephson junction [15.176]. Establishing a temperature gradient across the junction barrier, oscillations of the phase difference between the wave functions of the two electrodes are only effected by the supercurrent counterflow, even in the presence of an exactly

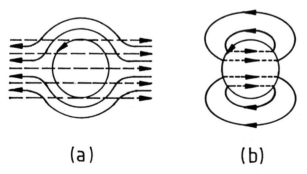

Fig. 15.10. Extended counterflow model of Ginzburg. In a temperature gradient oriented horizontally from right to left the counterflow of J_n (*dashed lines*) and J_S (*solid lines*) near a vortex line is shown. (**a**) T close to T_c. The quasiparticle concentration is still high everywhere. (**b**) $T \ll T_c$. Away from the vortex line the quasiparticle concentration is negligible

compensating (or even highly overcompensating) quasiparticle current in the opposite direction. For details see [15.174, 15.176].

The electric field E_x generated by the supercurrent density J_S is $E_x = \varrho_m(T) J_S$, where $\varrho_m(T)$ is the resistivity due to vortex motion in the mixed state. Using (15.40) and noting that $J_n = -J_S$ we obtain the Seebeck coefficient $s_m(T)$ in the mixed state [15.168, 15.174, 15.177]

$$s_m(T) = \frac{E_x}{\mathrm{grad}_x T} = \frac{\varrho_m(T)}{\varrho_n(T)} s_n(T) . \tag{15.41}$$

A discussion along similar lines has been presented by Zavaritzky et al. [15.178]. The result (15.41) has also been found by Maki from the time-dependent Ginzburg–Landau theory [10.12, 10.13]. The proportionality to $\varrho_m(T)$ indicates that the Seebeck effect in the mixed state exactly mirrors the broadening of the resistive transition. The result (15.41) has been well confirmed experimentally for epitaxial c-axis oriented films of $YBa_2Cu_3O_{7-\delta}$ [15.166, 15.167, 15.175, 15.179] and $Bi_2Sr_2CaCu_2O_{8+x}$ [15.167], and a $Tl_2Ba_2CaCu_2O_x$ single crystal [15.134]. In these experiments the magnetic field was oriented along the c-axis.

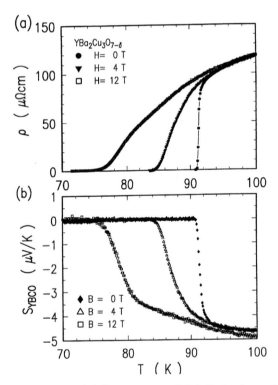

Fig. 15.11. (a) Resistivity and (b) Seebeck coefficient versus temperature of an epitaxial c-axis oriented film of $YBa_2Cu_3O_{7-\delta}$ for different magnetic fields oriented in the c-direction

In Fig. 15.11 we show typical results for $YBa_2Cu_3O_{7-\delta}$ at different magnetic fields. On the low-temperature side $\varrho_m(T)$ and $s_m(T)$ vanish because of flux pinning.

15.5.2 Thermal Diffusion of the Magnetic Flux Quanta

The thermal force $-S_\varphi \operatorname{grad} T$ driving the vortex structure down the temperature gradient results in the Nernst electric field oriented perpendicular to $\operatorname{grad} T$ and B. Taking $\operatorname{grad} T$ in the x-direction and B in the z-direction and neglecting the Hall force and flux pinning, we obtain from (9.2) and (7.9)

$$E_y = \frac{c\varrho_m(T)}{\varphi_0} \cdot S_\varphi(T) \cdot \operatorname{grad}_x T \,. \tag{15.42}$$

Again, the factor $\varrho_m(T)$ introduces the broadening of the resistive transition into the Nernst electric field E_y. In Fig. 15.12 we show typical results for the normalized Nernst electric field $E_y/\operatorname{grad}_x T$ and the resistivity as a function of temperature for an epitaxial c-axis-oriented $YBa_2Cu_3O_{7-\delta}$ film at different magnetic fields oriented along the c-axis. On the low-temperature end of the curves the resistivity and the Nernst field E_y vanish because of flux pinning. Detailed measurements of the Nernst effect in the mixed state have been performed for different high-temperature superconductors, and references can be found in [15.167, 15.169, 15.180].

In the presence of flux pinning the TAFF concept, discussed in Sect. 15.4.1 in combination with the Lorentz force, also applies to the case of the thermal force $-S_\varphi \operatorname{grad} T$. In the TAFF regime, the Nernst field E_y shows exact proportionality to $\operatorname{grad}_x T$, as has also been confirmed experimentally [15.177].

The important quantity to be extracted from measurements of the Nernst effect is the transport entropy S_φ per unit length of vortex line (see Sect. 9.3). An expression for the transport energy TS_φ can be obtained from simple physical arguments in the following way. From the magnetic energy density $W_m = BH/2$ one finds the energy per unit length of flux line $W_m/n = \varphi_0 H/2$ using $B = n\varphi_0$. To obtain the transport energy we must subtract the contribution from the uniform background field B/μ_0, yielding $TS_\varphi = \frac{1}{2}\varphi_0[H - (B/\mu_0)] = -\frac{1}{2}\varphi_0\langle M\rangle$. Here $\langle M\rangle$ is the spatially averaged magnetization. In the high-field limit ($H_{c2} - H \ll H_{c2}$) the following expression has been calculated from the time-dependent Ginzburg–Landau theory [10.12, 10.13, 10.34, 10.35]

$$TS_\varphi(T) = \frac{\varphi_0}{4\pi} \frac{H_{c2}(T) - H}{1.16(2\kappa^2 - 1) + 1} L_D(T) \tag{15.43}$$
$$= -\varphi_0 \langle M \rangle L_D(T) \,. \tag{15.44}$$

Here $L_D(T)$ is a numerical function that drops monotonically from unity to zero between $T = T_c$ and $T = 0$. Except for the function $L_D(T)$ the result (15.44) is very similar to that found from our simple argument.

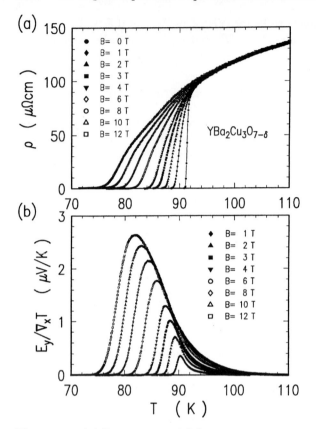

Fig. 15.12. (a) Resistivity and (b) normalized Nernst electric field versus temperature of an epitaxial c-axis-oriented film of $YBa_2Cu_3O_{7-\delta}$ for different magnetic fields oriented in the c-direction

For determining the transport entropy S_φ from the Nernst electric field E_y using (15.42), the resistivity $\varrho_m(T)$ due to vortex motion must be known exactly, and other possible resistive mechanisms must be accounted for. An additional complication in the analysis of the experimental data can arise from the fact that the theoretical expressions such as (15.43) are often obtained in the high-field limit, whereas the experiments do not reach this regime. Taking a typical value $dH_{c2}/dT = -2{,}5\,\mathrm{T/K}$ as an example, only 10 K below T_c we have $H_{c2} = 25\,\mathrm{T}$ which is difficult to approach in many laboratories. In this case the interpolation scheme of Hao and Clem [15.55, 15.56] discussed in Sect. 15.3.1 can greatly improve the data analysis [15.167–15.169].

We see from Fig. 15.12 that the Nernst effect clearly extends to temperatures well above the critical temperature T_c at zero magnetic field because of *fluctuation effects*. Furthermore, with increasing magnetic field these fluctuation effects become stronger. They can be measured with high accuracy, since the Nernst effect is negligible in the normal state, and a special subtraction procedure is unnecessary. In the cuprate superconductors, because of their small coherence length, high critical temperature, and large anisotropy, fluctuations are expected to play a much more prominent role than in the classical superconductors (see Sect. 15.4.2). The theoretical analysis of the fluctuation effects in the cuprate superconductors by Ullah and Dorsey [15.94, 15.95] yields a universal scaling function for the transport energy TS_φ which can be adjusted to the experimental values using a fitting procedure. From this fitting procedure the values of the derivative dH_{c2}/dT can be obtained. The Ullah–Dorsey scaling function discriminates between two-dimensional and three-dimensional systems. Such a scaling analysis has been performed for $YBa_2Cu_3O_{7-\delta}$ and $Bi_2Sr_2CaCu_2O_{8+x}$, indicating three-dimensional (two-dimensional) scaling for the former (latter) material [15.167, 15.169].

In concluding this section we note that the thermal diffusion process of magnetic flux quanta in the cuprate superconductors appears well understood. There exists consistency between the different reported experiments, including consistency between the Nernst and Ettinghausen effect in the cases where both were measured in the same material. Experimental values of the transport entropy S_φ and of the temperature derivatives $d(TS_\varphi)/dT$ and dH_{c2}/dT for $YBa_2Cu_3O_{7-\delta}$ and $Bi_2Sr_2CaCu_2O_{8+x}$ with $B \parallel c$ are listed in [15.167, 15.169, 15.180]. Similar information on $Nd_{2-x}Ce_xCuO_{4\pm y}$ can be found in [15.92].

15.5.3 Hall Angle

We have seen that in the mixed state of the cuprate superconductors the longitudinal (Seebeck effect) and the transverse (Nernst effect) electric field generated in a temperature gradient reach about the same order of magnitude. However, since the two electric fields originate from the thermal diffusion of the two different species existing together in the mixed state, quasiparticles and magnetic flux quanta, this result is not simply due to a large Hall angle. On the other hand, a finite Hall angle is associated with the thermal diffusion of each species separately, as we briefly discuss in the following.

In order to include the Hall angles in the thermal diffusion of the quasiparticles and the vortices, (15.41) and (15.42) must be extended, yielding the following equations [15.168, 15.169, 15.175]

$$\frac{E_x}{\mathrm{grad}_x T} = \frac{\varrho_m}{\varrho_n} s_n (1 + \delta_1 \delta_2) + \frac{cS_\varphi \varrho_m}{\varphi_0} \delta_1 \,, \tag{15.45}$$

$$\frac{E_y}{\mathrm{grad}_x T} = \frac{cS_\varphi \varrho_m}{\varphi_0} + \frac{\varrho_m}{\varrho_n} s_n (\delta_2 - \delta_1) \,. \tag{15.46}$$

Here $\delta_1 \equiv -a/\eta = \tan \theta_\varphi$ (see (15.22)) and $\delta_2 = \tan \theta_{\mathrm{qp}}$, where θ_φ and θ_{qp} are the Hall angles associated with the thermal diffusion of the vortices and the quasiparticles, respectively. Formally, both Hall angles are now indicated by a different notation. Expressions similar to (15.45) and (15.46) were derived by Maki [10.12, 10.13] from the time-dependent Ginzburg–Landau theory and were also obtained by Samoilov et al. [15.181]. We see that the longitudinal electric field E_x (Seebeck effect) contains a first-order contribution from the Hall angle of the thermal diffusion of the vortices, whereas the transverse electric field E_y (Nernst effect) only contains higher-order contributions from the Hall angle of the thermal diffusion of the quasiparticles. The difference $(\delta_2 - \delta_1)$ appears on the r.h.s. of (15.46), since the Hall angles of the vortices and of the quasiparticles are in opposite directions and partly cancel each other. (This compensation of the Hall angles originates from the inversion of the flow direction of J_n and J_S in the counterflow process.) Within the Bardeen–Stephen model [10.1] we have $\delta_1 = \delta_2$, and the second term in (15.46) vanishes.

Measurements of the Seebeck and Nernst effect in epitaxial $YBa_2Cu_3O_{7-\delta}$ films confirmed that the second term in (15.46) is, indeed, negligible [15.175]. An estimate from the experimental data yielded $(\delta_2 - \delta_1) < 0{,}005$. For determining the flux-flow Hall coefficient δ_1 from the measurements a careful treatment of the temperature-dependent electrothermal conductivity $s_n(T)/\varrho_n(T)$ below T_c is required. In the cuprate superconductors electron–electron scattering instead of electron–phonon scattering appears to dominate in the normal state. Therefore, the opening of the superconducting energy gap below T_c strongly affects the quasiparticle transport properties. A simple extrapolation of $s_n(T)/\varrho_n(T)$ from the normal state to temperatures below T_c is clearly inadequate and can lead to an overestimate of the Hall coefficient δ_1. A detailed discussion is given in [15.182].

15.6 Flux Pinning

The subject of flux pinning in high-temperature superconductors has already turned up in our discussion of vortex matter and TAFF. Because of its importance for technological applications, it has received much attention, and continues to represent a challenge to materials science and technology. Flux pinning is caused by spatial inhomogeneity of the superconducting material resulting in local depressions in the Gibbs free energy density of the magnetic flux structure. Because of the short coherence length in the high-temperature superconductors inhomogeneities, even on an atomic scale, can act as pinning centers. Here deviations from stoichiometry, oxygen vacancies in the CuO_2 planes, and twin boundaries are important examples. This point was already emphasized in an early paper by Deutscher and Müller [15.183]. The separation of a flux line into individual pancake vortices also promotes flux pinning effected by atomic size defects.

An illuminating early discussion of pinning due to atomic defects in the superconducting CuO_2 planes (in this case oxygen vacancies) has been given by Kes et al. [15.184, 15.185]. The elementary pinning interaction of vortices with the oxygen vacancies was calculated, and the vacancy concentration was related to the critical current density. The different structural defects in high-temperature superconductors acting as pinning centers were discussed in a review by Wördenweber [15.186]. The influence of flux pinning on magnetic relaxation in high-temperature superconductors has been reviewed in detail by Yeshurun et al. [15.187]. The statistical summation of pinning forces, briefly discussed in Sect. 11.2, continues to be a crucial issue. Here an important advance was achieved by the famous theory of *collective pinning* by Larkin and Ovchinnikov [15.188]. In this theory the elastic distortion of the vortex lattice in the presence of a random spatial distribution of pinning centers plays a central role, and the increase of the elastic energy is balanced against the energy gained by passing the flux lines through favorable pinning sites. A summarizing treatment emphasizing the main physical ideas is given in [15.7].

15.6.1 Vortex Glass and Universal Scaling

In Sect. 15.4.1 we have seen that according to the TAFF model in the limit $T \to 0$ the resistivity is exponentially small but remains finite for $I \to 0$ (see (15.16)). Hence, we have to deal with the question of the existence of superconductivity in the mixed state with *zero resistivity* for $I \to 0$. It turns out that superconductivity is recovered if the Anderson–Kim effective pinning potential $U_0(1 - J/J_c)$ of (15.17) is replaced by the function $U_0(J_c/J)^\mu$, with $\mu \leq 1$, obtained in the Larkin–Ovchinnikov collective pinning theory [15.188] and yielding $\varrho \to 0$ for $I \to 0$. From the latter form of the pinning potential one finds for the electric field

$$E \sim \exp\left[-\frac{U_0}{k_B T}\left(\frac{J_c}{J}\right)^\mu\right], \tag{15.47}$$

i.e., for $J \ll J_c$ a highly nonlinear glasslike response of the vortex motion to the electric current. μ is referred to as the vortex glass exponent. The possibility of a *vortex glass phase transition* with a well-defined glass-melting temperature T_g was proposed by Fisher [15.189] and further discussed in [15.190].

This phase transition was treated by Fisher et al. using scaling arguments along the following lines. A vortex glass phase correlation length $\xi_g \sim |T - T_g|^{-\nu}$ and a relaxation time $\tau_g \sim \xi_g^z \sim |T - T_g|^{-z\nu}$ is assumed, where the exponents ν and z describe the spatial divergence and the critical slowing down at the glass-transition temperature T_g, respectively. The vortex-glass model predicts scaling laws: the electric field should scale as $E\xi_g^{z+1} = \varepsilon_\pm(J\xi_g^{D-1})$, where $z \approx 4$, D is the spatial dimension, and $\varepsilon_\pm(x)$ are scaling functions for the regions above (+) and below (−) T_g. For $x \to 0$ $\varepsilon_-(x)$ should be proportional to $\exp(-x^{-\mu})$ so that the electric field is of the form (15.47). At $T = T_g$ we have $\xi_g \to \infty$, leading to the power law $E \sim J^{(z+1)/(D-1)}$. At $T > T_g$ one expects ohmic behavior $E \sim J$ for sufficiently low currents. In a plot of $\log E$ versus $\log J$ the curves above (below) T_g are expected to show positive (negative) curvature, whereas at $T = T_g$ a straight line will appear. This latter result can be used for empirically identifying T_g. Scaling functions for the scaled nonlinear resistivity ϱ versus the scaled current density J can be defined such that the measured values of ϱ and J for different temperatures at constant magnetic field collapse on two universal curves for $T > T_g$ and $T < T_g$, respectively.

Early experiments providing strong support for these predictions were performed by Koch et al. [15.86, 15.191] with epitaxial films and by Gammel et al. [15.192] with single crystals of $YBa_2Cu_3O_{7-\delta}$. They obtained similar values for the exponents: $z \approx 4 \pm 1$ and $\nu \approx 2$. Subsequently, vortex-glass scaling has been studied extensively by various groups including ac resistivity measurements [15.90]. Further details and references can be found in [15.7, 15.14, 15.50, 15.51]. A comprehensive review of the theory of the vortex-glass phases in type-II superconductors has been presented recently by Nattermann and Scheidl [15.193].

15.6.2 Nanostructuring and Confined Geometries

The recent advances in micro-fabrication technology have also been utilized for the preparation of spatial pinning structures on the sub-µm level. As an extension of the early work of Hebard et al. [11.23, 11.30] (see Sect. 11.4), spatial configurations of many antidots are now being investigated as prominent examples of artificial flux pinning structures. Here the possible existence of multiple flux quanta in single antidots and their spatial configuration in blind antidots represent interesting subjects. Further studies have investigated the stable vortex lattices at rational matching fields for given antidot lattices. Confinement geometries where the spatial structure of the vortex system is strongly influenced by the shielding supercurrent along the outer boundary are being studied experimentally and theoretically (mesoscopic effects on the vortex Gibbs free energy). An excellent summary of this rapidly expanding field has been prepared by Moshchalkov et al. [15.194]. Another important area to have emerged concerns columnar defects produced by nuclear tracts and acting as pinning sites.

Supplementary to these experimental studies, large-scale molecular-dynamics simulations of vortex systems have recently been carried out, dealing with up to 10^6 pinning sites. Here the work of Nori et al. on periodic arrays of pinning sites [15.195, 15.196], vortex avalanches [15.197], and the effect of twinning boundaries [15.198] represents a prominent example.

15.7 Vortex Imaging

Among the spatially averaging techniques for the observation of magnetic flux structures *small angle neutron scattering* and *muon spin rotation* continue to find important applications for high-temperature superconductors. References are given in [15.14].

As one of the important spatially resolving methods *Bitter decoration* has been applied to high-temperature superconductors by various groups [15.14].

An important advance in *magneto-optical flux detection* has been achieved by utilizing iron garnet films (developed for the bubble memory device technology) [15.199]. For magnetic flux detection the ferrimagnetic garnet films can be operated up to temperatures above 500 K. Hence, they are well-suited for the temperature range to be explored in high-temperature superconductors. Whereas in the initial experiments the magnetic anisotropy (easy magnetization axis) was perpendicular to the garnet film plane, the method has been developed further by the introduction of ferrimagnetic garnet films with in-plane anisotropy. With the garnet films a magnetic resolution of 10 µT is obtained. The spatial resolution can reach values less than the garnet film thickness, if the magnetic field gradients are sufficiently strong. A resolution of about 0,4 µm has been reported for imaging with a garnet film of 2 µm thickness. These recent advances in magneto-optical imaging were summarized in reviews by Koblischka and Wijngaarden [15.200], Vlasko-Vlasov et al. [15.201], and Polyanskii et al. [15.202]. The subjects which were studied include magnetic flux penetration into single crystals and thin films, and the effect of pinning defect structures. Another type of experiment dealt with the spatial distribution of the supercurrent density in multi-filamentary BSCCO tapes developed for power applications of high-temperature superconductors. In combination with a digital camera and an image-processing system, time-resolved observations are possible. In this way transient vortex states in $Bi_2Sr_2CaCu_2O_{8+\delta}$ crystals with 40 ms time resolution and up to $10^3 \times 10^3$ pixels per image were studied recently [15.203].

The *microscopic Hall probe array* developed by Zeldov et al. represents a promising novel technique for measuring the local magnetic field component perpendicular to the sample surface [15.53, 15.204]. The microscopic two-dimensional-electron-gas Hall sensors are etched in a GaAs/AlGaAs heterostructure. The active area of the Hall elements was initially $10 \times 10\,\mu m^2$ or larger, and was subsequently reduced to $3 \times 3\,\mu m^2$. Arrays of up to 20 Hall elements have been fabricated. The magnetic field sensitivity is better than 10 µT. In addition to studies of the vortex-lattice melting transition [15.53], this technique has been used for local magnetic relaxation measurements in crystals of $YBa_2Cu_3O_{7-\delta}$ [15.205] and $Nd_{1.85}Ce_{0.15}CuO_{4-\delta}$ [15.206].

The rapid growth of the field of *scanning probe microscopy* following the pioneering work of Binnig and Rohrer [15.207] has also had a strong impact on vortex imaging. A summary of the studies of high-temperature superconductors has been presented by de Lozanne [15.208].

The *scanning SQUID microscope* developed by Kirtley et al. is an instrument with the highest flux sensitivity [15.209]. It consists of a mechanical x-y scanning system combined with an integrated miniature SQUID magnetometer. The SQUID is made of 1 μm Nb–AlO$_x$–Nb junctions and an octagonal pick-up loop of 10 μm diameter integrated on a silicon substrate with a sharpened tip. The tip is in direct contact with the sample during the scan and serves for distance regulation. The spatial resolution is limited by the size of the pick-up loop to about 10 μm. The SQUID microscope has become famous because of its application for detecting half-integer magnetic flux quanta in geometries containing an effective π-junction (see Sect. 15.2.3).

By extending *scanning tunneling microscopy* to low temperatures one can measure the spatially varying quasiparticle density of states, yielding important information on the electronic properties of vortices and vortex structures. Pioneering experiments along these lines were performed by Hess ct al. on NbSc$_2$ crystals [15.210, 15.211]. NbSe$_2$ was chosen because of its excellent surface quality. In addition to imaging the Abrikosov vortex lattice, spectral images as a function of the distance from the center of an individual vortex were obtained. At the vortex center the spectral images clearly showed a conductance peak at zero bias. With increasing distance from the vortex center the zero-bias peak was found to split into two peaks below the superconducting energy gap. Apparently, this spatially resolved spectral behavior is due to the bound quasiparticle states in the vortex core (see Sect. 15.2.2). The splitting appears to be just another manifestation of the relationship between energy and radius of the bound state wave functions. A theoretical analysis by Dorsey et al. based on the Bogoliubov–de Gennes equations [15.212] or on the quasiclassical Eilenberger equations [15.213] confirmed this interpretation. A star-shaped conductance pattern around a vortex observed by Hess et al. was subsequently explained theoretically [15.214, 15.215]. Studying the vortex lattice in an inclined magnetic field, at high tilt angles structural instabilities and the evolution of buckled and disordered lattices were observed [15.216]. A useful summary was given by Hess in [15.217]. Scanning tunneling spectroscopy performed in the alloy system Nb$_{1-x}$Ta$_x$Se$_2$ revealed that the zero-bias conductance peak at the vortex center is very sensitive to disorder and gradually disappears as the mean free path is decreased by substitutional alloying. For $x = 0.2$ the density of states in the vortex center was found to be equal to that in the normal state [15.218].

Scanning tunneling spectroscopy in the mixed state of high-temperature superconductors is difficult because of the problem of the surface quality and the stringent requirements for highly reproducible tunneling conditions. The first successful experiments were reported by Maggio-Aprile et al. for single crystals of YBa$_2$Cu$_3$O$_{7-\delta}$ with B oriented along the c-axis [15.219]. In the differential tunneling conductivity they observed two maxima at $\pm 5,5$ meV which did not shift in energy with increasing distance from the vortex center. These two low energy peaks inside the vortex core were taken as the signature of a bound quasiparticle state corresponding to a minigap $\Delta^2/E_\mathrm{F} = 11$ meV. The vortex cores appear to be in an extreme quantum limit. Extending these experiments to single crystals of $\mathrm{Bi_2Sr_2CaCu_2O_{8+\delta}}$ brought the surprising result that the spectra inside the vortices were not much different from the zero-field spectra, in contrast to the situation in $\mathrm{NbSe_2}$ and YBaCuO [15.220]. Therefore, the detection of the vortices in BiSrCaCuO was much more difficult than in $\mathrm{NbSe_2}$ and YBaCuO. It was found that the spatial dependence of the spectra at 4,2 K through the vortex core corresponds to the temperature dependence of the spectra in zero magnetic field through T_c, if appropriate thermal smearing is taken into account. It appears that the local density of states measured at the vortex center is the *normal state pseudogap structure* and that the pseudogap structure measured above T_c is essentially the thermally smeared low-temperature pseudogap structure. Such behavior is consistent with the size of the vortex defined as the distance from the center to the point where the conductivity peaks of the superconducting state are fully restored. This size, given by the coherence length $\xi \sim \Delta^{-1}$, is at least a factor of two smaller in BiSrCaCuO than in YBaCuO.

The scanning tunneling spectroscopy studies discussed above clearly established this technique as a highly promising tool for investigating the electronic vortex properties in high-temperature superconductors. However, the surface quality of the samples continues to represent a challenging and complex issue.

Magnetic force microscopy measuring the force between a magnetized tip and the sample has been used for imaging vortices and for studying their interaction with defects in thin films of high-temperature superconductors [15.221] and in a $\mathrm{NbSe_2}$ crystal [15.222].

Abrikosov- and Josephson-vortices trapped in thin film $\mathrm{YBa_2Cu_3O_7}$ washer dc SQUIDS have been imaged using *low-temperature scanning electron microscopy* at liquid nitrogen temperature [15.223]. During scanning of the sample surface with the electron beam, the signal is generated due to the beam-induced local displacement of the vortices, resulting in a change of the flux coupled from the trapped vortices into the SQUID loop. In this way $1/f$ noise sources in the device have been identified. The spatial resolution of this method is about 1 μm.

Impressive advances in the application of *Lorentz microscopy* for vortex imaging have been achieved by Tonomura and co-workers. Using a high-voltage field emission electron microscope generating a coherent electron beam, the electron phase shift due to vortices in a tilted specimen is detected in an appropriately defocused image [15.224]. The experiments were performed with niobium films and the vortex images were obtained with a time resolution of 0,03 s. Subsequently, these studies were extended to observe the dynamics of vortices interacting with defect structures produced in the niobium films by 30 keV $\mathrm{Ga^+}$ ion beam irradiation [15.225, 15.226]. Extending this work to high-temperature superconductors, Tonomura's group used Lorentz microscopy to investigate the vortex lattice in (cleaved) single crystals of $Bi_2Sr_2CaCu_2O_{8+\delta}$ [15.227], including the imaging of vortex motion [15.228].

The coherent electron beam technology developed by Tonomura's group has also been used for the observation of vortex structures by *electron holography*. The samples were superconducting lead films [15.229] and thin niobium foils [15.230]. An electron biprism served to generate the hologram. In the first case, one of the divided electron waves passed near the lead film surface. In the second case, it was transmitted through the tilted niobium foil, allowing the reconstruction of the two-dimensional magnetic flux distribution in the interference micrograph. The phase distribution of the electron wave transmitted through the sample region/specimen film was quantitatively measured. The vortices were identified as tiny regions where the phase distribution changed rapidly.

List of Symbols

The same symbol is used for different quantities in some cases where the possibility of confusion is negligible.

Symbols Used in Chaps. 1–14

a	lattice parameter of the flux-line lattice
a	periodicity length of laminar domain structure
a_0	maximum flux-line lattice parameter due to attractive vortex interaction
a_{hk}	flux-line spacing between the (hk) lattice planes
a_n	width of normal domain for laminar domain structure
a_s	width of superconducting domain for laminar domain structure
a_{spot}	equilibrium spacing of flux spots
c	velocity of light
d	thickness of plate or film
d_c	critical film thickness separating type-I and type-II behavior
d_s	critical sample thickness separating branching and nonbranching domain pattern
$d\underline{s}$	path element of integration
$d\underline{S}$	area element of integration
$d\sigma$	surface element of integration
e	elementary charge
$\underline{e}$	local electric field
e^*	$2e$
f	normalized wave function ψ/ψ_0
f	free-energy density
f	factor describing the contribution of the Magnus force
$f(\tilde{h})$	sum $f_1(\tilde{h}) + f_2(\tilde{h})$
f_{kin}	kinetic energy term in GL expression of the free-energy density
$\underline{f}_L$	Lorentz force per unit length of vortex line
f_n	free-energy density of normal phase
$\underline{f}_p$	pinning force per unit length of vortex line or flux tube

$f_{p\,core}$	core contribution to the pinning force per unit length of vortex line
$f_{p\,mag}$	magnetic contribution to the pinning force per unit length of vortex line
f_s	free-energy density of superconducting phase
f_{th}	thermal force per unit length of vortex line or flux tube
$f_1(\tilde{h})$	numerical function in the contribution F_2 to the intermediate-state energy
$f_2(\tilde{h})$	numerical function in the contribution F_3 to the intermediate-state energy
f_{2x}	force on line 2 in x-direction due to interaction between two vortex lines
f_∇	$|\nabla\Psi|$-term in GL expression of the free-energy density
g	Gibbs free-energy density
h	Planck's constant
$\hbar$	$h/2\pi$
$\underline{h}$	local magnetic flux density
$\bar{h}$	Fourier transform of flux density h
$\tilde{h}$	reduced magnetic field H/H_c or H/H_{c2}
$\underline{h}_s$	local flux density arising from supercurrents
$\bar{h}_o$	Fourier transform of flux density for vortex line centered at the origin
k	wave number
k_B	Boltzmann constant
k_c	logarithmic creep-rate constant
k_o	constant describing coupling between amplitude and lifetime of voltage pulses
ℓ	electron mean free path
ℓ_b	electron mean free path of bulk material
ℓ_{12}	distance between voltage probes
m	electron mass
m^*	2m
$\underline{m}$	magnetic moment of particles in the Träuble-Essmann decoration method
n	vortex-line density
n	number of flux quanta per flux tube
n	integer
n_s	single-electron number density of superconducting electrons

n_s^*	$n_s/2$
p	electron momentum
p	pinned fraction of vortex lines not taking part in flux flow
p_s	generalized particle momentum
q	wave number describing periodicity of solution of GL equations
$\underline{q}$	reciprocal lattice vector
r	radial coordinate
r	radius of superconducting domain
$\underline{r}$	spatial coordinate
$\underline{r}_\ell$	spatial coordinate of vortex line
t	reduced temperature T/T_c
t	thickness of magneto-optical layer
t	time
v	velocity
v_F	Fermi velocity
v_n	velocity of the neutron beam
$\underline{v}_s$	velocity of superconducting electrons
$\underline{v}_\varphi$	flux-flow velocity
$w(\omega)$	power spectrum
w_s	sample width
$\hat{y}$	unit vector in y-direction
$y(t)$	$(1 - t^4)^{-\frac{1}{2}}$
$\underline{A}$	vector potential
$\underline{B}$	magnetic flux density
B_m	mean flux density
B_o	minimum flux density due to attractive vortex interaction
C	specific heat per unit volume
D	demagnetization coefficient
D	normal-state diffusion constant, Eq.(10.29)
D_{spot}	flux-spot diameter
$\underline{E}$	electric field
E_{eff}	effective activation energy
E_F	Fermi energy
E_n	energy eigenvalues
E_1	energy per unit length of vortex line
E_{1c}	core energy per unit length of vortex line
E_{12}	interaction energy per unit length between two vortex lines

F	free energy
$F(q)$	form factor
$\overline{F}_c$	area of the two-dimensional unit cell of the vortex lattice
$\underline{F}_p$	pinning force density
F_p	pinning energy associated with a small inclusion
F_1	wall-energy contribution to the intermediate-state energy, Eq.(2.36)
F_2	contribution of field-inhomogeneity to the intermediate-state energy, Eq.(2.37)
F_3	contribution of domain broadening near the surface to the intermediate-state energy, Eq.(2.38)
G	Gibbs free energy
H	spatially averaged magnetic field
H_c	critical field
H_{c1}	lower critical field
H_{c2}	upper critical field
H_{c3}	critical field of surface superconductivity
$H_{c\perp}$	perpendicular critical field (for thin films)
H_e	applied external magnetic field
H_{en}	critical entry field
H_i	magnetic field inside hollow tube
H_φ	azimuthal component of magnetic field
$H_\parallel$	parallel magnetic field
$H_\perp$	perpendicular magnetic field
I	electric current
I_c	critical current
I_{hk}	intensity of the (hk) reflection of the neutron beam
I_0	intensity of incoming light in magneto-optical flux detection
I_1	intensity of reflected light in magneto-optical flux detection
I_2	intensity of light behind the analyzer in magneto-optical flux detection
$\underline{J}$	electric current density
J_c	critical current density
$\underline{J}_n$	normal current density
$\underline{J}_s$	supercurrent density
K	heat conductivity
K^*	coefficient describing coupling between both sides of Josephson junction
K_n	heat conductivity of the normal phase

K_p	pinning force caused by a small inclusion
K_s	heat conductivity of the superconducting phase
K_0	Hankel function of order zero
K_1	Hankel function of order one
L	length of vortex line
$L_D(T)$	numerical function defined in Eq.(10.41)
L_h	healing length of a phase-slip center
L_s	sample length
$\underline{M}$	magnetization
N	number of turns of induction coil
N_T	number of vortex lines with short-range correlation during flux flow
Q	heat input per unit volume
Q_J	Joule heat
R_c	critical radius of a current-carrying wire
R_0	radius of disk or cylinder
S	(cross-sectional) area
S_c	contribution of the core of the vortex line or flux tube to the transport entropy
S_i	incremental entropy per unit length of vortex line
S_n	entropy density in the normal phase
S_s	entropy density in the superconducting phase
S_φ	transport entropy
S_∇	contribution of local temperature gradients to the transport entropy
T	temperature
T_b	bath temperature
T_c	critical temperature
$\underline{U}$	heat current density
V	voltage
V	Verdet constant
$V(x)$	potential
W	energy dissipation per unit length of vortex line
W_{core}	energy dissipation per unit length of vortex core
α	expansion coefficient in the Ginzburg-Landau expression for the free-energy density
α	wall energy per unit area
α^*	normalized slope $(H_{c2}/\rho_n)(\partial \rho_f/\partial H)_{H_{c2}}$

α_h	heat transfer coefficient
β	angle between applied magnetic field and large sample face
β	absorption coefficient of magneto-optical film
β	expansion coefficient in the Ginzburg-Landau expression for the free-energy density
γ	expansion coefficient in the Ginzburg-Landau expression for the free-energy density
γ	ratio of the magnetic moment of the neutron and the nuclear magneton
γ	transport coefficient; Eq.(10.16)
δ	wall-energy parameter
$\delta_2(r)$	two-dimensional delta function
ϵ	Ettinghausen coefficient
ϵ_1	correction term in the logarithmic creep-rate constant
η	viscosity coefficient in the damping force on the moving vortex line or flux tube
η_n	characteristic length $(K_n d/\alpha_h)^{1/2}$
η_s	characteristic length $(K_s d/\alpha_h)^{1/2}$
ϑ_{hk}	neutron scattering angle
κ	Ginzburg-Landau parameter
λ	penetration depth
λ	wavelength of light
λ_{eff}	field-dependent penetration depth
λ_L	London penetration depth
λ_n	wavelength of neutron beam
λ_{TF}	Thomas-Fermi static-charge screening length
μ	chemical potential
μ_1, μ_2	energy on both sides of the Josephson junction
ν	Nernst coefficient
ν	jump rate during flux creep
ν^*	flux-tube nucleation frequency
ν_p	average pulse rate of statistical voltage pulses
$\xi(T)$	temperature-dependent coherence length
ξ_c	radius of vortex core
ξ_v	variational core-radius parameter
ξ_o	Pippard coherence length
ρ	charge density
$\tilde{\rho}$	quantity defined in Eq.(10.42)
ρ_f	flux-flow resistivity

ρ_n	electrical resistivity of the normal phase
σ	electrical conductivity
σ_n	electric conductivity of the normal phase
τ	electron relaxation time
τ	relaxation time of flux penetration
τ^*	transit time of vortex motion
τ_s	spin-flip scattering time
τ_0	lifetime of voltage pulse
τ_2	inelastic scattering time
φ	phase of order parameter
φ	magnetic flux
φ'	fluxoid
φ_0	flux quantum
χ	autocorrelation function
$\chi(\underline{r},t)$	function effecting the gauge transformation of the potentials $\underline{A}$ and $\tilde{\Phi}$
ω	frequency
ω_c	cyclotron frequency
Δ	energy gap
θ	Hall angle
θ	angle of Faraday rotation in magneto-optical flux detection
$\underline{\theta}$	vector in azimuthal direction
Λ	quasiparticle diffusion length
Φ	difference of phase of the complex order parameter
Φ	electrical potential
$\tilde{\Phi}$	effective potential
Φ_n	unscattered neutron flux behind the sample
Ψ	Ginzburg-Landau complex order parameter
$\Psi^{(1)}(x)$	first derivative of the di-gamma function
$\Psi^{(2)}(x)$	second derivative of the di-gamma function
Ψ_L	solution of linearized Ginzburg-Landau equations
Ω	volume (macroscopic)

Additional Symbols for Section 15

s	layer spacing
s	jumping distance of flux line
m_{ab}	effective Cooper pair mass in ab plane
m_c	effective Cooper pair mass in c-direction
μ_0	vacuum permeability
Λ	screening length for interaction between pancake vortices
ε_n	energy levels of quasiparticles in vortex core
$\delta\varepsilon$	energy smearing due to electronic scattering time
k_F	Fermi wave vector
c_L	Lindemann number
u	thermal displacement of position
T_m	vortex lattice melting temperature
U_0	depth of pinning potential well
R_j	flux-line jumping rate
ν_0	attempt frequency
l_c	flux line correlation length in magnetic field direction
ϱ_c	electric resistivity at critical current density
$\boldsymbol{n}$	unit vector in magnetic field direction
α	damping coefficient of the Hall force
n^*	charge carrier density
η_{eff}	effective damping coefficient
τ_s	electronic scattering time in the superconducting mixed state
τ_n	electronic scattering time in the normal state
ϱ_{xx}	longitudinal resistivity
ϱ_{xy}	Hall resistivity
v_φ^*	vortex velocity at instability
E^*	electric field at instability
$\zeta(x)$	Riemann zeta function
τ_ε	quasiparticle energy relaxation time
ℓ_ε	quasiparticle energy relaxation length
$\tilde{\varrho}$	core resistivity
T^*	effective quasiparticle temperature
J_0	pair breaking current density
J^*	current density at which $k_B T^* = \Delta$
τ_B	cycle time of Bloch oscillation
ω_B	Bloch frequency
ω_J	Josephson frequency
S_E^{eff}	effective Euclidean action of tunneling process
s_n	normal-state Seebeck coefficient
S_q	transport entropy of charge carriers
q	electric charge
ϱ_m	resistivity in the mixed state
s_m	Seebeck coefficient in the mixed state

θ_φ	Hall angle of thermal diffusion of vortices
θ_{qp}	Hall angle of thermal diffusion of quasiparticles
δ_1	$\tan\theta_\varphi$
δ_2	$\tan\theta_{\mathrm{qp}}$
μ	vortex glass exponent
T_{g}	vortex glass melting temperature
ξ_{g}	vortex glass correlation length
τ_{g}	vortex glass relaxation time
D	spatial dimension
ν	spatial critical exponent of vortex glass
z	temporal critical exponent of vortex glass

Bibliography

In the following we list a number of textbooks on the subject of superconductivity. Further we indicate various reviews and collections dealing with magnetic flux structures in superconductors. Completeness of this list is not attempted.

Introductory Books on Superconductivity

Buckel, W.: *Supraleitung, Grundlagen und Anwendungen*, 2nd ed. (Physik Verlag, Weinheim 1977)
De Gennes, P.G.: *Superconductivity of Metals and Alloys* (W.A. Benjamin, Inc., New York 1966)
Kuper, C.G.: *Introduction to the Theory of Superconductivity* (Clarendon Press, Oxford 1968)
London, F.: *Superfluids*, Vol. I (Dover Publications, Inc., New York 1961)
Lynton, E.A.: *Superconductivity* (Chapman & Hall, London 1971)
Rickayzen, G.: *Theory of Superconductivity* (John Wiley & Sons, Inc., New York 1965)
Saint-James, D., Sarma, G., Thomas, E.J.: *Type-II Superconductivity* (Pergamon Press, New York 1969)
Schrieffer, J.R.: *Theory of Superconductivity* (W.A. Benjamin, Inc., New York 1964)
Tinkham, M.: *Introduction to Superconductivity* (McGraw-Hill, Inc., New York 1975)

Reviews and Collections

Campbell, A.M., Evetts, J.E.: Flux vortices and transport currents in type II superconductors. Adv. Phys. *21*, 199 (1972)
Cyrot, M.: Ginzburg-Landau theory for superconductors. Rep. Prog. Phys. *36*, 103 (1973)
Goodman, B.B.: Type II Superconductors. Rep. Prog. Phys. *29*, 445 (1966)
Haasen, P., Freyhardt, H.C. (eds.): *International Discussion Meeting on Flux Pinning in Superconductors* (Akademie der Wissenschaften in Göttingen 1975)
Huebener, R.P.: Dynamics of magnetic flux structures in superconductors. Phys. Repts. (Section C of Phys. Lett.) *13*, 143 (1974)
Huebener, R.P., Clem, J.R.: Magnetic flux structures in superconductors - a conference summary. Rev. Mod. Phys. *46*, 409 (1974)
Parks, R.D. (ed.): *Superconductivity*, Vols. I and II (Marcel Dekker, Inc., New York 1969)
Ullmaier, H.: *Irreversible Properties of Type II Superconductors*, Springer Tracts in Modern Physics, Vol. 76 (Springer, Berlin, Heidelberg, New York 1975)
Weber, H.W. (ed.): *Anisotropy Effects in Superconductors* (Plenum Press, New York 1977)

References

Chapter 1. Introduction

1.1 W. Meissner, R. Ochsenfeld: Naturwiss. 21, 787 (1933)
1.2 L.D. Landau: Zh. Eksp. Teor. Fiz. 7, 371 (1937)
1.3 A.A. Abrikosov: Zh. Eksp. Teor. Fiz. 32, 1442
    [Sov. Phys.-JETP 5, 1174 (1957)]
1.4 V.L. Ginzburg, L.D. Landau: Zh. Eksp. Teor. Fiz. 20, 1064 (1950)
    [English translation in *Men of Physics*: L.D. Landau, Vol. I, ed. by
    D. Ter Haar (Pergamon Press, New York 1965)]
1.5 L.P. Gor'kov: Zh. Eksp. Teor. Fiz. 34, 735 [Sov. Phys.-JETP 7, 505
    (1958)]
1.6 L.P. Gor'kov: Zh. Eksp. Teor. Fiz. 36, 1918 [Sov. Phys.-JETP 9, 1364
    (1959)]
1.7 A.L. Fetter, P.C. Hohenberg: In *Superconductivity*, ed. by R.D. Parks
    (Marcel Dekker, New York 1969) p. 817

Chapter 2. Magnetic Properties of Type-I Superconductors

2.1 C.J. Gorter, H.B.G. Casimir: Physica 1, 306 (1934)
2.2 F. London, H. London: Proc. Roy. Soc. (London) A 149, 71 (1935)
2.3 J. Bardeen, L.N. Cooper, J.R. Schrieffer: Phys. Rev. 108, 1175 (1957)
2.4 A.L. Schawlow, G.E. Devlin: Phys. Rev. 113, 120 (1959)
2.5 D. Shoenberg: *Superconductivity* (Cambridge University Press, Cambridge,
    1952)
2.6 J.R. Waldram: Adv. Phys. 13, 1 (1964)
2.7 A.B. Pippard: Proc. Roy. Soc. A 203, 210 (1950)
2.8 A.B. Pippard: Proc. Roy. Soc. (London) A 216, 547 (1953)
2.9 L.D. Landau, E.M. Lifshitz: *Electrodynamics of Continuous Media*
    (Pergamon Press, Oxford 1960) p. 182
2.10 E.M. Lifshitz, Yu.V. Sharvin: Dokl. Akad. Nauk 79, 783 (1951)
2.11 F. Haenssler, L. Rinderer: Helv. Phys. Acta 40, 659 (1967)
2.12 L.D. Landau: Nature, London 141, 688 (1938)
2.13 L.D. Landau: J. Phys. U.S.S.R. 7, 99 (1943)
2.14 C.G. Kuper: Philos. Mag. 42, 961 (1951)
2.14a E.R. Andrew: Proc. Roy. Soc. (London) A 194, 98 (1948)
2.15 A. Hubert: Phys. Status Solidi 24, 669 (1967)
2.16 H. Kirchner: Siemens Forsch.- u. Entwickl. Ber. 1, 39 (1971)
2.17 D.E. Farrell, R.P. Huebener, R.T. Kampwirth: J. Low Temp. Phys. 19,
     99 (1975)
2.18 R.N. Goren, M. Tinkham: J. Low Temp. Phys. 5, 465 (1971)
2.19 J.D. Livingston, W. De Sorbo: In *Superconductivity*, ed. by R.D. Parks
     (Marcel Dekker, New York 1969) Chap.21
2.20 H. Kirchner: Phys. Status Solidi (a) 4, 531 (1971)
2.21 A. Bodmer, U. Essmann, H. Träuble: Phys. Status Solidi (a) 13, 471
     (1972)
2.22 R.P. Huebener, R.T. Kampwirth, V.A. Rowe: Cryogenics 12, 100 (1972)

2.23 R.P. Huebener, R.T. Kampwirth: Phys. Status Solidi (a) *13*, 255 (1972)
2.24 A. Bodmer: Phys. Status Solidi (a) *19*, 513 (1973)
2.25 U. Kunze, B. Lischke, W. Rodewald: Phys. Status Solidi (b) *62*, 377 (1974)
2.26 A.L. Schawlow: Phys. Rev. *101*, 573 (1956)
2.27 A.L. Schawlow, G.E. Devlin: Phys. Rev. *110*, 1011 (1958)
2.28 T.E. Faber: Proc. Roy. Soc. (London) A *248*, 460 (1958)
2.29 Yu.V. Sharvin: Zh. Eksp. Teor. Fiz. *33*, 1341 (1957) [Sov. Phys.-JETP *6*, 1031 (1958)]
2.30 Yu.V. Sharvin: Sov. Phys.-JETP *11*, 216 (1960)
2.31 Yu.V. Sharvin, V.F. Gantmakher: Zh. Eksp. Teor. Fiz. *38*, 1456 [Sov. Phys.-JETP *11*, 1052 (1960)]
2.32 W. DeSorbo, W.A. Healy: Cryogenics *4*, 257 (1964)
2.33 F. Haenssler, L. Rinderer: Helv. Phys. Acta *38*, 448 (1965)
2.34 H. Träuble, U. Essmann: Phys. Status Solidi *18*, 813 (1966)
2.35 D.E. Farrell, R.P. Huebener, R.T. Kampwirth: Phys. Status Solidi (a) *20*, 419 (1973)
2.36 D.E. Farrell, R.P. Huebener, R.T. Kampwirth: Solid State Commun. *11*, 1647 (1972)
2.37 H. Träuble, U. Essmann: Phys. Status Solidi *25*, 395 (1968)
2.38 R.P. Huebener, R.T. Kampwirth: J. Low Temp. Phys. *15*, 47 (1974)
2.39 G.J. Dolan: J. Low Temp. Phys. *15*, 111 (1974)
2.40 A. Kiendl, H. Kirchner: J. Low Temp. Phys. *14*, 349 (1974)
2.41 H. Kirchner, A. Kiendl: Phys. Lett. *39*A, 293 (1972)

Chapter 3. Ginzburg-Landau Theory

3.1 H. Haken: In *Synergetics, Cooperative Phenomena in Multi-Component Systems*, ed. by H. Haken, B.G. Teubner (Stuttgart 1973)
3.2 P.G. DeGennes: *Superconductivity of Metals and Alloys* (W.A. Benjamin, New York 1966)
3.3 J.E. Zimmermann, J.E. Mercereau: Phys. Rev. Lett. *14*, 887 (1965)
3.4 J. Bardeen: Rev. Mod. Phys. *34*, 667 (1962)
3.5 F. London: *Superfluids*, Vol. I (John Wiley & Sons, New York 1950)
3.6 R. Doll, M. Näbauer: Phys. Rev. Lett. *7*, 51 (1961)
3.7 B.S. Deaver, W.M. Fairbank: Phys. Rev. Lett. *7*, 43 (1961)
3.8 W.A. Little, R.D. Parks: Phys. Rev. Lett. *9*, 9 (1962)
3.9 W.A. Little, R.D. Parks: Phys. Rev. *133*, A97 (1964)
3.10 J. Feder, D.S. McLachlan: Phys. Rev. *177*, 763 (1969)
3.11 F.W. Smith, A. Baratoff, M. Cardona: Phys. Kondens. Mater. *12*, 145 (1970)
3.12 D. Saint-James, P.G. DeGennes: Phys. Lett. *7*, 306 (1963)
3.13 W.H. Kleiner, L.M. Roth, S.H. Autler: Phys. Rev. *133*, A1226 (1964)

Chapter 4. Magnetic Properties of Type-II Superconductors

4.1 L.V. Shubnikov, V.I. Khotkevich, Yu.D. Shepelev, Yu.N. Riabinin: Zh. Eksp. Teor. Fiz. *7*, 221 (1937)
4.2 B.B. Goodman: Rep. Progr. Phys. *29*, 445 (1966)
4.3 B. Serin: In *Superconductivity*, ed. by R.D. Parks (Marcel Dekker, New York 1969) p.925
4.4 U. Essmann, H. Träuble: Phys. Lett. *24*A, 526 (1967)
4.5 D. Cribier, B. Jacrot, L.M. Rao, B. Farnoux: Phys. Lett. *9*, 106 (1964)
4.6 J. Schelten, H. Ullmaier, W. Schmatz: Phys. Status Solidi (b) *48*, 619 (1971)
4.7 J. Schelten, H. Ullmaier, G. Lippmann: Z. Physik *253*, 219 (1972)
4.8 H.W. Weber, J. Schelten, G. Lippmann: Phys. Status Solidi (b) *57*, 515 (1973)

4.9   J. Schelten, G. Lippmann, H. Ullmaier: J. Low Temp. Phys. *14*, 213 (1974)
4.10 A.G. Redfield: Phys. Rev. *162*, 367 (1967)
4.11 A. Kung: Phys. Rev. Lett. *25*, 1006 (1970)
4.12 J.M. Delrieu: J. Low Temp. Phys. *6*, 197 (1972)
4.13 J.R. Clem: J. Low Temp. Phys. *18*, 427 (1975)
4.14 J.R. Clem: Proc. 14th Intern. Conf. Low Temp. Phys., eds. M. Krusius, M. Vuorio (North-Holland, Amsterdam 1975) Vol. 2, p.285
4.15 P.M. Morse, H. Feshbach: *Methods of Theoretical Physics* (McGraw Hill, New York 1953) Chap.10
4.16 L. Neumann, L. Tewordt: Z. Phys. *189*, 55 (1966)
4.17 N.R. Werthamer: *Superconductivity*, ed. by R.D. Parks (Marcel Dekker, New York 1969) p.321
4.18 E.H. Brandt: J. Low Temp. Phys. *24*, 409 (1976)
4.19a L. Tewordt: Z. Phys. *180*, 385 (1964)
4.19b L. Tewordt: Phys. Rev. *137*, A1745 (1965)
4.20 L. Neumann, L. Tewordt: Z. Phys. *191*, 73 (1966)
4.21 G. Eilenberger: Z. Phys. *214*, 195 (1968)
4.22 K. Usadel: Phys. Rev. Lett. *25*, 507 (1970)
4.23 K. Usadel: Phys. Rev. B *4*, 99 (1971)
4.24 E.H. Brandt: J. Low Temp. Phys. *24*, 427 (1976)
4.25 W. Pesch, L. Kramer: J. Low Temp. Phys. *15*, 367 (1974)
4.26 L. Kramer, W. Pesch, R.J. Watts-Tobin: J. Low Temp. Phys. *14*, 29 (1974)
4.27 L. Kramer, W. Pesch, R.J. Watts-Tobin: Solid State Commun. *14*, 1251 (1974)
4.28 R.J. Watts-Tobin, L. Kramer, W. Pesch: J. Low Temp. Phys. *17*, 71 (1974)
4.29 W. Pesch: Z. Phys. B *21*, 263 (1975)
4.30 E.H. Brandt: Phys. Status Solidi (b) *77*, 105 (1976)
4.31 C. Caroli, P.G. DeGennes, J. Matricon: Phys. Lett. *9*, 307 (1964)
4.32 C. Caroli, J. Matricon: Phys. Kondens. Mater. *3*, 380 (1965)
4.33 J. Bardeen, R. Kümmel, A.E. Jacobs, L. Tewordt: Phys. Rev. *187*, 556 (1969)
4.34 L. Kramer, W. Pesch: Z. Phys. *269*, 59 (1974)
4.35 L. Kramer: Proc. 14th Intern. Conf. Low Temp. Phys., ed. by M. Krusius, M. Vuorio (North-Holland, Amsterdam 1975) Vol. 2, p.281
4.36 B. Obst: Phys. Lett *28A*, 662 (1969)
4.37 B. Obst: Phys. Status Solidi (b) *45*, 467 (1971)
4.38 R. Kahn, G. Parette: Solid State Commun. *13*, 1839 (1973)
4.39 P. Thorel, R. Kahn, Y. Simon, D. Cribier: J. Phys. *34*, 447 (1973)
4.40 B. Obst: *Anisotropy Effects in Superconductors*, ed. by H.W. Weber (Plenum Press, New York 1977) p.139
4.41 J. Schelten: *Anisotropy Effects in Superconductors*, ed. by H.W. Weber (Plenum Press, New York 1977) p.113
4.42 H. Ullmaier, R. Zeller, P.H. Dederichs: Phys. Lett *44A*, 331 (1973)
4.43 K. Takanaka: Prog. Theor. Phys. *49*, 64 (1973)
4.44 K. Takanaka: Prog. Theor. Phys. *50*, 365 (1973)
4.45 K. Takanaka: Phys. Status Solidi (b) *68*, 623 (1975)
4.46 K. Takanaka: *Anisotropy Effects in Superconductors*, ed. by H.W. Weber (Plenum Press, New York 1977) p.93
4.47 H. Teichler: Philos. Mag. *30*, 1209 (1974)
4.48 H. Teichler: Philos. Mag. *31*, 775 (1975)
4.49 H. Teichler: Philos. Mag. *31*, 789 (1975)
4.50 H. Teichler: *Anisotropy Effects in Superconductors*, ed. by H.W. Weber (Plenum Press, New York 1977) p.7
4.51 H. Träuble, U. Essmann: Phys. Status Solidi *25*, 373 (1968)
4.52 H. Träuble, U. Essmann: J. Appl. Phys. *39*, 4052 (1968)
4.53 U. Essmann, H. Träuble: Phys. Status Solidi *32*, 337 (1969)

4.54 R. Labusch: Phys. Lett. 22, 9 (1966)
4.55 E.H. Brandt: Phys. Status Solidi 35, 1027 (1969)
4.56 E.H. Brandt: Phys. Status Solidi 36, 371 (1969)
4.57 D.C. Hill, D.D. Morrison, R.M. Rose: J. Appl. Phys. 40, 5160 (1969)
4.58 E.H. Brandt: Phys. Status Solidi 36, 381 (1969)
4.59 E.H. Brandt: Phys. Status Solidi 36, 393 (1969)
4.60 E.H. Brandt: J. Low Temp. Phys. 26, 709 (1977)
4.61 E.H. Brandt: J. Low Temp. Phys. 26, 735 (1977)
4.62 E.H. Brandt: J. Low Temp. Phys. 28, 263 (1977)
4.63 E.H. Brandt: J. Low Temp. Phys. 28, 291 (1977)
4.64 C.P. Bean, J.D. Livingston: Phys. Rev. Lett. 12, 14 (1964)
4.65 J.R. Clem, R.P. Huebener, D.E. Gallus: J. Low Temp. Phys. 12, 449 (1973)
4.66 B.L. Walton, B. Rosenblum: Proc. 13th Intern. Conf. Low Temp. Phys., ed. by K.D. Timmerhaus, W.J. O'Sullivan, E.F. Hammel (Plenum Press, New York 1972) Vol. 3, p.172
4.67 B.L. Walton, B. Rosenblum, F. Bridges: Phys. Lett. 43A, 263 (1973)
4.68 L. Kramer: Z. Phys. 259, 333 (1973)
4.69 H. Träuble, U. Essmann: Phys. Status Solidi 20, 95 (1967)
4.70 N.V. Sarma: Philos. Mag. 18, 171 (1968)
4.71 U. Krägeloh: Phys. Status Solidi 42, 559 (1970)
4.72 U. Kumpf: Phys. Status Solidi (b) 44, 829 (1971)
4.73 D.R. Aston, L.W. Dubeck, F. Rothwarf: Phys. Rev. B 3, 2231 (1971)
4.74 D.K. Finnemore, J.R. Clem, T.F. Stromberg: Phys. Rev. B 6, 1056 (1972)
4.75 J.J. Wollan, K.W. Haas, J.R. Clem, D.K. Finnemore: Phys. Rev. B 10, 1874 (1974)
4.76 U. Essmann, R. Schmucker: Phys. Status Solidi (b) 64, 605 (1974)
4.77 J. Auer, H. Ullmaier: Phys. Rev. B 7, 136 (1973)
4.78 G. Eilenberger, H. Büttner: Z. Phys. 224, 335 (1969)
4.79 K. Dichtel: Phys. Lett. 35A, 285 (1971)
4.80 E.H. Brandt: Phys. Lett. 39A, 193 (1972)
4.81 R.M. Cleary: Phys. Rev. Lett. 24, 940 (1970)
4.82 A.E. Jacobs: Phys. Rev. B 4, 3016 (1971)
4.83 A.E. Jacobs: Phys. Rev. B 4, 3022 (1971)
4.84 A.E. Jacobs: Phys. Rev. B 4, 3029 (1971)
4.85 A.E. Jacobs: J. Low Temp. Phys. 10, 137 (1973)
4.86 M.C. Leung: J. Low Temp. Phys. 12, 215 (1973)
4.87 M.C. Leung, A.E. Jacobs: Z. Phys. 253, 89 (1972)
4.88 M.C. Leung, A.E. Jacobs: J. Low Temp. Phys. 11, 395 (1973)
4.89 M.C. Leung, A.E. Jacobs: Low Temperature Physics - Lt 13, Vol. 3, ed. by K.D. Timmerhaus, W.J. O'Sullivan, E.F. Hammel (Plenum Press, New York 1974) p.46
4.90 A. Hubert: Phys. Status Solidi (b) 53, 147 (1972)
4.91 L. Kramer: Phys. Rev. B 3, 3821 (1971)
4.92 S. Grossmann, Ch. Wissel: Z. Phys. 252, 74 (1972)
4.93 L. Kramer: Z. Phys. 258, 367 (1973)

Chapter 5. Thin Films

5.1 M. Tinkham: Phys. Rev. 129, 2413 (1963)
5.2 H. Boersch, U. Kunze, B. Lischke, W. Rodewald: Phys. Lett. 44A, 273 (1973)
5.3 B. Lischke, W. Rodewald: Phys. Status Solidi (b) 63, 97 (1974)
5.4 W. Rodewald: Phys. Lett. 55A, 135 (1975)
5.5 G.J. Dolan, J. Silcox: Proc. Intern. Conf. Low Temp. Phys. LT 13, 1972, Vol. 3, ed. by K.D. Timmerhaus, W.J. O'Sullivan, E.F. Hammel (Plenum Press, New York 1974) p.147

5.6 G.J. Dolan, J. Silcox: Phys. Rev. Lett. *30*, 603 (1973)
5.7 G.J. Dolan: J. Low Temp. Phys. *15*, 111 (1974)
5.8 G.J. Dolan: J. Low Temp. Phys. *15*, 133 (1974)
5.9 E. Guyon, C. Caroli, A. Martinet: J. Phys. Radium *25*, 683 (1964)
5.10 J. Pearl: Low Temperature Physics - Lt 9, ed. by J.G. Daunt, D.V. Edwards, F.J. Milford, M. Yaqub (Plenum Press, New York 1965) part B, p.566
5.11 J. Pearl: J. Appl. Phys. *37*, 4139 (1966)
5.12 K. Maki: Ann. Phys. (N.Y.) *34*, 363 (1965)
5.13 G. Lasher: Phys. Rev. *154*, 345 (1967)
5.14 A.L. Fetter, P.C. Hohenberg: Phys. Rev. *159*, 330 (1967)
5.16 J.P. Burger, G. Deutscher, E. Guyon, A. Martinet: Phys. Rev. *137*, A853 (1965)
     853 (1965)
5.17 G.D. Cody, R.E. Miller: Phys. Rev. *173*, 481 (1968)
5.18 G.D. Cody, R.E. Miller: Phys. Rev. B *5*, 1834 (1972)
5.19 R.E. Miller, G.D. Cody: Phys. Rev. *173*, 494 (1968)
5.20 B.L. Brandt, R.D. Parks, R.D. Chaudhari: J. Low Temp. Phys. *4*, 41 (1971)
5.21 M.D. Maloney, F. De la Cruz, M. Cardona: Phys. Rev. B *5*, 3558 (1972)
5.22 K.E. Gray: J. Low Temp. Phys. *15*, 335 (1974)

Chapter 6. Experimental Techniques

6.1 H. Kirchner: Proc. Intern. Conf. Magnetic Structures in Superconductors, Argonne National Laboratory, Report ANL-8054 (1973)
6.2 F. Bitter: Phys. Rev. *38*, 1903 (1931)
6.3 B.M. Balashova, Yu.V. Sharvin: Zh. Eksp. Teor. Fiz. *31*, 40 (1956) [Sov. Phys.-JETP *4*, 54 (1957)]
6.4 A.L. Schawlow, G.E. Devlin, J.K. Hulm: Phys. Rev. *116*, 626 (1959)
6.5 P.B. Alers: Phys. Rev. *105*, 104 (1957)
6.6 W. DeSorbo: Phys. Rev. Lett. *4*, 406 (1960)
6.7 H. Kirchner: Phys. Lett. *26*A, 651 (1968)
6.8 G. Güntherodt: Phys. cond. Matter *18*, 37 (1974)
6.9 P. Laeng, L. Rinderer: Cryogenics *12*, 315 (1972)
6.10 H. Kirchner: Phys. Lett. *30*A, 437 (1969)
6.11 A. Meshkovsky, A. Shalnikov: Zh. Eksp. Teor. Fiz. *11*, 1 (1947)
6.12 A. Meshkovsky, A. Shalnikov: Zh. Eksp. Teor. Fiz. *34*, 312 (1958)
6.13 R.F. Broom, E.H. Rhoderick: Proc. Phys. Soc. (London) *79*, 586 (1962)
6.14 H.W. Weber, R. Riegler: Solid State Commun. *12*, 121 (1973)
6.15 J. Schelten: Kerntechnik *14*, 86 (1972)
6.16 J. Schelten, H. Ullmaier, G. Lippmann, W. Schmatz: Low Temp. Physics-LT 13, Vol. 3, ed. by K.D. Timmerhaus, W.J. O'Sullivan, E.F. Hammel (Plenum Press, New York 1973) p.54
6.17 G. Lippmann, J. Schelten, R.W. Hendricks, W. Schmatz: Phys. Status Solidi (b) *58*, 633 (1973)
6.18 P. Thorel, R. Kahn: Proceed. Intern. Conf. Magnetic Structures in Superconductors, Argonne National Laboratory Report ANL 8054, p.85 (1973)
6.19 S. Foner: Rev. Sci. Instr. *30*, 548 (1959)
6.20 P. Pincus, A.C. Gossard, V. Jaccarino, J.H. Wernick: Phys. Lett. *13*, 21 (1964)
6.21 W. Fite, A.G. Redfield: Phys. Rev. Lett. *17*, 381 (1966)
6.22 J.M. Delrieu, J.M. Winter: Solid State Commun. *4*, 545 (1966)
6.23 D. Rossier, D.E. MacLaughlin: Phys. Kondens. Mater. *11*, 66 (1970)
6.24 T.R. Brown, J.G. King: Phys. Rev. Lett. *26*, 969 (1971)
6.25 A.T. Fiory, D.E. Murnick, M. Leventhal, W.J. Kossler: Phys. Rev. Lett. *33*, 969 (1974)

6.26 L. Reimer: *Elektronenmikroskopische Untersuchungs- und Präparationsmethoden*, 2. Auflage (Springer, Berlin, Heidelberg, New York 1967)
6.27 D.E. Newbury, H. Yakowitz: In *Practical Scanning Electron Microscopy*, ed. by J.I. Goldstein, H. Yakowitz (Plenum Press, New York 1975) p.149
6.28 J. Philibert, R. Tixier: Micron *1*, 174 (1969)
6.29 D.J. Fathers, J.P. Jakubovics: Philos. Mag. *27*, 765 (1973)
6.30 D.J. Fathers, D.C. Joy, J.P. Jakubovics: Proc. 8th Intern. Congress on Electron Microscopy, ed. by J.V. Sanders, D.J. Goodchild (1974)
6.31 T. Ikuta, R. Shimizu: Phys. Status Solidi (a) *23*, 605 (1974)
6.32 J.R. Dorsey: Proc. 1st Nat. Conf. on Electron Probe Microanalysis, Maryland, USA (1966)
6.33 M.E. Barnett, W.C. Nixon: J. Sci. Instr. *44*, 893 (1967)
6.34 I.R. Banbury, W.C. Nixon: J. Sci. Instr. *44*, 889 (1967)
6.35 I.R. Banbury, W.C. Nixon: J. Phys. E: Sci. Instr. *2*, 1055 (1969)
6.36 P.W. Hawkes, U. Valdrè: J. Phys. E: Sci. Instr. *10*, 309 (1977)

Chapter 7. Lorentz Force and Flux Motion

7.1 C.J. Gorter: Physica *23*, 45 (1957)
7.2 C.J. Gorter: Phys. Lett. *1*, 69 (1962)
7.3 C.J. Gorter: Phys. Lett. *2*, 26 (1962)
7.4 P.W. Anderson: Phys. Rev. Lett. *9*, 309 (1962)
7.5 Y.B. Kim, C.F. Hempstead, A.R. Strnad: Phys. Rev. *131*, 2486 (1963)
7.6 Y.B. Kim, C.F. Hempstead, A.R. Strnad: Phys. Rev. *129*, 528 (1963)
7.7 Y.B. Kim, C.F. Hempstead, A.R. Strnad: Phys. Rev. *139*, A1163 (1965)
7.8 Y.B. Kim, M.J. Stephen: In *Superconductivity*, ed. by R.D. Parks (Marcel Dekker, New York 1969) p.1107
7.9 Y.B. Kim: Proc. 12th Intern. Conf. Low Temp. Phys., Kyoto, 1970, ed. by E. Kanda (Academic Press of Japan, Kyoto 1971) p.231
7.10 A.M. Campbell, J.E. Evetts: Adv. in Phys. *21*, 199 (1972)
7.11 R.P. Huebener: *Solid State Physics*, Vol. 27, eds. H. Ehrenreich, F. Seitz, D. Turnbull (Academic Press, New York 1972) p.63
7.12 R.P. Huebener: Phys. Reports C *13*, 143 (1974)
7.13 P.G. DeGennes, J. Matricon: Rev. Mod. Phys. *36*, 45 (1964)
7.14 A.G. Van Vijfeijken, A.K. Niessen: Philips Res. Rep. *20*, 505 (1965)
7.15 A.G. Van Vijfeijken, A.K. Niessen: Phys. Lett. *16*, 23 (1965)
7.16 A.G. Van Vijfeijken: Philips Res. Rep. Suppl. *8*, 1 (1968)
7.17 P.R. Solomon: Phys. Rev. *179*, 475 (1969)
7.18 J. Schelten, H. Ullmaier, G. Lippmann: Phys. Rev. B *12*, 1772 (1975)
7.19 R.P. Huebener, R.T. Kampwirth, A. Seher: J. Low Temp. Phys. *2*, 113 (1970)
7.19a E.J. Kramer: Proc. Intern. Discussion Meeting on Flux Pinning in Superconductors, ed. by P. Haasen, H.C. Freyhardt (Akad. d. Wiss., Göttingen 1975)
7.20 R.P. Huebener, G. Kostorz, V.A. Rowe: J. Low Temp. Phys. *4*, 73 (1971)
7.21 B. Rosenblum, M. Cardona: Phys. Rev. Lett. *12*, 657 (1964)
7.22 J. Le G. Gilchrist, P. Monceau: J. Phys. C: Solid State Phys. *3*, 1399 (1970)
7.23 J. Le G. Gilchrist, P. Monceau: J. Phys. Chem. Sol. *32*, 2101 (1971)
7.24 R.J. Pedersen, Y.B. Kim, R.S. Thompson: Phys. Rev. B *7*, 982 (1973)
7.25 J.W. Ekin, B. Serin, J.R. Clem: Phys. Rev. B *9*, 912 (1974)
7.26 J.W. Ekin: Phys. Rev. B *12*, 2676 (1975)
7.27 A.M. Clogston: Phys. Rev. Lett. *9*, 266 (1962)
7.28 B.S. Chandrasekhar: Appl. Phys. Lett. *1*, 7 (1962)
7.29 C.J. Axt, W.C.H. Joiner: Phys. Rev. Lett. *21*, 1168 (1968)
7.30 W.C.H. Joiner, J. Thompson: Solid State Commun. *11*, 1393 (1972)

7.31   J.R. Clem: Phys. Rev. Lett. *20*, 735 (1968)
7.32   K. Yamafuji, F. Irie: Phys. Lett. *25A*, 387 (1967)
7.33   J. Lowell: J. Phys. C: Solid State Phys. *3*, 712 (1970)
7.34   J.A. Good, E.J. Kramer: Philos. Mag. *22*, 329 (1970)
7.35   D.E. Farrell: Phys. Rev. Lett. *28*, 154 (1972)
7.36   D.E. Farrell, L. Isett: Phys. Rev. B *5*, 3523 (1972)
7.37   R.P. Huebener, R.T. Kampwirth, D.E. Farrell: Phys. Lett. *41A*, 105 (1972)
7.38   H. Takayama, T. Ogushi, Y. Shibuya: J. Phys. Soc. Jpn. *30*, 1083 (1971)
7.39   B. König, H. Kirchner: Phys. Status Solidi (a) *28*, 467 (1975)
7.40   R.P. Huebener: Phys. Rev. B *2*, 3540 (1970)
7.41   R.P. Huebener, R.T. Kampwirth: J. Low Temp. Phys. *7*, 229 (1972)
7.42   P.R. Solomon, R.E. Harris: Proc. 12th Intern. Conf. Low Temp. Phys., Kyoto, 1970, ed. by E. Kanda (Academic Press of Japan, 1971) p.475
7.43   P.R. Solomon, R.E. Harris: Phys. Rev. B *3*, 2969 (1971)
7.44   A.B. Pippard: Philos. Mag. *41*, 243 (1950)
7.45   E.M. Lifshitz: Zh. Eksp. Teor. Fiz. *9*, 834 (1950)
7.46   T.E. Faber: Proc. Roy. Soc. (London) A *223*, 174 (1954)
7.47   P. Laeng, F. Haenssler, L. Rinderer: J. Low Temp. Phys. *4*, 533 (1971)
7.48   R.P. Huebener, L.G. Stafford, F.E. Aspen: Phys. Rev. B *5*, 3581 (1972)
7.49   D.E. Stevens, R.P. Huebener: Phys. Rev. B *6*, 3547 (1972)
7.50   W.A. Reed, E. Fawcett, Y.B. Kim: Phys. Rev. Lett. *14*, 790 (1965)
7.51   A.T. Fiory, B. Serin: Phys. Rev. Lett. *21*, 359 (1968)
7.52   K. Noto, Y. Muto: Proc. 12th Intern. Conf. Low Temp. Phys., Kyoto, 1970, ed. by E. Kanda (Academic Press of Japan, 1971) p.399
7.53   J. Le G. Gilchrist, J.C. Vallier: Phys. Rev. B *3*, 3878 (1971)
7.54   K. Noto, S. Shinzawa, Y. Muto: Solid State Commun. *18*, 1081 (1976)
7.55   N. Usui, T. Ogasawara, K. Yasukochi, S. Tomoda: J. Phys. Soc. Jpn. *27*, 574 (1969)
7.56   H. Van Beelen, J.P. Van Braam Houckgeest, H.M. Thomas, C. Stolk, R. De Bruyn Ouboter: Physica *36*, 241 (1967)
7.57   H. Ullmaier: Solid State Commun. *7*, 1565 (1969)
7.58   R.R. Hake: Phys. Rev. *168*, 442 (1968)
7.59   B. Byrnak, F.B. Rasmussen: Proc. Intern. Conf. Science of Superconductivity, Stanford, 1969, ed. by F. Chilton (North-Holland, Amsterdam 1971) p.357
7.60   C.H. Weijsenfeld: Physica *45*, 241 (1969)
7.61   E. Hering, E. Cruceanu: Phys. Lett. *38A*, 431 (1972)
7.62   P.R. Solomon, F.A. Otter Jr.: Phys. Rev. *164*, 608 (1967)
7.63   J. Lowell, J.S. Munoz, J.B. Sousa: Phys. Rev. *183*, 497 (1969)
7.64   Y. Muto, K. Noto, K. Mori: Proc. 12th Intern. Conf. Low Temp. Phys., Kyoto, 1970, ed. by E. Kanda (Academic Press of Japan, 1971) p.401
7.65   V.A. Rowe, R.P. Huebener: Phys. Rev. B *2*, 4489 (1970)
7.66   F. Vidal: Phys. Rev. B *8*, 1982 (1973)
7.67   B.D. Josephson: Phys. Lett. *16*, 242 (1965)
7.68   B.D. Josephson: Phys. Lett. *1*, 251 (1962)
7.69   J.R. Clem: Phys. Rev. B *1*, 2140 (1970)
7.70   M. Sugahara: Phys. Rev. B *6*, 130 (1972)
7.71   R.S. Thompson, C.-R. Hu: Phys. Rev. Lett. *31*, 883 (1973)
7.72   P. Tholfsen, H. Meissner: Phys. Rev. *185*, 653 (1969)
7.73   H. Meissner, P. Tholfsen: J. Low Temp. Phys. *4*, 141 (1971)
7.74   H. Meissner: J. Low Temp. Phys. *2*, 267 (1970)
7.75   P. Monceau: Phys. Lett. *47A*, 193 (1974)
7.76   K.E. Gray: J. Low Temp. Phys. *23*, 679 (1976)
7.77   H. Meissner: Phys. Rev. *97*, 1627 (1955)
7.78   H. Meissner: Phys. Rev. *101*, 31 (1956)
7.79   S.T. Sekula, R.W. Boom, C.J. Bergeron: Appl. Phys. Lett. *2*, 102 (1963)
7.80   C.J. Bergeron: Appl. Phys. Lett. *3*, 63 (1963)

7.81 W.E. Timms, D.G. Walmsley: J. Phys. F: Metal Phys. *5*, 287 (1975)
7.82 J.R. Clem: Phys. Rev. Lett. *38*, 1425 (1977)

Chapter 8. Special Experiments

8.1 J.T. Hanlon, J.F. Dillon: J. Appl. Phys. *36*, 1269 (1965)
8.2 B.B. Goodman, M.R. Wertheimer: Phys. Lett. *18*, 236 (1965)
8.3 B.B. Goodman, A. Lacaze, M.R. Wertheimer: C.R. Acad. Sci. *262*, B 12 (1966)
8.4 D.C. Baird, L.S. Wright: J. Low Temp. Phys. *8*, 177 (1972)
8.5 R.B. Harrison, M.R. Wertheimer, L.S. Wright: Proc. 13th Intern. Conf. Low Temp. Phys., Vol. 3, ed. by K.D. Timmerhaus, W.J. O'Sullivan, E.F. Hammel (Plenum Press, New York 1974) p.79
8.6 D.E. Chimenti, R.P. Huebener: Solid State Commun. *21*, 467 (1977)
8.7 J.M. Delrieu: J. Phys. F: Metal Phys. *3*, 893 (1973)
8.8 F. Mezei: Z. Phys. *255*, 146 (1972)
8.9 Y. Simon, P. Thorel: Phys. Lett. *35A*, 450 (1971)
8.10 D.M. Kroeger, J. Schelten: J. Low Temp. Phys. *25*, 369 (1976)
8.11 Yu.V. Sharvin: Zh. Eksp. Teor. Fiz. Pis. Red.*2*, 287 (1965) [JETP Letters *2*, 183 (1965)]
8.12 Yu.V. Sharvin, I.L. Landau: Zh. Eksp. Teor. Fiz. *58*, 1943 (1970) [Sov. Phys.-JETP *31*, 1047 (1970)]
8.13 A.K. Gupta, D.E. Farrell: Phys. Rev. B *7*, 3037 (1973)
8.14 S. Wolf, D. Gubser, D.E. Farrell: Solid State Commun. *14*, 457 (1974)
8.15 M.K. Chien, D.E. Farrell: Phys. Rev. B *9*, 2902 (1974)
8.16 I. Giaever : Phys. Rev. Lett. *15*, 825 (1965)
8.17 I. Giaever: Phys. Rev. Lett. *16*, 460 (1966)
8.18 P.R. Solomon: Phys. Rev. Lett. *16*, 50 (1966)
8.19 P.E. Cladis, R.D. Parks, J.M. Daniels: Phys. Rev. Lett. *21*, 1521 (1968)
8.20 J.R. Clem: Phys. Rev. B *9*, 898 (1974)
8.21 J.R. Clem: Phys. Rev. B *12*, 1742 (1975)
8.22 J.W. Ekin, J.R. Clem: Phys. Rev. B *12*, 1753 (1975)
8.23 M.D. Sherrill: Phys. Rev. B *7*, 1908 (1973)
8.24 J.R. Manson, M.D. Sherrill: Phys. Rev. B *11*, 1066 (1975)
8.25 M.D. Sherrill, W.A. Lindstrom: Phys. Rev. B *11*, 1125 (1975)
8.26 K.P. Selig, D.E. Chimenti, R.P. Huebener: Z. Phys. B *29*, 33 (1978)
8.27 A.C. Rose-Innes, E.A. Stangham: Cryogenics *9*, 456 (1969)
8.28 P.H. Melville, M.T. Taylor: Cryogenics *10*, 491 (1970)
8.29 J.P. Pye, A.C. Rose-Innes: Cryogenics *13*, 216 (1973)
8.30 H. Meissner, P. Crowley: Cryogenics *12*, 91 (1972)
8.31 H. Meissner: Cryogenics *14*, 36 (1974)

Chapter 9. Thermal Force and Flux Motion

9.1 R.P. Huebener, A. Seher: Phys. Rev. *181*, 701 (1969)
9.2 S.R. De Groot: *Thermodynamics of Irreversible Processes* (North-Holland, Amsterdam 1951)
9.3 V.A. Rowe, R.P. Huebener: Phys. Rev. *185*, 666 (1969)
9.4 J. Lowell, J.S. Munoz, J.B. Sousa: Phys. Lett. *24A*, 376 (1967)
9.5 R.P. Huebener: Phys. Lett. *24A*, 651 (1967)
9.6 R.P. Huebener: Phys. Lett. *25A*, 588 (1967)
9.7 R.P. Huebener, A. Seher: Phys. Rev. *181*, 710 (1969)
9.8 L. Rinderer: Proc. Intern. Conf. on Magnetic Structures in Superconductors, Argonne, Argonne Report ANL-8054 (1973)
9.9 R.P. Huebener: Phys. Lett. *28A*, 383 (1968)
9.10 R.P. Huebener, V.A. Rowe: J. Vac. Sci. Technol. *6*, 677 (1969)

9.11 C.H. Stephan, B.W. Maxfield: J. Low Temp. Phys. *10*, 185 (1973)
9.12 E.K. Sichel, B. Serin: J. Low Temp. Phys. *3*, 635 (1970)
9.13 E.K. Sichel, B. Serin: Phys. Lett. *37*A, 123 (1971)
9.14 E.K. Sichel, B. Serin: J. Low Temp. Phys. *24*, 145 (1976)
9.15 J.R. Clem: Phys. Rev. *176*, 531 (1968)
9.16 A.F. Andreev, Yu.K. Dzhikhaev: Zh. Eksp. Teor. Fiz. *60*, 298 (1971) [Sov. Phys.-JETP *33*, 163 (1971)]
9.17 O.L. De Lange, F.A. Otter: Phys. Lett. *38*A, 13 (1972)
9.18 O.L. De Lange, F.A. Otter: J. Phys. Chem. Solids *33*, 1571 (1972)
9.19 M.J. Stephen: Phys. Rev. Lett. *16*, 801 (1966)
9.20 A.G. Van Vijfeijken: Phys. Lett. *23*, 65 (1966)
9.21 A.T. Fiory, B. Serin: Proc. Intern. Conf. Science of Supercond., Stanford, Physica *55*, 73 (1971)
9.22 F.A. Otter, G.B. Yntema: Proc. Low Temp. Calorimetry Conf., Helsinki, 1966; Ann. Acad. Sci. Fennicae *210*, 98 (1966)
9.23 D.C. Hopkins, R.R. Rice, J.M. Carter, J.D. Hayes: Phys. Rev. *183*, 516 (1969)
9.24 R. Ehrat, L. Rinderer: Phys. Lett. *30*A, 95 (1969)
9.25 R. Ehrat, L. Rinderer: J. Low Temp. Phys. *7*, 533 (1972)

Chapter 10. Time-Dependent Theories

10.1 J. Bardeen, M.J. Stephen: Phys. Rev. *140*, A1197 (1965)
10.2 P. Noziėres, W.F. Vinen: Philos. Mag. *14*, 667 (1966)
10.3 J. Bardeen, R.D. Sherman: Phys. Rev. B *12*, 2634 (1975)
10.4 A.I. Larkin, Yu.V. Ovchinnikov: Pis'ma Zh. Eksp. Teor. Fiz. *23*, 210 (1976) [JETP Letters *23*, 187 (1976)]
10.5 N.B. Kopnin, V.E. Kravtsov: Pis'ma Zh. Eksp. Teor. Fiz. *23*, 631 (1976) [JETP Letters *23*, 578 (1976)]
10.6 L.P. Gor'kov, G.M. Eliashberg: Zh. Eksp. Teor. Fiz. *54*, 612 (1968) [Sov. Phys.-JETP *27*, 328 (1968)]
10.7 E. Abrahams, T. Tsuneto: Phys. Rev. *152*, 416 (1966)
10.8 A. Schmid: Phys. Kondens. Mater. *5*, 302 (1966)
10.9 C. Caroli, K. Maki: Phys. Rev. *159*, 306 (1967)
10.10 C. Caroli, K. Maki: Phys. Rev. *159*, 316 (1967)
10.11 C. Caroli, K. Maki: Phys. Rev. *164*, 591 (1967)
10.12 K. Maki: J. Low Temp. Phys. *1*, 45 (1969)
10.13 K. Maki: Prog. Theor. Phys. (Jpn.) *41*, 902 (1969)
10.14 K. Maki: Proc. Intern. Conf. Science of Supercond., Stanford, 1969, ed. by F. Chilton (North-Holland, Amsterdam 1971) p.124
10.15 R.S. Thompson: Phys. Rev. B *1*, 327 (1970)
10.16 H. Takayama, H. Ebisawa: Prog. Theor. Phys. (Jpn.) *44*, 1450 (1970)
10.17 H. Ebisawa: J. Low Temp. Phys. *9*, 11 (1972)
10.18 R.S. Thompson, C.-R. Hu: Phys. Rev. Lett. *27*, 1352 (1971)
10.19 R.S. Thompson, C.-R. Hu: Proc. 13th Intern. Conf. Low Temp. Phys., Boulder, Vol. 3, ed. by K.D. Timmerhaus, W.J. O'Sullivan, E.F. Hammel (Plenum Press, New York 1974) p.163
10.20 C.-R. Hu, R.S. Thompson: Phys. Rev. B *6*, 110 (1972)
10.21 C.-R. Hu, R.S. Thompson: Phys. Rev. Lett. *31*, 217 (1973)
10.22 R.S. Thompson, C.-R. Hu, T.I. Smith: Phys. Rev. B *6*, 2044 (1972)
10.23 M. Cyrot: Rept. Progr. Phys. *36*, 103 (1973)
10.24 L.D. Landau, I.M. Khalatnikov: Dokl. Akad. Nauk SSSR *96*, 469 (1954) [Engl. translation: Collected Papers of L.D. Landau, ed. by D. Ter Haar (Pergamon Press, Oxford) p. 626]
10.25 K. Maki: Physics 1, *21*, 127 (1964)
10.26 S. Imai: Prog. Theor. Phys. (Jpn.) *54*, 624 (1975)
10.27 P. Hagmann, R. Meier-Hirmer, H. Winter: Z. Phys. B *26*, 233 (1977)

10.28 Y. Muto, K. Mori, K. Noto: Proc. Intern. Conf. Science of Supercond., Stanford, 1969, ed. by F. Chilton (North-Holland, Amsterdam 1971) p.362
10.29 C.J. Axt, W.C.H. Joiner: Phys. Rev. *171*, 461 (1968)
10.30 G.E. Kuhl, M.C. Ohmer: Phys. Rev. B *2*, 1264 (1970)
10.31 Y. Muto, K. Noto, M. Hongo, K. Mori: Phys. Lett. *30*A, 480 (1969)
10.32 H. Ebisawa: Proc. Intern. Conf. Low Temp. Phys. - LT 13, Vol. 3 (1972), ed. by K.D. Timmerhaus, W.J. O'Sullivan, E.F. Hammel (Plenum Press, New York 1974) p.177
10.33 A. Houghton, K. Maki: Phys. Rev. B *3*, 1625 (1971)
10.34 C.-R. Hu: Phys. Rev. B *13*, 4780 (1976)
10.35 C.-R. Hu: Phys. Rev. B *14*, 4834 (1976)

Chapter 11. Flux Pinning

11.1 H. Ullmaier: *Irreversible Properties of Type II Superconductors*, Springer Tracts in Modern Physics, Vol. 76 (Springer, Berlin, Heidelberg, New York 1975)
11.2 J. Friedel, P.G. De Gennes, J. Matricon: Appl. Phys. Lett. *2*, 119 (1963)
11.3 C.P. Bean: Phys. Rev. Lett. *8*, 250 (1962)
11.4 Y.B. Kim, C.F. Hempstead, A.R. Strnad: Phys. Rev. Lett. *9*, 306 (1962)
11.5 C.P. Bean: Rev. Mod. Phys. *36*, 36 (1964)
11.6 H. Ullmaier: Phys. Status Solidi *17*, 631 (1966)
11.7 A.M. Campbell: J. Phys. C. *2*, 1492 (1969)
11.8 R.W. Rollins, H. Küpfer, W. Gey: J. Appl. Phys. *45*, 5392 (1974)
11.9 J.R. Clem, H.R. Kerchner, S.T. Sekula: Phys. Rev. B *14*, 1893 (1976)
11.10 H. Koppe: Phys. Status Solidi *17*, K 229 (1966)
11.11 W.T. Norris: J. Phys. D *3*, 489 (1970)
11.12 W.T. Norris: J. Phys. D *4*, 1358 (1971)
11.13 H. Ullmaier, R.H. Kernohan: Phys. Status Solidi *17*, K 233 (1966)
11.14 A. Migliori, R.J. Bartlett, R.D. Taylor: J. Appl. Phys. *47*, 3266 (1976)
11.15 A. Migliori, R.D. Taylor, R.J. Bartlett: IEEE Trans. Magnetics, MAG *13*, 198 (1977)
11.16 E. Nembach: Phys. Status Solidi *13*, 543 (1966)
11.17 T.H. Alden, J.D. Livingston: J. Appl. Phys. *37*, 3551 (1966)
11.18 R. Labusch: Cryst. Lattice Defects *1*, 1 (1969)
11.19 R. Labusch: Phys. Status Solidi *32*, 439 (1969)
11.20 J. Lowell: J. Phys. F *2*, 547 (1972)
11.21 J.A. Good, E.J. Kramer: Philos. Mag. *24*, 339 (1971)
11.22 D. Dew-Hughes: Philos. Mag. *30*, 293 (1974)
11.23 A.F. Hebard, A.T. Fiory, S. Somekh: IEEE Trans. Mag. *13*, 589 (1977)
11.24 D. Saint-James, G. Sarma, E.J. Thomas: *Type II Superconductivity* (Pergamon Press, Oxford 1969)
11.25 D.D. Morrison, R.M. Rose: Phys. Rev. Lett. *25*, 356 (1970)
11.26 P.S. Swartz, H.R. Hart: Phys. Rev. *137*, 818 (1965)
11.27 O. Daldini, P. Martinoli, J.L. Olsen, G. Berner: Phys. Rev. Lett. *32*, 218 (1974)
11.28 P. Martinoli, O. Daldini, C. Leemann, B. Van Den Brandt: Phys. Rev. Lett. *36*, 382 (1976)
11.29 P. Martinoli: Phys. Rev. B. *17*, 1175 (1978)
11.30 A.T. Fiory, A.F. Hebard, S. Somekh: Appl. Phys. Lett. *32*, 73 (1978)
11.31 H. Raffy, J.C. Renard, E. Guyon: Solid State Commun. *11*, 1679 (1972)
11.32 H. Raffy, E. Guyon, J.C. Renard: Solid State Commun. *14*, 427 (1974)
11.33 H. Raffy, E. Guyon, J.C. Renard: Solid State Commun. *14*, 431 (1974)
11.33a G. Antesberger, H. Ullmaier: Philos. Mag. *29*, 1101 (1974)

11.34 C.P. Herring: Phys. Lett. *47*A, 105 (1974)
11.35 W. Rodewald: Proceed. 6th Intern. Cryogenic Engineering Conf., ed.
      by K. Mendelssohn (IPC Science and Technology Press 1977) p.400
11.36 B. Lischke, W. Rodewald: Proc. Intern. Discussion Meeting on Flux
      Pinning in Superconductors, ed. by P. Haasen, H.C. Freyhardt
      (Akademie der Wissenschaften, Göttingen 1974) p.323

Chapter 12. Flux Creep and Flux Jumps

12.1 R.D. Dunlap, C.F. Hempstead, Y.B. Kim: J. Appl. Phys. *34*, 3147 (1963)
12.2 M.R. Beasley, R. Labusch, W.W. Webb: Phys. Rev. *181*, 682 (1969)
12.3 H. Boersch, B. Lischke, H. Söllig: Phys. Status Solidi (b) *61*, 215
     (1974)

Chapter 13. Electrical Noise Power

13.1 L. Van Hove: Phys. Rev. *95*, 249 (1954)
13.2 L. Van Hove: Phys. Rev. *95*, 1374 (1954)
13.3 D.K.C. MacDonald: *Noise and Fluctuations* (Wiley and Sons, New York
     1962)
13.4 C. Heiden: Phys. Rev. *188*, 319 (1969)
13.5 C. Heiden: Habilitationsschrift, University of Münster (1971)
13.6 A.T. Fiory: Phys. Rev. Lett. *27*, 501 (1971)
13.7 A.T. Fiory: Phys. Rev. B *7*, 1881 (1973)
13.8 J.G. Park: J. Phys. F: Metal Phys. *2*, 957 (1972)
13.9 P. Jarvis, J.G. Park: J. Phys. F: Metal Phys. *4*, 1238 (1974)
13.10 G.J. Van Gurp: Phys. Rev. *166*, 436 (1968)
13.11 G.J. Van Gurp: Phys. Rev. *178*, 650 (1969)
13.12 D.J. Van Ooijen, G.J. Van Gurp: Philips Res. Repts. *21*, 343 (1966)
13.13 G. Fournet, J. Baixeras: Phys. Lett. *25*A, 552 (1967)
13.14 J.M.A. Wade: Philos. Mag. *23*, 1029 (1971)
13.15 C. Heiden: Proc. 13th Intern. Conf. Low Temp. Phys., Boulder, 1972,
      ed. by K.D. Timmerhaus, W.J. O'Sullivan, E.F. Hammel (Plenum Press,
      New York 1974) p.75
13.16 J.B. Kruger, C.Heiden: Proc. 12th Intern. Conf. Low Temp. Phys.,
      Kyoto, ed. by E. Kanda (Academic Press of Japan, 1970) p.415
13.17 C. Heiden, H.P. Friedrich: Solid State Commun. *9*, 323 (1971)
13.18 C. Heiden, D. Kohake: Phys. Status Solidi (b) *64*, K 83 (1974)
13.19 C. Heiden, D. Kohake, W. Krings, L. Ratke: J. Low Temp. Phys. *27*, 1
      (1977)
13.20 P. Jarvis, J.G. Park: J. Phys. F: Metal Phys. *5*, 1573 (1975)
13.21 J. Thompson, W.C.H. Joiner: Solid State Commun. *16*, 849 (1975)
13.22 F. Habbal, W.C.H. Joiner: Phys. Lett. *60*A, 434 (1977)
13.23 F. Habbal, W.C.H. Joiner: J. Low Temp. Phys. *28*, 83 (1977)

Chapter 14. Current-Induced Resistive State

14.1 F. London: *Une Conception Nouvelle de la Superconductibilité*
     (Hermann, Paris 1937)
14.2 H.L. Watson, R.P. Huebener: Phys. Rev. B *10*, 4577 (1974)
14.3 C.J. Gorter, M.L. Potters: Physica *24*, 169 (1958)
14.4 H. Meissner: Phys. Rev. *113*, 1183 (1959)
14.5 A.F. Andreev, Yu.V. Sharvin: Zh. Eksp. Teor. Fiz. *53*, 1499 (1967)
     [Sov. Phys.-JETP *26*, 865 (1968)]
14.6 A.F. Andreev: Zh. Eksp. Teor. Fiz. *54*, 1510 (1968) [Sov. Phys.-JETP
     27, 809 (1968)]
14.7 D.C. Baird, B.K. Mukherjee: Phys. Rev. Lett. *21*, 996 (1968)

14.8 D.C. Baird, B.K. Mukherjee: Phys. Rev. B 3, 1043 (1971)
14.9 L. Rinderer: Helv. Phys. Acta 29, 339 (1956)
14.10 H.D. Wiederick, B.K. Mukherjee, D.C. Baird: J. Phys. D 4, 1365 (1971)
14.11 B.K. Mukherjee: J. Low Temp. Phys. 12, 181 (1973)
14.12 L. Rinderer: Helv. Phys. Acta 32, 320 (1959)
14.13 E. Posada, L. Rinderer: J. Low Temp. Phys. 9, 33 (1972)
14.14 H.D. Wiederick, B.K. Mukherjee, D.C. Baird: J. Low Temp. Phys. 21, 257 (1975)
14.15 M.Ya. Azbel: JETP Lett. 10, 351 (1969)
14.16 B. Lalevic: Phys. Rev. 128, 1070 (1962)
14.17 B. Lalevic: Phys. Status Solidi 9, 63 (1965)
14.18 I.L. Landau, Yu.V. Sharvin: Zh. Eksp. Teor. Fiz. Pis. Red. 10, 192 (1969) [JETP Letters 10, 121 (1969)]
14.19 I.L. Landau, Yu.V. Sharvin: Zh. Eksp. Teor. Fiz. Pis. Red. 15, 88 (1972) [JETP Letters 15, 59 (1972)]
14.20 W. Bestgen: Habilitationsschrift, Universität Marburg (1974)
14.21 P. Tekiel: Zh. Eksp. Teor. Fiz. 61, 1691 (1971) [Sov. Phys.-JETP 34, 902 (1972)]
14.22 A.F. Andreev, P. Tekiel: Zh. Eksp. Teor. Fiz. 62, 1540 (1972) [Sov. Phys.-JETP 35, 807 (1972)]
14.23 A.F. Andreev, W. Bestgen: Zh. Eksp. Teor. Fiz. 64, 1865 (1973) [Sov. Phys.-JETP 37, 942 (1973)]
14.24 W. Bestgen: Zh. Eksp. Teor. Fiz. 65, 2097 (1973) [Sov. Phys.-JETP 38, 1048 (1974)]
14.25 W. Bestgen: Z. Phys. 269, 73 (1974)
14.26 H. Meissner: J. Low Temp. Phys. 3, 563 (1970)
14.27 H.L. Phillips, H. Meissner: Phys. Rev. B 5, 3572 (1972)
14.28 R.P. Huebener, D.E. Gallus: Phys. Rev. B 7, 4089 (1973)
14.29 E.H. Rhoderick, E.M. Wilson: Nature (London) 194, 1167 (1962)
14.30 R.P. Huebener, R.T. Kampwirth, J.R. Clem: J. Low Temp. Phys. 6, 275 (1972)
14.31 R.P. Huebener, R.T. Kampwirth: Proc. 1972 Applied Supercond. Conf., ed. by H.M. Long, W.F. Gauster (IEEE, New York 1972) p.422
14.32 R.P. Huebener, H.L. Watson: Phys. Rev. B 9, 3725 (1974)
14.33 R.P. Huebener, D.E. Gallus: Phys. Lett. 44A, 443 (1973)
14.34 G.E. Churilov, V.M. Dmitriev, A.P. Beskorsyi: Zh. Eksp. Teor. Fiz. Pis. Red. 10, 231 (1969) [JETP Letters 10, 146 (1969)]
14.35 V.I. Galaiko, V.M. Dmitriev, G.E. Churilov: Zh. Eksp. Teor. Fiz. Pis. Red. 18, 362 (1973) [JETP Letters 18, 213 (1973)]
14.36 V.N. Gubankov, V.P. Koshelets, K.K. Likharev, G.A. Ovsyannikov: Zh. Eksp. Teor. Fiz. Pis. Red. 18, 292 (1973) [JETP Letters 18, 171 (1973)]
14.37 V.N. Gubankov, V.P. Koshelets, G.A. Ovsyannikov: Zh. Eksp. Teor. Fiz. 71, 348 (1976) [Sov. Phys.-JETP 44, 181 (1976)]
14.38 L.E. Musirenko, V.I. Shnyrkov, V.G. Volotskaya, I.M. Dmitrenko: Fiz. Nizkikh Temp. 1, 413 (1975) [Sov. J. Low Temp. Phys. 1, 205 (1975)]
14.39 K.K. Likharev: Zh. Eksp. Teor. Fiz. 61, 1700 (1971) [Sov. Phys.-JETP 34, 906 (1972)]
14.40 L.G. Aslamazov, A.I. Larkin: Zh. Eksp. Teor. Fiz. 68, 766 (1975) [Sov. Phys.-JETP 41, 381 (1975)]
14.41 R.P. Huebener, D.E. Gallus: Appl. Phys. Lett. 22, 597 (1973)
14.42 W.F. Brown: *Magnetostatic Principles in Ferromagnetism* (North-Holland, Amsterdam 1962)
14.43 D.E. Chimenti, H.L. Watson, R.P. Huebener: J. Low Temp. Phys. 23, 303 (1976)
14.44 D.E. Chimenti: Phys. Lett. 64A, 81 (1977)

14.45 M.C.L. Orlowski, W. Buck, R.P. Huebener: J. Low Temp. Phys. *27*, 159 (1977)
14.46 I.O. Kulik, I.K. Yanson: *The Josephson Effect in Superconductive Tunneling Structures* (Israel Program for Scientific Translations 1972)
14.47 L. Solymar: *Superconductive Tunneling and Applications* (Chapman and Hall, London 1972)
14.48 A.H. Silver, J.E. Zimmermann: *Applied Superconductivity*, Vol. 1, ed. by V.E. Newhouse (Academic Press, New York 1975) p.1
14.49 H.M. Long, W.F. Gauster: Proc. 1972 Appl. Supercond. Conf., Annapolis, IEEE Pub. No. 72CH0682-5-TABSC (1972)
14.50 J.M. Rowell: Proc. 1974 Appl. Supercond. Conference, Argonne, Illinois IEEE Trans. Mag. *11*, No. 2 (1975)
14.51 B.P. Strauss (ed.): Proc. 1976 Appl. Supercond. Conf., Stanford, California, IEEE Trans. Mag. *13*, No. 1 (1977)
14.52 G.I. Rochlin: Proc. Conf. on Fluct. in Supercond., ed. by W.S. Goree, F. Chilton (Stanford Research Institute, Palo Alto 1968)
14.53 W.H.-G. Müller: Thesis, Universität Karlsruhe (1973)
14.54 T.M. Klapwijk, T.B. Veenstra: Phys. Lett. *47*A, 351 (1974)
14.55 C. Guthmann, J. Maurer, M., Belin, J. Bok, A. Libchaber: Phys. Rev. B *11*, 1909 (1975)
14.56 I.K. Yanson: Fiz. Nizk. Temp. *1*, 141 (1975) [Sov. J. Low Temp. Phys. *1*, 67 (1975)]
14.57 V.P. Galaiko, V.M. Dmitriev, G.E. Churilov: Fiz. Nizk. Temp. *2*, 299 (1976) [Sov. J. Low Temp. Phys. *2*, 148 (1976)]
14.58 V.M. Dmitriev, E.V. Khristenko, L.V. Esichko: Fiz. Nizk. Temp. *2*, 318 (1976) [Sov. J. Low Temp. Phys. *2*, 159 (1976)]
14.59 W.W. Webb, R.J. Warburton: Phys. Rev. Lett. *20*, 461 (1968)
14.60 J.D. Meyer, G. Von Minnigerode: Phys. Lett. *38*A, 529 (1972)
14.61 J.D. Meyer: Appl. Phys. *2*, 303 (1973)
14.62 W.J. Skocpol, M.R. Beasley, M. Tinkham: J. Low Temp. Phys. *16*, 145 (1974)
14.63 J.D. Meyer, R. Tidecks: Solid State Commun. *24*, 639 (1977)
14.64 A.B. Pippard, J.G. Shepherd, D.A. Tindall: Proc. Roy. Soc. (London) A *324*, 17 (1971)
14.65 J.D. Meyer, R. Tidecks: Solid State Commun. *18*, 305 (1976)
14.66 T.M. Klapwijk, J.E. Mooij: Phys. Lett. *57*A, 97 (1976)
14.67 G.J. Dolan, L.D. Jackel: Phys. Rev. Lett. *39*, 1628 (1977)
14.68 L. Kramer, A. Baratoff: Phys. Rev. Lett. *38*, 518 (1977)
14.69 V. Ambegaokar: Phys. Rev. Lett. *39*, 235 (1977)
14.70 W.J. Skocpol, M.R. Beasley, M. Tinkham: J. Appl. Phys. *45*, 4054 (1974)
14.71 R.P. Huebener: J. Appl. Phys. *46*, 4982 (1975)

Chapter 15. High Temperature Superconductors

15.1 J.G. Bednorz, K.A. Müller: Z. Phys. B – Condensed Matter **64**, 189 (1986)
15.2 J.G. Bednorz, K.A. Müller: Rev. Mod. Phys. **60**, 585 (1988)
15.3 A. Schilling, M. Cantori, J.D. Guo, H.R. Ott: Nature **363**, 56 (1993)
15.4 J.G. Bednorz, M. Takashige, K.A. Müller: Europhys. Lett. **3**, 379 (1987)
15.5 J.R. Waldram: *Superconductivity of Metals and Cuprates* (Institute of Physics, London 1996)
15.6 C.P. Poole, H.A. Farach, R.J. Creswick: *Superconductivity* (Academic Press, San Diego 1995)
15.7 M. Tinkham: *Introduction to Superconductivity*, Second Edition (McGraw-Hill, New York 1996)

15.8   M. Tinkham, C.J. Lobb: In Solid State Physics, Vol. 42, ed. by H. Ehrenreich, D. Turnbull (Academic Press, Boston 1989) p. 91
15.9   D.J. Scalapino: Physics Reports **250**, 329 (1995)
15.10 C.C. Tsuei, J.R. Kirtley, C.C. Chi, Lock See Yu-Jahnes, A. Gupta, T. Shaw, J.Z. Sun, M.B. Ketchen: Phys. Rev. Lett. **73**, 593 (1994)
15.11 C.C. Tsuei, J.R. Kirtley: Scient. American **275**, 50 (1996)
15.12 D.J. Van Harlingen: Rev. Mod. Phys. **67**, 515 (1995)
15.13 W.E. Lawrence, S. Doniach: Proc. 12th Int. Conf. Low Temperature Physics LT12, E. Kanda, ed. (Academic Press of Japan, Kyoto 1971, p. 361)
15.14 E.H. Brandt: Rep. Prog. Phys. **58**, 1465 (1995)
15.15 J.R. Clem: Phys. Rev. B **43**, 7837 (1991)
15.16 K.H. Fischer: Physica C **178**, 161 (1991)
15.17 C. Caroli, P.G. De Gennes, J. Matricon: Phys. Lett. **9**, 307 (1964)
15.18 J. Bardeen, R. Kümmel, A.E. Jacobs, L. Tewordt: Phys. Rev. **187**, 556 (1969)
15.19 Chr. Bruder: Phys. Rev. B **41**, 4017 (1990)
15.20 C.W.J. Beenakker: Rev. Mod. Phys. **69**, 731 (1997)
15.21 P.G. De Gennes: *Superconductivity of Metals and Alloys* (W.A. Benjamin, New York 1966)
15.22 J.B. Ketterson, S.N. Song: *Superconductivity* (Cambridge University Press 1999)
15.23 D. Rainer, J.A. Sauls, D. Waxman: Phys. Rev. B **54**, 10094 (1996)
15.24 L. Kramer, W. Pesch: Z. Phys. **269**, 59 (1974)
15.25 W. Pesch, L. Kramer: J. Low Temp. Phys. **15**, 367 (1973)
15.26 J. Bardeen, R. Sherman: Phys. Rev. B **12**, 2634 (1975)
15.27 F. Gygi, M. Schlüter: Phys. Rev. B **43**, 7609 (1991)
15.28 N. Hayashi, T. Isoshima, M. Ichioka, K. Machida: Phys. Rev. Lett. **80**, 2921 (1998)
15.29 Yong Ren, Ji-Hai Xu, C.S. Ting: Phys. Rev. Lett. **74**, 3680 (1995)
15.30 J.H. Xu, Y. Ren, C.S. Ting: Phys. Rev. B **53**, R2991 (1996)
15.31 M. Franz, C. Kallin, P.I. Soininen, A.J. Berlinsky, A.L. Fetter: Phys. Rev. B **53**, 5795 (1996)
15.32 N. Enomoto, M. Ichioka, K. Machida: J. Phys. Soc. Japan **66**, 204 (1997)
15.33 N. Schopohl, K. Maki: Phys. Rev. B **52**, 490 (1995)
15.34 M. Ichioka, N. Hayashi, N. Enomoto, K. Machida: Phys. Rev. B **53**, 15316 (1996)
15.35 N. Hayashi, M. Ichioka, K. Machida: Phys. Rev. B **56**, 9052 (1997)
15.36 I. Knezevic, Z. Radovic: in Superconducting and Related Oxides: Physics and Nanoengineering III, SPIE Proceedings (Bellingham 1998, p. 106)
15.37 P.I. Soininen, C. Kallin, A.J. Berlinsky: Phys. Rev. B **50**, 13883 (1994)
15.38 Y. Morita, M. Kohmoto, K. Maki: Phys. Rev. Lett. **78**, 4841 (1997); Europhys. Lett. **40**, 207 (1997)
15.39 M. Franz, Z. Tesanovic: Phys. Rev. Lett. **80**, 4763 (1998)
15.40 Y. Morita, M. Kohmoto, K. Maki: Int.J. Mod. Phys. B **12**, 989 (1998)
15.41 R.P. Huebener, N. Schopohl, G.E. Volovik (eds.): *Microscopic Structure and Dynamics of Vortices in Unconventional Superconductors and Superfluids* (Springer, Berlin 2001)
15.42 D. Xu, S.K. Yip, J.A. Sauls: Phys. Rev. B **51**, 16233 (1995)
15.43 M. Sigrist, T.M. Rice: Rev. Mod. Phys. **67**, 503 (1995)
15.44 J.R. Kirtley, C.C. Tsuei, J.Z. Sun, C.C. Chi, Lock-See Yu-Jahnes, A. Gupta, M. Rupp, M.B. Ketchen: Nature **373**, 225 (1995)
15.45 C.C. Tsuei, J.R. Kirtley, M. Rupp, J.Z. Sun, Lock-See Yu-Jahnes, C.C. Chi, A. Gupta, M.B. Ketchen: J. Phys. Chem. Solids **56**, 1787 (1995)

15.46	C.C. Tsuei, J.R. Kirtley: Rev. Mod. Phys. **72**, 969 (2000)
15.47	K.A. Müller, M. Takashige, J.G. Bednorz: Phys. Rev. Lett. **58**, 1143 (1987)
15.48	Y. Yeshurun, A.P. Malozemoff: Phys. Rev. Lett. **60**, 2202 (1988)
15.49	E.H. Brandt: Int.J. Mod. Phys. B **5**, 751 (1991)
15.50	G. Blatter, M.V. Feigel'man, V.B. Geshkenbein, A.I. Larkin, V.M. Vinokur: Rev. Mod. Phys. **66**, 1125 (1994)
15.51	K.H. Fischer: Superconductivity Review **1**, 153 (1995)
15.52	G.W. Crabtree, W.K. Kwok, U. Welp, D. Lopez, J.A. Fendrich: Proceed. NATO Advanced Study Institute on the Physics and Materials Science of Vortex States, Flux Pinning and Dynamics, S. Bose and R. Kossowski, eds. (Kluwer Academic Publishers 1999)
15.53	E. Zeldov, D. Majer, M. Konczykowski, V.B. Geshkenbein, V.M. Vinokur, H. Shtrikman: Nature **375**, 373 (1995)
15.54	G.W. Crabtree, D.R. Nelson: Physics Today **50**, 38 (1997)
15.55	Z. Hao, J.R. Clem, M.W. Mc Elfresh, L. Civale, A.P. Malozemoff, F. Holtzberg: Phys. Rev. B **43**, 2844 (1991)
15.56	Z. Hao, J.R. Clem: Phys. Rev. Lett. **67**, 2371 (1991)
15.57	A.L. Fetter, P.C. Hohenberg, in *Superconductivity*, R.D. Parks, ed. Vol. 2 (Marcel Dekker, New York 1969, p. 817)
15.58	M. Cyrot: Physik Kondens. Mater. **3**, 374 (1965)
15.59	U. Brandt, W. Pesch, L. Tewordt: Z. Phys. **201**, 209 (1967)
15.60	K. Maki: Phys. Rev. **156**, 437 (1967)
15.61	E. Canel: Phys. Lett. **16**, 101 (1965)
15.62	A.P. Van Gelder: Phys. Rev. **181**, 787 (1969)
15.63	R. Kümmel: Phys. Rev. B **3**, 3787 (1971)
15.64	P. Pöttinger, U. Klein: Phys. Rev. Lett. **70**, 2806 (1993)
15.65	G.E. Volovik: JEPT Lett. **58**, 469 (1993); **65**, 491 (1997)
15.66	S.H. Simon, P.A. Lee: Phys. Rev. Lett. **78**, 1548 (1997)
15.67	K.A. Moler, D.J. Baar, J.S. Urbach, R. Liang, W.N. Hardy, A. Kapitulnik: Phys. Rev. Lett. **73**, 2744 (1994)
15.68	K.A. Moler, D.L. Sisson, J.S. Urbach, M.R. Beasley, A. Kapitulnik, D.J. Baar, R. Liang, W.N. Hardy: Phys. Rev. B **55**, 3954 (1997)
15.69	R.A. Fischer, J.E. Gordon, S.F. Reklis, D.A. Wright, J.P. Emerson, B.F. Woodfield, E.M. Mc Carron III, N.E. Phillips: Physica C **252**, 237 (1995)
15.70	D.A. Wright et al.: J. Low Temp. Phys. **105**, 897 (1996)
15.71	B. Revaz, J.Y. Genoud, A. Junod, K. Neumeier, A. Erb, E. Walker: Phys. Rev. Lett. **80**, 3364 (1998)
15.72	M. Ichioka, N. Enomoto, K. Machida: J. Phys. Soc,. Japan **66**, 3928 (1997)
15.73	M. Ichioka, N. Hayashi, K. Machida: Phys. Rev. B **55**, 6565 (1997)
15.74	M. Ichioka, A. Hasegawa, K. Machida: Phys. Rev. B **59**, 184 (1999)
15.75	M. Ichioka, A. Hasegawa, K. Machida: Phys. Rev. B **59**, 8902 (1999)
15.76	A.S. Mel'nikov: J. Phys.: Condens. Matter **11**, 4219 (1999)
15.77	A.S. Mel'nikov: JETP Lett. **71**, 327 (2000)
15.78	K. Yasui, T. Kita: Phys. Rev. Lett. **83**, 4168 (1999)
15.79	P.W. Anderson, Y.B. Kim: Rev. Mod. Phys. **36**, 39 (1964)
15.80	P.H. Kes, J. Aarts, J. van den Berg, C.J. van der Beek, J.A. Mydosh: Supercond. Sci. Technol. **1**, 242 (1989)
15.81	E.H. Brandt: Physica B **169**, 91 (1991)
15.82	A. Freimuth: in *Selected Topics in Superconductivity*, ed. by L.C. Gupta, M.S. Multani (World Scientific, Singapore 1993) p. 393
15.83	T.T.M. Palstra, B. Batlogg, L.F. Schneemeyer, J.V. Waszczak: Phys. Rev. Lett. **61**, 1662 (1988)

15.84 T.T.M. Palstra, B. Batlogg, R.B. van Dover, L.F. Schneemeyer, J.V. Waszczak: Appl. Phys. Lett. **54**, 763 (1989)
15.85 T.T.M. Palstra, B. Batlogg, R.B. van Dover, L.F. Schneemeyer, J.V. Waszczak: Phys. Rev. B **41**, 6621 (1990)
15.86 R.H. Koch, V. Foglietti, W.J. Gallagher, G. Koren, A. Gupta, M.P.A. Fisher: Phys. Rev. Lett. **63**, 1511 (1989)
15.87 E. Zeldov, N.M. Amer, G. Koren, A. Gupta: Appl. Phys. Lett. **56**, 1700 (1990)
15.88 R. Griessen: Phys. Rev. Lett. **64**, 1674 (1990)
15.89 J. van den Berg, C.J. van der Beek, P.H. Kes, J.A. Mydosh, M.J.V. Menken, A.A. Menovsky: Supercond. Sci. Technol. **1**, 249 (1989)
15.90 Ph. Seng, R. Gross, U. Baier, M. Rupp, D. Koelle, R.P. Huebener, P. Schmitt, G. Saemann-Ischenko, L. Schulz: Physica C **192**, 403 (1992)
15.91 T.T.M. Palstra, B. Batlogg, L.F. Schneemeyer, J.V. Waszcak: Phys. Rev. Lett. **64**, 3090 (1990)
15.92 F. Gollnik, M. Naito: Phys. Rev. B **58**, 11734 (1998)
15.93 W.J. Skocpol, M. Tinkham: Rep. Prog. Phys. **38**, 1049 (1975)
15.94 S. Ullah, A.T. Dorsey: Phys. Rev. Lett. **65**, 2066 (1990)
15.95 S. Ullah, A.T. Dorsey: Phys. Rev. B **44**, 262 (1991)
15.96 U. Welp, S. Fleshler, W.K. Kwok, R.A. Klemm, V.M. Vinokur, J. Downey, B. Veal, G.W. Crabtree: Phys. Rev. Lett. **67**, 3180 (1991)
15.97 W.F. Vinen, A.C. Warren: Proc. Phys. Soc. London **91**, 409 (1967)
15.98 N.B. Kopnin, V.E. Kravtsov: Zh. Eksp. Teor. Fiz. **71**, 1644 (1976) [Sov. Phys. JETP **44**, 861 (1976)]
15.99 N.B. Kopnin, V.E. Kravtsov: Zh. Eksp. Teor. Fiz. Pis'ma Red. **23**, 631 (1976) [JETP Lett. **23**, 578 (1976)]
15.100 N.B. Kopnin, M.M. Salomaa: Phys. Rev. B **44**, 9667 (1991)
15.101 G. Blatter, B.I. Ivlev: Phys. Rev. B **50**, 10272 (1994)
15.102 A. van Otterlo, M. Feigelman, V. Geshkenbein, G. Blatter: Phys. Rev. Lett. **75**, 3736 (1995)
15.103 M. Golosovsky, M. Tsindlekht, D. Davidov: Supercond. Sci. Technol. **9**, 1 (1996)
15.104 Y. Matsuda, N.P. Ong, Y.F. Yan, J.M. Harris, J.B. Peterson: Phys. Rev. B **49**, 4380 (1994)
15.105 J.M. Harris, Y.F. Yan, O.K.C. Tsui, Y. Matsuda, N.P. Ong: Phys. Rev. Lett. **73**, 1711 (1994)
15.106 N.B. Kopnin, V.M. Vinokur: Phys. Rev. Lett. **83**, 4864 (1999)
15.107 A. Virosztek, J. Ruvalds: Phys. Rev. B **42**, 4064 (1990); **45**, 347 (1992)
15.108 C.C. Tsuei, A. Gupta, G. Koren: Physica C **161**, 415 (1989)
15.109 D.N. Newns, C.C. Tsuei, P.C. Pattnaik, C.L. Kane: Comments Condens. Matter Phys. **15**, 273 (1992)
15.110 C.T. Rieck, W.A. Little, J. Ruvalds, A. Virosztek: Phys. Rev. B **51**, 3772 (1995)
15.111 O.M. Stoll, R.P. Huebener, S. Kaiser, M. Naito: J. Low Temp. Phys. **118**, 59 (2000)
15.112 R.P. Huebener, O.M. Stoll, A. Wehner, M. Naito: Physica C **338**, 221 (2000)
15.113 G.E. Volovik: Zh. Eksp. Teor. Fiz. **104**, 3070 (1993) [Sov. Phys. JETP **77**, 435 (1993)]; Pis'ma Zh. Eksp. Teor. Fiz. **57**, 233 (1993) [JETP Lett. **57**, 244 (1993)]
15.114 N.B. Kopnin, G.E. Volovik, Ü. Parts: Europhys. Lett. **32**, 651 (1995)
15.115 M. Stone: Phys. Rev. B **54**, 13222 (1996)
15.116 G. Blatter, V.B. Geshkenbein, N.B. Kopnin: Phys. Rev. B **59**, 14663 (1999)

15.117 N.B. Kopnin: Physica B280, 231 (2000)
15.118 G.E. Volovik: Pis'ma Zh. Eksp. Teor. Fiz. **62**, 58 (1995) [JETP Lett. **62**, 65 (1995)]
15.119 E.B. Sonin: Phys. Rev. B **55**, 485 (1997)
15.120 S.J. Hagen, C.J. Lobb, R.L. Greene, M.G. Forrester, J.H. Kang: Phys. Rev. B **41**, 11630 (1990)
15.121 S.J. Hagen, C.J. Lobb, R.L. Greene, M. Eddy: Phys. Rev. B **43**, 6246 (1991)
15.122 S.J. Hagen, A.W. Smith, M. Rajeswari, J.L. Peng, Z.Y. Li, R.L. Greene, S.N. Mao, X.X. Xi, S. Bhattacharya, Qi Li, C.J. Lobb: Phys. Rev. B **47**, 1064 (1993)
15.123 M.N. Kunchur, D.K. Kristen, C.E. Klabunde, J.M. Phillips: Phys. Rev. Lett. **72**, 2259 (1994)
15.124 D.A. Beam, N.-C. Yeh, R.P. Vasquez: Phys. Rev. B **60**, 601 (1999)
15.125 T. Nagaoka, Y. Matsuda, H. Obara, A. Sawa, T. Terashima, I. Chong, M. Takano, M. Suzuki: Phys. Rev. Lett. **80**, 3594 (1998)
15.126 A.T. Dorsey: Phys. Rev. B **46**, 8376 (1992)
15.127 J. Luo, T.P. Orlando, J.M. Graybeal, X.D. Wu, R. Muenchausen: Phys. Rev. Lett. **68**, 690 (1992)
15.128 W.N. Kang, Wan-Seon Kim, S.J. Oh, Sung-Ik Lee, D.H. Kim, C.H. Choi, H.-C. Ri, C.W. Chu: cond-mat/9905201 preprint
15.129 A.T. Dorsey, M.P.A. Fisher: Phys. Rev. Lett. **68**, 694 (1992)
15.130 V.M. Vinokur, V.B. Geshkenbein, M.V. Feigel'man, G. Blatter: Phys. Rev. Lett. **71**, 1242 (1993)
15.131 R.P. Huebener, F. Gollnik: Phys. Rev. B **57**, 13393 (1998)
15.132 R.P. Huebener: Physica C **168**, 605 (1990)
15.133 G.Yu. Logvenov, V.V. Ryazanov, A.V. Ustinov, R.P. Huebener: Physica C **175**, 179 (1991)
15.134 G.Yu. Logvenov, M. Hartmann, R.P. Huebener: Phys. Rev. B **46**, 11102 (1992)
15.135 A.I. Larkin, Yu.N. Ovchinnikov: Z. Eksp. Teor. Fiz. **68**, 1915 (1975) [Sov. Phys. JETP **41**, 960 (1976)]
15.136 L.E. Musienko, I.M. Dmitrenko, V.G. Volotskaya: Pis'ma Zh. Eksp. Teor. Fiz. **31**, 603 (1980) [JETP Lett. **31**, 567 (1980)]
15.137 W. Klein, R.P. Huebener, S. Gauss, J. Parisi: J. Low Temp. Phys. **61**, 413 (1985)
15.138 S.G. Doettinger, R.P. Huebener, R. Gerdemann, A. Kühle, S. Anders, T.G. Träuble, J.C. Villegier: Phys. Rev. Lett. **73**, 1691 (1994)
15.139 S.G. Doettinger, S. Kittelberger, R.P. Huebener, C.C. Tsuei: Phys. Rev. B **56**, 14157 (1997)
15.140 Z.L. Xiao, P. Ziemann: Phys. Rev. B **53**, 15265 (1996)
15.141 Z.L. Xiao, E.Y. Andrei, P. Ziemann: Phys. Rev. B **58**, 11185 (1998)
15.142 Z.L. Xiao, P. Voss-de Haan, G. Jakob, Th. Kluge, P. Haibach, H. Adrian, E.Y. Andrei: Phys. Rev. B **59**, 1481 (1999)
15.143 S.G. Doettinger, R.P. Huebener, A. Kühle: Physica C **251**, 285 (1995)
15.144 S.G. Doettinger, R.P. Huebener: Chinese J. Phys. **34**, 527 (1996)
15.145 A.I. Larkin, Yu.N. Ovchinnikov: Pis'ma Zh. Eksp. Teor. Fiz. **23**, 210 (1976) [JETP Lett. **23**, 187 (1976)]
15.146 A.I. Larkin, Yu.N. Ovchinnikov: Zh. Eksp. Teor. Fiz. **73**, 299 (1977) [Sov. Phys. JETP **46**, 155 (1977)]
15.147 O.M. Stoll, S. Kaiser, R.P. Huebener, M. Naito: Phys. Rev. Lett. **81**, 2994 (1998)

15.148 O.M. Stoll, R.P. Huebener, S. Kaiser, M. Naito: Phys. Rev. B **60**, 12424 (1999)
15.149 R.P. Huebener, S. Kaiser, O.M. Stoll: Europhys. Lett. **44**, 772 (1998)
15.150 U. Yaron, P.L. Gammel, D.A. Huse, R.N. Kleiman, C.S. Oglesby, E. Bucher, B. Batlogg, D.J. Bishop, K. Mortensen, K. Clausen, C.A. Bolle, F. De La Cruz: Phys. Rev. Lett. **73**, 2748 (1994)
15.151 A. Duarte, E.F. Righi, C.A. Bolle, F. De La Cruz, P.L. Gammel, C.S. Oglesby, B. Bucher, B. Batlogg, D.J. Bishop: Phys. Rev. B **53**, 11336 (1996)
15.152 F. Pardo, F. De La Cruz, P.L. Gammel, C.S. Oglesby, E. Bucher, B. Batlogg, D.J. Bishop: Phys. Rev. Lett. **78**, 4633 (1997)
15.153 A. Van Otterlo, R.T. Scalettar, G.T. Zimanyi, R. Olsson, A. Petrean, W. Kwok, V.M. Vinokur: Phys. Rev. Lett. 84, 2493 (2000)
15.154 D. Lopez, W.K. Kwok, H. Safar, R.J. Olsson, A.M. Petrean, L. Paulius, G.W. Crabtree: Phys. Rev. Lett. **82**, 1277 (1999)
15.155 G.W. Crabtree, D. Lopez, W.K. Kwok, H. Safar, L.M. Paulius: J. Low Temp. Phys. **117**, 1313 (1999)
15.156 A.C. Mota, A. Pollini, P. Visani, K.A. Müller, J.G. Bednorz: Phys. Rev. B **36**, 4011 (1987)
15.157 A.C. Mota, G. Juri, P. Visani, A. Pollini, T. Teruzzi, K. Aupke, B. Hilty: Physica C **185**, 343 (1991)
15.158 L. Fruchter, A.P. Malozemoff, I.A. Campbell, J. Sanchez, M. Konczykowski, R. Griessen, F. Holtzberg: Phys. Rev. B **43**, 8709 (1991)
15.159 A.C. Mota: in *The Vortex State*, N. Bontemps et al. (eds.) (Kluwer Academic Publishers 1994) p. 265
15.160 L. Fruchter, D. Prost, I.A. Campbell: Physica 235-40C, 249 (1994)
15.161 A.O. Caldeira, A.J. Leggett: Phys. Rev. Lett. **46**, 211 (1981); Ann. Phys. (N.Y.) **149**, 374 (1983)
15.162 A.I. Larkin, Yu.N. Ovchinnikov: JETP Lett. **37**, 382 (1983)
15.163 G. Blatter, V.B. Geshkenbein, V.M. Vinokur: Phys. Rev. Lett. **66**, 3297 (1991)
15.164 A.F.Th. Hoekstra, R. Griessen, A.M. Testa, J. el Fattahi, M. Brinkmann, K. Westerholt, W.K. Kwok, G.W. Crabtree: Phys. Rev. Lett. **80**, 4293 (1998)
15.165 A.F.Th. Hoekstra, A.M. Testa, G. Doornbos, J.C. Martinez, B. Dam, R. Griessen, B.I. Ivlev, M. Brinkmann, K. Westerholt, W.K. Kwok, G.W. Crabtree: Phys. Rev. B **59**, 7222 (1999)
15.166 R.P. Huebener, F. Kober, H.-C. Ri, K. Knorr, C.C. Tsuei, C.C. Chi, M.R. Scheuermann: Physica C **181**, 345 (1991)
15.167 H.-C. Ri, R. Gross, F. Gollnik, A. Beck, R.P. Huebener, P. Wagner, H. Adrian: Phys. Rev. B **50**, 3312 (1994)
15.168 R.P. Huebener, R. Gross, H.-C. Ri, F. Gollnik: Physica B **197**, 588 (1994)
15.169 R.P. Huebener: Supercond. Sci. Technol. **8**, 189 (1995)
15.170 V.L. Ginzburg: Zh. Eksp. Teor. Fiz. **14**, 177 (1944) [J. Phys. USSR **8**, 148 (1944)]
15.171 V.L. Ginzburg: JETP Lett. **49**, 58 (1989)
15.172 D.J. Van Harlingen: Physica B **109–110**, 1710 (1982)
15.173 A.V. Ustinov, M. Hartmann, R.P. Huebener: Europhys. Lett. **13**, 175 (1990)
15.174 R.P. Huebener, A.V. Ustinov, V.K. Kaplunenko: Phys. Rev. B **42**, 4831 (1990)
15.175 H.-C. Ri, F. Kober, A. Beck, L. Alff, R. Gross, R.P. Huebener: Phys. Rev. B **47**, 12312 (1993)

15.176 V.K. Kaplunenko, V.V. Ryazanov, V.V. Schmidt: Zh. Eksp. Teor. Fiz. **89**, 1389 (1985) [Sov. Phys. JETP **62**, 804 (1985)]
15.177 R.P. Huebener, F. Kober, R. Gross, H.-C. Ri: Physica C **185–189**, 349 (1991)
15.178 N.V. Zavaritsky, A.V. Samoilov, A.A. Yurgens: Physica C **180**, 417 (1991)
15.179 H.-C. Ri, F. Kober, R. Gross, R.P. Huebener, A. Gupta: Phys. Rev. B **43**, 13739 (1991)
15.180 F. Kober, H.-C. Ri, R. Gross, D. Koelle, R.P. Huebener, A. Gupta: Phys. Rev. B **44**, 11951 (1991)
15.181 A.V. Samoilov, A.A. Yurgens, N.V. Zavaritsky: Phys. Rev. B **46**, 6643 (1992)
15.182 F. Gollnik, R.P. Huebener: Phys. Rev. B **57**, 7495 (1998)
15.183 G. Deutscher, K.A. Müller: Phys. Rev. Lett. **59**, 1745 (1987)
15.184 C.J. van der Beek, P.H. Kes: Phys. Rev. B **43**, 13032 (1991)
15.185 P.H. Kes: Physica C **185–189**, 288 (1991)
15.186 R. Wördenweber: Rep. Prog. Phys. **62**, 187 (1999)
15.187 Y. Yeshurun, A.P. Malozemoff, A. Shaulov: Rev. Mod. Phys. **68**, 911 (1996)
15.188 A.I. Larkin, Yu.N. Ovchinnikov: J. Low Temp. Phys. **34**, 409 (1979)
15.189 M.P.A. Fisher: Phys. Rev. Lett. **62**, 1415 (1989)
15.190 D.S. Fisher, M.P.A. Fisher, D.A. Huse: Phys. Rev. B **43**, 130 (1991)
15.191 R.H. Koch, V. Foglietti, M.P.A. Fisher: Phys. Rev. Lett. **64**, 2586 (1990)
15.192 P.L. Gammel, L.F. Schneemeyer, D.J. Bishop: Phys. Rev. Lett. **66**, 953 (1991)
15.193 T. Nattermann, S. Scheidl: Adv. in Phys. **49**, 607 (2000)
15.194 V.V. Moshchalkov, V. Bruyndoncx, L. Van Look, M.J. Van Bael, Y. Bruynseraede, A. Tonomura: *Handbook of Nanostructured Materials and Nanotechnology*, ed. by H.S. Nalwa (Academic Press, San Diego, 2000) p. 451
15.195 C. Reichhardt, C.J. Olson, F. Nori: Phys. Rev. B **58**, 6534 (1998)
15.196 C. Reichhardt, J. Groth, C.J. Olson, S.B. Field, F. Nori: Phys. Rev. B **54**, 16108 (1996)
15.197 C.H. Olson, C. Reichhardt, F. Nori: Phys. Rev. B **56**, 6175 (1997)
15.198 C. Reichhardt, C.J. Olson, F. Nori: Phys. Rev. B **61**, 3665 (2000)
15.199 M.V. Indenbom, V.I. Nikitenko, A.A. Polyanskii, V.K. Vlasko-Vlasov: Cryogenics **30**, 747 (1990)
15.200 M.R. Koblischka, R.J. Wijngaarden: Supercond. Sci. Technol. **8**, 199 (1995)
15.201 V.K. Vlasko-Vlasov, G.W. Crabtree, U. Welp, V.I. Nikitenko: in *Physics and Materials Science of Vortex States, Flux Pinning and Dynamics*, ed. by R. Kossowsky et al., NATO ASI, Ser. E, Vol. 356 (Kluwer, Dordrecht 1999), p. 205
15.202 A.A. Polyanskii, X.Y. Cai, D.M. Feldmann, D.C. Larbalestier: in *NATO Science Series 3. High Technology*, Vol. 72, eds., I. Nedkov, M. Ausloos (Kluwer Academic, Netherlands 1999) p. 353
15.203 D. Giller, A. Shaulov, T. Tamegai, Y. Yeshurun: Phys. Rev. lett. 84, 3698 (2000)
15.204 E. Zeldov, A.I. Larkin, V.B. Geshkenbein, M. Konczykowski, D. Majer, B. Khaykovich, V.M. Vinokur, H. Shtrikman: Phys. Rev. Lett. **73**, 1428 (1994)
15.205 Y. Abulafia, A. Shaulov, Y. Wolfus, R. Prozorov, L. Burlachkov, Y. Yeshurun, D. Majer, E. Zeldov, V.M. Vinokur: Phys. Rev. Lett. **75**, 2404 (1995)

15.206 D. Giller, A. Shaulov, R. Prozorov, Y. Abulafia, Y. Wolfus, L. Burlachkov, Y. Yeshurun, E. Zeldov, V.M. Vinokur, J.L. Peng, R.L. Greene: Phys. Rev. Lett. **79**, 2542 (1997)
15.207 G. Binnig, H. Rohrer: Rev. Mod. Phys. **59**, 615 (1987)
15.208 A. de Lozanne: Supercond. Sci. Technol. **12**, R43 (1999)
15.209 J.R. Kirtley, M.B. Ketchen, K.G. Stawiasz, J.Z. Sun, W.J. Gallagher, S.H. Blanton, S.J. Wind: Appl. Phys. Lett. **66**, 1138 (1995)
15.210 H.F. Hess, R.B. Robinson, R.C. Dynes, J.M. Valles, J.V. Waszczak: Phys. Rev. Lett. **62**, 214 (1989)
15.211 H.F. Hess, R.B. Robinson, J.V. Waszczak: Phys. Rev. Lett. **64**, 2711 (1990)
15.212 J.D. Shore, M. Huang, A.T. Dorsey, J.P. Sethna: Phys. Rev. Lett. **62**, 3089 (1989)
15.213 S. Ullah, A.T. Dorsey, L.J. Buchholtz: Phys. Rev. B **42**, 9950 (1990)
15.214 F. Gygi, M. Schlüter: Phys. Rev. Lett. **65**, 1820 (1990)
15.215 N. Hayashi, M. Ichioka, K. Machida: Phys. Rev. Lett. **77**, 4074 (1996)
15.216 H.F. Hess, C.A. Murray, J.V. Waszczak: Phys. Rev. Lett. **69**, 2138 (1992); Phys. Rev. B **50**, 16528 (1994)
15.217 H.F. Hess: in *Methods of Experimental Physics*, ed. by J.A. Stroscio, W.A. Kaiser (Academic Press, New York 1993), Vol. 27, p. 427
15.218 Ch. Renner, A.D. Kent, Ph. Niedermann, O. Fischer, F. Levy: Phys. Rev. Lett. **67**, 1650 (1991)
15.219 I. Maggio-Aprile, Ch. Renner, A. Erb, E. Walker, O. Fischer: Phys. Rev. Lett. **75**, 2754 (1995)
15.220 Ch. Renner, B. Revaz, K. Kadowaki, I. Maggio-Aprile, O. Fischer: Phys. Rev. Lett. **80**, 3606 (1998)
15.221 C.W. Yuan, Z. Zheng, A.L. de Lozanne, M. Tortonese, D.A. Rudman, J.N. Eckstein: J. Vac. Sci. Technol. B **14**, 1210 (1996)
15.222 A. Volodin, K. Temst, C. Van Haesendonck, Y. Bruynseraede: Appl. Phys. Lett. **73**, 1134 (1998)
15.223 S. Keil, R. Straub, R. Gerber, R.P. Huebener, D. Koelle, R. Gross, K. Barthel: IEEE Trans. Appl. Supercond. **9**, 2961 (1999)
15.224 K. Harada, T. Matsuda, J. Bonevich, M. Igarashi, S. Kondo, G. Pozzi, U. Kawabe, A. Tonomura: Nature **360**, 51 (1992)
15.225 T. Matsuda, K. Harada, H. Kasai, O. Kamimura, A. Tonomura: Science **271**, 1393 (1996)
15.226 K. Harada, O. Kamimura, H. Kasai, T. Matsuda, A. Tonomura, V.V. Moshchalkov: Science **274**, 1167 (1996)
15.227 K. Harada, T. Matsuda, H. Kasai, J.E. Bonevich, T. Yoshida, U. Kawabe, A. Tonomura: Phys. Rev. Lett. **71**, 3371 (1993)
15.228 A. Tonomura, H. Kasai, O. Kamimura, T. Matsuda, K. Harada, J. Shimoyama, K. Kishio, K. Kitazawa: Nature **397**, 308 (1999)
15.229 S. Hasegawa, T. Matsuda, J. Endo, N. Osakabe, M. Igarashi, T. Kobayashi, M. Naito, A. Tonomura, R. Aoki: Phys. Rev. B **43**, 7631 (1991)
15.230 J.E. Bonevich, K. Harada, T. Matsuda, H. Kasai, T. Yoshida, G. Pozzi, A. Tonomura: Phys. Rev. Lett. **70**, 2952 (1993)

Index

Abrikosov solution of the GL equations, 50–55
Absorption coefficient of the magneto-optical material, 104,105
Andreev bound states, 238, 244
Anisotropy of the domain structure, 26
Anisotropy of the superconductor,
– influence on the flux-line lattice, 75–78
Anomalous skin effect, 11
Antidots, 270
Atomic beam technique, 119
Attractive interaction between flux lines, 87–93
Autocorrelation function, 200–205, 226

Bardeen–Stephen model, 165, 252
Bean model, 180, 181
Bitter method, 59, 100
Bloch frequency, 259
Bloch oscillation of the quasiparticles, 259
Bohm–Aharonov effect, 196
Bohr–Sommerfeld quantum condition, 45
Branching in the intermediate-mixed state, 88
Branching model, 18
Bridgman relation, 160
Broadening of the resistive transition, 249

Clem model, 69–72
Cluster precipitation, 193
Coherence length, 9
– dependence on electron mean free path, 12
– temperature dependence, 38
– experimental values, 10, 235
Collective pinning, 269
Columnar defects, 270

Composite conductor, 199
Condensation energy, 4
Constricted geometry, 152, 221–227
Cooper pair, 36, 139
Corbino disk geometry, 128, 208, 260
Core resistivity, 257
Correlation between vortex lines, 205
Coupling parameter, 149
Creep rate, 195
– constant, 197
Critical current, 125, 127, 221
– enhancement, 221
– enhancement due to surface energy barrier, 85
– for solid cylinder, 183, 184, 211
– from GL theory, 41, 228
Critical current density, 246
Critical entry field, 84, 220
Critical field H_c (thermodynamic critical field), 5, 58
– H_{c1}, 58, 65
– H_{c2}, 47, 58
– H_{c3}, for surface superconductivity, 49
Critical perpendicular field $H_{c\perp}$, 96, 97
Critical radius of vortex-line ring, 183
Critical state, 146, 179–181
Critical temperature gradient, 156, 158
Critical thickness for the vortex
– state of thin films, 96–98
– experimental values, 98
Critical thickness for branching and nonbranching behavior, 19
Critical vortex velocity, 255, 256
Cross correlation, 210
Crystal lattice, influence on flux-line lattice, 75–78, 88, 113
Current-induced resistive state, 152, 211–233
– thin film geometry, 215–227
– wire geometry, 211–215
Cyclotron frequency, 46, 168

$d_{x^2-y^2}$ pairing symmetry, 236
d-wave symmetry, 239, 240, 244
Damping coefficient, 255
Damping force, 122, 123, 155, 251
DC transformer, 149
Defects in the flux-line lattice, 78–82
Demagnetization, 13, 131, 132, 175, 220, 221
– coefficient, 14
Diamagnetism, 5
Diffusion constant, 173
"Dirty" limit, 239
Dislocation in the flux-line lattice, 80–82
Dissipation of energy, 167
Distribution of supercurrent density, 158
Domain pattern, path dependence, 26, 30
Domain patterns, 21–32
Domain structure, 13
Doppler shift, 119, 146
Dynamic correlation, 244, 260
Dynamic pinning force, 129
Dynamic structure factor, 201

Eddy current damping, 123, 131, 138
– screening, 132
Edge dislocation in the flux-line lattice, 80, 81
Effective damping coefficient, 252
Effective quasiparticle temperature, 257
Eilenberger equations, 74, 239
Einstein relation, 139
Elastic distortions of the flux-line lattice, 129, 185
Electron doped cuprate superconductor $Nd_{2-x}Ce_xCuO_{4\pm y}$, 249, 254
Electron–electron interaction, 252
Electron holography, 274
Electron-optical method for flux detection, 119, 120
Electrothermal conductivity, 268
Energy barrier of the flux line, 83–86, 187, 220
Ettinghausen coefficient, 159
Ettinghausen effect, 136, 158, 249
Europium compounds for magneto-optical flux detection, 103, 104

Faraday rotation, 102, 103
Faraday's law, 124, 138
First order phase transition, 87, 88, 91

Floating magnet model, 154
Fluctuation effects, 249, 266
Flux creep, 195–198, 247
Flux flow, 114, 125, 158, 247
– experimental methods, 144–154
Flux-flow instabilities, 254–260
Flux flow noise, 209, 210
Flux-flow resistance, 124–128, 167
– (from TDGL), 174
Flux-flow resistivity minimum, 128
Flux jump, 198, 199
Flux-line lattice, 75–82
– anisotropy, 75–78
– orientation, 75–78, 113
– symmetry, 75–78
Fluxoid, 44
Fluxoid quantum, 45
Flux pinning, 268
Flux quantum, 15
Flux spots, 20, 25
Flux-tube lattice, 26
Flux-tube train, 216, 217
Force-free configuration, 143
Form factor, 71, 112
Fountain effect, 262
Free-energy-density, 4
Free energy expansion, 34, 35
Frequency modulated nucleation rate, 226

Gap parameter, 34, 57
Gapless superconductor, 170
Gauge invariance, 172
Generalized force, 171
Ginzburg–Landau equations, 37
Ginzburg–Landau parameter, 39, 174, 235
Ginzburg–Landau theory, 33–57
GLAG theory, 57, 73
– extensions, 73
Goren–Tinkham model, 20, 25
Granular aluminum film, 151, 152, 191, 192
Growth process, 30

Half-integer magnetic flux quanta, 240
Hall angle, 122, 125, 135, 138, 155, 157, 168, 169, 252, 267
– (from TDGL), 177
Hall effect, 134–136, 251–254, 267
Hall probe, 110
Hall probe array, 271
Hall sensors, 243
Hall voltage, 124

Hao–Clem interpolation scheme, 243, 249, 266
Harmonic oscillator, 48
Healing length, 229, 230
Heat transfer coefficient, 231, 232
Helicon resonance, 134
High-temperature superconductors, 235
Honeycomb domain structure, 24
Hopping process, 210
Hotspot model, 231–233
Hysteresis of the electrical resistance, 129, 130, 258

Incremental entropy, 163
Inelastic scattering time, 230
Inhomogeneous state, 142
Instability
 – of current-density distribution, 214, 219
 – of the flux-flow state, 142, 254–260
 – thermal, 141, 198
Intermediate state, 13–16
Intermediate-mixed state, 87–93
Interstitial flux line, 78–80
Irreversibility line, 241
Isolated vortex line, 61

Josephson frequency, 259
Josephson junction, 139, 192
Josephson relation, 138, 204, 218, 226
Josephson steps, 226, 227
Josephson strings, 237

Kim model, 180, 181
Kinetic energy of the supercurrents, 36, 94
Kink instability, 215
Kramer–Pesch effect, 239, 257, 258
Kutta–Joukowski relation, 123

Laminar Landau model, 16, 24
Landau domain theory, 16–20
Landau level, 46
Larkin–Ovchinnikov theory, 255, 256
Lindemann criterion, 242
Lindemann number, 242
Little–Parks experiment, 46
London equations, 7
London model, 61–68
London theory, 5–7
Lorentz force, 66, 122, 245
Lorentz microscopy, 273
Low-temperature scanning electron microscopy, 273

Macroscopic superlattice, 88
Magnetic coupling, 148–152, 216
Magnetic force microscopy, 273
Magnetization
 – ballistic method, 116
 – electronic integration, 116, 117
 – London model, 67
 – measurement, 114–118, 181
Magneto-optical flux detection, 102–109, 144, 271
 – interference condition, 106, 107
 – time resolution, 144
Magnetoresistive field probe, 110, 152, 153, 223
Magnus force, 122, 123, 134, 155, 253
Mass tensor, 236
Matching condition, 187, 191–193
Maxwell stress, 122
Meandering domain structure, 24
Meissner effect, 4
Melting of the vortex lattice, 243, 248
Metastable resistance level, 214
Microbridge, 228–231
Micro field probe, 152
Microstructure, metallurgical, 21, 191, 217, 223
Microwave surface resistance, 86, 127, 134
Minigap, 238
Mixed state, 58
Moderately clean limit, 239
Modulation
 – of composition, 192
 – of the surface, 191
 – of thickness, 191
Monocrystalline samples, 26–29, 75, 88
Mosaic spread of the flux-line lattice, 113
Motion picture technique, 144
Multiple-quantum vortex, 192

Nascent vortices, 86
Negative differential resistance, 220, 232, 260
Nernst coefficient, 159
Nernst effect, 138, 156, 261, 265, 268
Neutron diffraction, 59, 72, 75, 91, 110–114, 125, 145–147
 – small-angle scattering apparatus, 111
Neutron spin-echo technique, 146
Noise power, 140, 200–210, 219
Nonbranching model, 18

Nonlocality of superconductivity, 9
Nuclear magnetic resonance, 59, 118, 145
Nucleation process, 30, 223
Nucleation of superconductivity
- at the surface, 48
- in the interior, 46
Nucleation of vortex lines at the surface, 86, 87
Nucleation rate of flux tubes, 152–154, 218, 226
Nucleation spectrum of flux tubes, 223–226

Onsager reciprocity relations, 138, 156
Order parameter, 33
Oxygen vacancies, 248, 268, 269

Pancake vortices, 236, 241
Paramagnetic effect, 128
Paramagnetic Meissner effect, 240
Paramagnetic moment (longitudinal), 143
Path dependence of domain pattern, 26, 29, 30
Peak effect, 127
Peltier effect, 136, 158, 254
Penetration depth, 5
- experimental values, 10, 236
- field dependence, 70
- London value, 6
- temperature dependence, 7, 8, 39
Periodicity length, 16–18, 24
Perovskites, 235
Phase of the order parameter, 36
Phase-slip center, 228–231
Phase slippage, 231
π-junction, 240
Pinning force, 122, 123, 129, 147, 155, 158, 179–194, 210
- core interaction, 186
- elastic interaction, 189
- from Ginzburg-Landau free energy, 189
- magnetic interaction, 187
- maximum value, 187
- summation, 184, 185
Pinning potential well, 245, 247
Polarized muons, spin precession, 119
Power spectrum, 200–205, 219, 225, 226
Pseudogap, 273
Pseudo wave function, 33, 139

Quantum tunneling of vortices, 261
Quasiclassical approximation, 240, 243
Quasiclassical limit, 253
Quasiparticle diffusion coefficient, 256
Quasiparticle diffusion length, 229, 230
Quasiparticle energy relaxation, 256
Quasiparticle energy relaxation length, 257
Quasiparticle scattering time, 252

Rectification effect, 191
Relaxation oscillation, 229, 232
Relaxation time for flux penetration, 131, 133
Righi-Leduc effect, 158
Rocking curve, 113–115, 146

s-wave pairing symmetry, 236, 239, 244
Scanning probe microscopy, 271
Scanning tunneling microscopy, 272
Schottky relation, 204
Second order phase transition, 47, 59
- Landau theory, 33, 35
Second-sound generation, 137
Seebeck coefficient, 254
Seebeck effect, 138, 156, 261, 262, 268
Sharvin geometry, 24, 129
Sharvin point contact, 147, 158
Shot noise, 202
Shubnikov phase, 58
Silsbee's rule, 85, 211
Simulation experiments, 154
Skin effect, 131, 133
Sound barrier of flux-flow speed, 141
Specific heat, 244
Spectral flow, 253
Spin-flip scattering time, 173
Square vortex lattice, 52, 53, 56
SQUID microscope, 240, 272
Stability criterion, 198, 199
Stroboscope for magneto-optical flux detection, 144, 145, 222–224
Subbands, 244, 252, 258
"Superclean" limit, 239
Superconducting domain flow, 130, 164
Supercooling, 47
Supercurrent backflow, 263
Supercurrent density, 6
Supercurrent velocity, 36
Surface resistance, 134
Surface superconductivity, 50
Synchronization of phase slippage, 222

Thermal diffusion of magnetic flux quanta, 265
Thermal diffusion of quasiparticles, 262
Thermal effects, 231
Thermal force, 155, 261, 265
Thermal instability, 141, 198
Thermally assisted flux flow, 245
Thin films, 94–99
– vortex state, 94–98
Time-dependent Ginzburg-Landau equations, 173
Time-dependent Ginzburg-Landau theory, 170–178
Topology of domains, 30
Transport energy, 265
Transport entropy, 136, 138, 156, 160–164, 265
– (from TDGL), 177
Transport equation, 171
Träuble-Essmann technique, 101, 193
Triangular vortex lattice, 52, 53, 56, 58
Two-dimensional mixed state, 214
Type-I superconductor, 15, 41, 47
– magnetic properties, 4–32
Type-II superconductor, 15, 41, 47, 58–93

Uncertainty principle, 9
Universal scaling, 269
Universal scaling function, 249, 267
Usadel equations, 74

Vacancy of flux-line, 78, 79
Verdet constant, 104
Vibrating-specimen magnetometer, 117, 118
Viscosity coefficient, 128, 167
Voltage criterion, 127
Voltage pulse, 140, 202–208
– dependence on geometry of contacts, 205–208
Voltage step, 226–228
Voltage, time-averaged value, 141
Vortx–antivortex pair, 237
Vortex core, local approximation, 165, 166
Vortex glass, 243, 248, 269
Vortex glass-melting temperature, 269
Vortex imaging, 271
Vortex lattice melting, 243, 248
Vortex line energy, 64
Vortex line interaction, 65
Vortex liquid, 243
Vortex matter, 241, 242
Vortex state, 50–57

Wall energy, 14
Wall-energy parameter, 15, 40
– experimental values, 10
Weak link, 139, 140, 228
Wiener–Khintchine theorem, 200
Wohlleben effect, 240

Zener breakdown, 260

Printing: Mercedes-Druck, Berlin
Binding: Stürtz AG, Würzburg

Springer Series in Solid-State Sciences
Editors: M. Cardona   P. Fulde   K. von Klitzing   H.-J. Queisser

1 **Principles of Magnetic Resonance**
  3rd Edition   By C. P. Slichter
2 **Introduction to Solid-State Theory**
  By O. Madelung
3 **Dynamical Scattering of X-Rays in Crystals**   By Z. G. Pinsker
4 **Inelastic Electron Tunneling Spectroscopy**
  Editor: T. Wolfram
5 **Fundamentals of Crystal Growth I**
  Macroscopic Equilibrium and Transport Concepts
  By F. E. Rosenberger
6 **Magnetic Flux Structures in Superconductors**
  2nd Edition   By R. P. Huebener
7 **Green's Functions in Quantum Physics**
  2nd Edition   By E. N. Economou
8 **Solitons and Condensed Matter Physics**
  Editors: A. R. Bishop and T. Schneider
9 **Photoferroelectrics**   By V. M. Fridkin
10 **Phonon Dispersion Relations in Insulators**   By H. Bilz and W. Kress
11 **Electron Transport in Compound Semiconductors**   By B. R. Nag
12 **The Physics of Elementary Excitations**
   By S. Nakajima, Y. Toyozawa, and R. Abe
13 **The Physics of Selenium and Tellurium**
   Editors: E. Gerlach and P. Grosse
14 **Magnetic Bubble Technology**   2nd Edition
   By A. H. Eschenfelder
15 **Modern Crystallography I**
   Fundamentals of Crystals
   Symmetry, and Methods of Structural Crystallography
   2nd Edition
   By B. K. Vainshtein
16 **Organic Molecular Crystals**
   Their Electronic States   By E. A. Silinsh
17 **The Theory of Magnetism I**
   Statics and Dynamics
   By D. C. Mattis
18 **Relaxation of Elementary Excitations**
   Editors: R. Kubo and E. Hanamura
19 **Solitons**   Mathematical Methods for Physicists
   By. G. Eilenberger
20 **Theory of Nonlinear Lattices**
   2nd Edition   By M. Toda
21 **Modern Crystallography II**
   Structure of Crystals   2nd Edition
   By B. K. Vainshtein, V. L. Indenbom, and V. M. Fridkin
22 **Point Defects in Semiconductors I**
   Theoretical Aspects
   By M. Lannoo and J. Bourgoin
23 **Physics in One Dimension**
   Editors: J. Bernasconi and T. Schneider
24 **Physics in High Magnetics Fields**
   Editors: S. Chikazumi and N. Miura
25 **Fundamental Physics of Amorphous Semiconductors**   Editor: F. Yonezawa
26 **Elastic Media with Microstructure I**
   One-Dimensional Models   By I. A. Kunin
27 **Superconductivity of Transition Metals**
   Their Alloys and Compounds
   By S. V. Vonsovsky, Yu. A. Izyumov, and E. Z. Kurmaev
28 **The Structure and Properties of Matter**
   Editor: T. Matsubara
29 **Electron Correlation and Magnetism in Narrow-Band Systems**   Editor: T. Moriya
30 **Statistical Physics I**   Equilibrium Statistical Mechanics   2nd Edition
   By M. Toda, R. Kubo, N. Saito
31 **Statistical Physics II**   Nonequilibrium Statistical Mechanics   2nd Edition
   By R. Kubo, M. Toda, N. Hashitsume
32 **Quantum Theory of Magnetism**
   2nd Edition   By R. M. White
33 **Mixed Crystals**   By A. I. Kitaigorodsky
34 **Phonons: Theory and Experiments I**
   Lattice Dynamics and Models of Interatomic Forces   By P. Brüesch
35 **Point Defects in Semiconductors II**
   Experimental Aspects
   By J. Bourgoin and M. Lannoo
36 **Modern Crystallography III**
   Crystal Growth
   By A. A. Chernov
37 **Modern Chrystallography IV**
   Physical Properties of Crystals
   Editor: L. A. Shuvalov
38 **Physics of Intercalation Compounds**
   Editors: L. Pietronero and E. Tosatti
39 **Anderson Localization**
   Editors: Y. Nagaoka and H. Fukuyama
40 **Semiconductor Physics**   An Introduction
   6th Edition   By K. Seeger
41 **The LMTO Method**
   Muffin-Tin Orbitals and Electronic Structure
   By H. L. Skriver
42 **Crystal Optics with Spatial Dispersion, and Excitons**   2nd Edition
   By V. M. Agranovich and V. L. Ginzburg
43 **Structure Analysis of Point Defects in Solids**
   An Introduction to Multiple Magnetic Resonance Spectroscopy
   By J.-M. Spaeth, J. R. Niklas, and R. H. Bartram
44 **Elastic Media with Microstructure II**
   Three-Dimensional Models   By I. A. Kunin
45 **Electronic Properties of Doped Semiconductors**
   By B. I. Shklovskii and A. L. Efros
46 **Topological Disorder in Condensed Matter**
   Editors: F. Yonezawa and T. Ninomiya

Springer Series in Solid-State Sciences
Editors: M. Cardona P. Fulde K. von Klitzing H.-J. Queisser

47 **Statics and Dynamics of Nonlinear Systems**
   Editors: G. Benedek, H. Bilz, and R. Zeyher
48 **Magnetic Phase Transitions**
   Editors: M. Ausloos and R. J. Elliott
49 **Organic Molecular Aggregates**
   Electronic Excitation and Interaction Processes
   Editors: P. Reineker, H. Haken, and H. C. Wolf
50 **Multiple Diffraction of X-Rays in Crystals**
   By Shih-Lin Chang
51 **Phonon Scattering in Condensed Matter**
   Editors: W. Eisenmenger, K. Laßmann,
   and S. Döttinger
52 **Superconductivity in Magnetic and Exotic Materials** Editors: T. Matsubara and A. Kotani
53 **Two-Dimensional Systems, Heterostructures, and Superlattices**
   Editors: G. Bauer, F. Kuchar, and H. Heinrich
54 **Magnetic Excitations and Fluctuations**
   Editors: S. W. Lovesey, U. Balucani, F. Borsa, and V. Tognetti
55 **The Theory of Magnetism II** Thermodynamics and Statistical Mechanics By D. C. Mattis
56 **Spin Fluctuations in Itinerant Electron Magnetism** By T. Moriya
57 **Polycrystalline Semiconductors**
   Physical Properties and Applications
   Editor: G. Harbeke
58 **The Recursion Method and Its Applications**
   Editors: D. G. Pettifor and D. L. Weaire
59 **Dynamical Processes and Ordering on Solid Surfaces** Editors: A. Yoshimori and M. Tsukada
60 **Excitonic Processes in Solids**
   By M. Ueta, H. Kanzaki, K. Kobayashi, Y. Toyozawa, and E. Hanamura
61 **Localization, Interaction, and Transport Phenomena** Editors: B. Kramer, G. Bergmann, and Y. Bruynseraede
62 **Theory of Heavy Fermions and Valence Fluctuations** Editors: T. Kasuya and T. Saso
63 **Electronic Properties of Polymers and Related Compounds**
   Editors: H. Kuzmany, M. Mehring, and S. Roth
64 **Symmetries in Physics** Group Theory Applied to Physical Problems 2nd Edition
   By W. Ludwig and C. Falter
65 **Phonons: Theory and Experiments II**
   Experiments and Interpretation of Experimental Results By P. Brüesch
66 **Phonons: Theory and Experiments III**
   Phenomena Related to Phonons
   By P. Brüesch
67 **Two-Dimensional Systems: Physics and New Devices**
   Editors: G. Bauer, F. Kuchar, and H. Heinrich

68 **Phonon Scattering in Condensed Matter V**
   Editors: A. C. Anderson and J. P. Wolfe
69 **Nonlinearity in Condensed Matter**
   Editors: A. R. Bishop, D. K. Campbell, P. Kumar, and S. E. Trullinger
70 **From Hamiltonians to Phase Diagrams**
   The Electronic and Statistical-Mechanical Theory of sp-Bonded Metals and Alloys By J. Hafner
71 **High Magnetic Fields in Semiconductor Physics**
   Editor: G. Landwehr
72 **One-Dimensional Conductors**
   By S. Kagoshima, H. Nagasawa, and T. Sambongi
73 **Quantum Solid-State Physics**
   Editors: S. V. Vonsovsky and M. I. Katsnelson
74 **Quantum Monte Carlo Methods in Equilibrium and Nonequilibrium Systems** Editor: M. Suzuki
75 **Electronic Structure and Optical Properties of Semiconductors** 2nd Edition
   By M. L. Cohen and J. R. Chelikowsky
76 **Electronic Properties of Conjugated Polymers**
   Editors: H. Kuzmany, M. Mehring, and S. Roth
77 **Fermi Surface Effects**
   Editors: J. Kondo and A. Yoshimori
78 **Group Theory and Its Applications in Physics**
   2nd Edition
   By T. Inui, Y. Tanabe, and Y. Onodera
79 **Elementary Excitations in Quantum Fluids**
   Editors: K. Ohbayashi and M. Watabe
80 **Monte Carlo Simulation in Statistical Physics**
   An Introduction 3rd Edition
   By K. Binder and D. W. Heermann
81 **Core-Level Spectroscopy in Condensed Systems**
   Editors: J. Kanamori and A. Kotani
82 **Photoelectron Spectroscopy**
   Principle and Applications 2nd Edition
   By S. Hüfner
83 **Physics and Technology of Submicron Structures**
   Editors: H. Heinrich, G. Bauer, and F. Kuchar
84 **Beyond the Crystalline State** An Emerging Perspective By G. Venkataraman, D. Sahoo, and V. Balakrishnan
85 **The Quantum Hall Effects**
   Fractional and Integral 2nd Edition
   By T. Chakraborty and P. Pietiläinen
86 **The Quantum Statistics of Dynamic Processes**
   By E. Fick and G. Sauermann
87 **High Magnetic Fields in Semiconductor Physics II**
   Transport and Optics Editor: G. Landwehr
88 **Organic Superconductors** 2nd Edition
   By T. Ishiguro, K. Yamaji, and G. Saito
89 **Strong Correlation and Superconductivity**
   Editors: H. Fukuyama, S. Maekawa, and A. P. Malozemoff

Springer Series in Solid-State Sciences
Editors: M. Cardona P. Fulde K. von Klitzing H.-J. Queisser

Managing Editor: H. K. V. Lotsch

90 **Earlier and Recent Aspects of Superconductivity**
Editors: J. G. Bednorz and K. A. Müller

91 **Electronic Properties of Conjugated Polymers III** Basic Models and Applications
Editors: H. Kuzmany, M. Mehring, and S. Roth

92 **Physics and Engineering Applications of Magnetism** Editors: Y. Ishikawa and N. Miura

93 **Quasicrystals** Editors: T. Fujiwara and T. Ogawa

94 **Electronic Conduction in Oxides** 2nd Edition
By N. Tsuda, K. Nasu, F. Atsushi, and K. Siratori

95 **Electronic Materials**
A New Era in Materials Science
Editors: J. R. Chelikowsky and A. Franciosi

96 **Electron Liquids** 2nd Edition By A. Isihara

97 **Localization and Confinement of Electrons in Semiconductors**
Editors: F. Kuchar, H. Heinrich, and G. Bauer

98 **Magnetism and the Electronic Structure of Crystals** By V. A. Gubanov, A. I. Liechtenstein, and A. V. Postnikov

99 **Electronic Properties of High-T_c Superconductors and Related Compounds**
Editors: H. Kuzmany, M. Mehring, and J. Fink

100 **Electron Correlations in Molecules and Solids** 3rd Edition By P. Fulde

101 **High Magnetic Fields in Semiconductor Physics III** Quantum Hall Effect, Transport and Optics By G. Landwehr

102 **Conjugated Conducting Polymers**
Editor: H. Kiess

103 **Molecular Dynamics Simulations**
Editor: F. Yonezawa

104 **Products of Random Matrices**
in Statistical Physics By A. Crisanti, G. Paladin, and A. Vulpiani

105 **Self-Trapped Excitons**
2nd Edition By K. S. Song and R. T. Williams

106 **Physics of High-Temperature Superconductors**
Editors: S. Maekawa and M. Sato

107 **Electronic Properties of Polymers**
Orientation and Dimensionality
of Conjugated Systems   Editors: H. Kuzmany, M. Mehring, and S. Roth

108 **Site Symmetry in Crystals**
Theory and Applications 2nd Edition
By R. A. Evarestov and V. P. Smirnov

109 **Transport Phenomena in Mesoscopic Systems** Editors: H. Fukuyama and T. Ando

110 **Superlattices and Other Heterostructures**
Symmetry and Optical Phenomena 2nd Edition
By E. L. Ivchenko and G. E. Pikus

111 **Low-Dimensional Electronic Systems**
New Concepts
Editors: G. Bauer, F. Kuchar, and H. Heinrich

112 **Phonon Scattering in Condensed Matter VII**
Editors: M. Meissner and R. O. Pohl

113 **Electronic Properties of High-T_c Superconductors**
Editors: H. Kuzmany, M. Mehring, and J. Fink

114 **Interatomic Potential and Structural Stability**
Editors: K. Terakura and H. Akai

115 **Ultrafast Spectroscopy of Semiconductors and Semiconductor Nanostructures**
2nd Edition By J. Shah

116 **Electron Spectrum of Gapless Semiconductors**
By J. M. Tsidilkovski

117 **Electronic Properties of Fullerenes**
Editors: H. Kuzmany, J. Fink, M. Mehring, and S. Roth

118 **Correlation Effects in Low-Dimensional Electron Systems**
Editors: A. Okiji and N. Kawakami

119 **Spectroscopy of Mott Insulators and Correlated Metals**
Editors: A. Fujimori and Y. Tokura

120 **Optical Properties of III–V Semiconductors**
The Influence of Multi-Valley Band Structures
By H. Kalt

121 **Elementary Processes in Excitations and Reactions on Solid Surfaces**
Editors: A. Okiji, H. Kasai, and K. Makoshi

122 **Theory of Magnetism**
By K. Yosida

123 **Quantum Kinetics in Transport and Optics of Semiconductors**
By H. Haug and A.-P. Jauho

124 **Relaxations of Excited States and Photo-Induced Structural Phase Transitions**
Editor: K. Nasu

125 **Physics and Chemistry of Transition-Metal Oxides**
Editors: H. Fukuyama and N. Nagaosa